NUCLEAR TECHNOLOGY DEMYSTIFIED

EVERYTHING YOU NEED TO KNOW ABOUT EVERYTHING NUCLEAR

William K. Terry

Published 2020

Printed in the United States of America
Print ISBN: 978-1-951490-06-5

Canoe Tree Press
4697 Main Street
Manchester, VT 05255

www.CanoeTreePress.com

ABOUT THE AUTHOR

William K. Terry earned his PhD in Nuclear Engineering at the University of Washington in 1980, specializing in fusion reactor plasma physics. From there, he spent five years as an Assistant Professor at Purdue University, where he launched the fusion program in the university's School of Nuclear Engineering.

After Purdue, he spent three years at the Department of Energy's Hanford Site, working on the site characterization plan for a high-level nuclear waste repository. He spent the last 20 years of his career at the Idaho National Laboratory, primarily studying advanced reactor concepts, most notably the pebble-bed reactor.

Outside of work, his interests have included aviation, fly-fishing, birdwatching, horses, and natural resource conservation. He is a private pilot with multi-engine and instrument ratings. He has chased trout and salmon from Norway to New Zealand, and he supports numerous conservation groups.

To Julie

TABLE OF CONTENTS

ACKNOWLEDGEMENTS

When I retired in 2007 from the Idaho National Laboratory, I moved from the relatively pro-nuclear community of Idaho Falls, Idaho, to the "People's Republic of Boulder," where nuclear energy is widely regarded as the spawn of Satan. Writers expressing that view in the Boulder *Daily Camera* reveal profound ignorance on the subject. This book was directly motivated by my desire to provide in-depth factual information on all aspects of nuclear technology for them and anyone else who is willing to look at it objectively. So in a way I have to be grateful to Boulder's anti-nukes who spurred me into writing.

I have had the privilege of working through the years with some brilliant scientists who have influenced my development as a researcher. In particular, I would like to express my gratitude to my dissertation adviser, George Vlases, and to my former colleagues Karl Ott and Abderrafi Ougouag.

I have drawn on many sources for the illustrations in the book. I am grateful to the many people who have made their work freely available on Wikimedia Commons and other Internet sources. I am particularly indebted to the following people for granting me permission, or working to obtain their institutions' permission, to use their copyrighted work: Nicola Scafetta of Duke University, Frances Marshall of the Idaho National Laboratory, Lynda Seaver of Lawrence Livermore National Laboratory, Jason Post of GE Power & Water, Marion Brünglinghaus of Informationskreis KernEnergie, Berlin, James Mellott of NuScale Power, Bonnifer Ballard of the American Nuclear Society, Jeanette L. Taplin of the Electric Power Research Institute, Inc., Mary R. Hale of Argonne National Laboratory, Gregory Perret of the Paul Scherrer Institute, Terry L. Schultz of Westinghouse, and Wim Uyttenhove of SCK-CEN. Uncredited illustrations are my own creations.

I am especially indebted to my dear friend and former colleague Chuck Wemple for his careful review of my first draft, and to my beloved wife Julie Devine for proofreading the final draft and for her constant encouragement and love. If any errors or typos remain, do not blame Chuck or Julie; they are my fault.

Finally, I would like to thank the people at Canoe Tree Press—Simona Meloni for converting my Word file into an actual book, and Suanne Laqueur for help with all aspects of the publication process. Thanks also to the illustrator at Canoe Tree Press for converting some of my crude sketches into presentable drawings.

INTRODUCTION

As the environmental consequences of burning fossil fuels become more and more apparent, citizens around the world are becoming increasingly interested in energy sources that do not contain carbon. One carbon-free form of energy is nuclear energy.

Nuclear technology is mysterious, even scary, to many people. Much of what has been claimed in the popular media is highly misleading, and I attempt to present a more informed viewpoint. Even if you are opposed to nuclear power, you must base your arguments on facts if you argue your position with integrity.

Many fine books have been written to present facts about nuclear technology, particularly about nuclear power plants, to help open-minded laypersons separate the truth from misleading and erroneous claims made by opponents of nuclear power. An excellent example is *Power to Save the World: The Truth About Nuclear Energy,* by Gwyneth Cravens, a former antinuclear activist. Because their primary objective is to defend nuclear power, such books devote much of their space to discussions of the allegations of the opponents and rebuttals based on facts. This emphasis leaves little room to present the fundamentals of nuclear technology. This book takes the opposite approach. Having worked in nuclear technology research and development for 32 years, I am an unabashed nuclear power advocate, but I don't intend to preach in this book. I think intelligent people can come to the right decision when they understand the facts. So this book is primarily intended to help people understand nuclear technology on a fairly sophisticated level, but without the mathematics required for practical application or true in-depth understanding. An intelligent person with a good high school education should be able to understand this book, although it will take some effort in places.

However, this is a book on science. As I learned after some initial struggles in high school, you can't read a science or math book like you read a book on history. At least, where basic principles of science are being discussed, you should proceed slowly, making sure that you understand every term introduced, and every logical or mathematical argument made, before going on. I present some complex mathematical expressions only to show you what they look like, but I do present a few simple mathematical thought processes, which you should try to follow.

If you have total math phobia, you can still understand much of the conceptual and descriptive material without struggling to understand the mathematical ideas, but the more of the mathematics you follow, the better you will understand everything else. In particular, if your eyes glaze over when you see numbers, you will miss some key arguments. For example, when I state that the average background radiation dose for Americans is 102 mrem per year, but that jet pilots can increase their radiation exposure by 468 mrem per year by making three transcontinental flights per week, the numerical comparison is the whole point. Try to understand such arguments even if it hurts.

Readers with little or no scientific or mathematics background may not be familiar with the shorthand notation for denoting large or small numbers. Rather than write the speed of light, for example, as 300,000,000 meters/second (or m/s), we write 3.0×10^8 m/s, where 10^8 means one followed by eight zeros. The notation 10^{-8} means one divided by 10^8. This notation becomes more and more convenient as numbers become very large or very small, but often it is used even for numbers closer to one. And we should also know that $10^0 = 1$.

There is some redundancy in the book. Sometimes I restate something that appears in an earlier chapter, such as the definition of a unit for a physical quantity like energy, so that the reader won't have to go hunting for it if memory fails, as it so often does. I also remind the reader where to find discussions of complicated concepts when they are called on in chapters after their first appearance. My goal in such redundancy is simply to help the reader out. For the most part, the book does not have to be read from front to back. Each chapter almost stands alone. Where concepts from previous chapters are invoked, the previous relevant material is identified.

Although my primary goal in this book is to enable the reader to understand the principles at work in nuclear technology, much of the discussion relies on specific examples of nuclear systems. Currently, the technology is changing quite rapidly, and some of my examples will be out of date even before the book is printed. The best I can hope to do is to keep updating future editions.

In a scholarly work, thorough documentation is given to enable the reader to verify claims. The most original references are preferred, which are usually found in archival journals available in technical libraries. My objective is to make it easy for the reader to find more information on topics of discussion, and the most convenient source of information on most topics is the Internet. So I make frequent references to websites, especially Wikipedia. The editors of Wikipedia try to check new input for accuracy, but since anyone can contribute to Wikipedia, the information contained there cannot be trusted completely. Nevertheless, I have found the technical articles in Wikipedia to be generally informative and accurate, and since they are so easy to access and they are likely to be available for a long time, I use them extensively in this book for the reader's convenience.

I would like to insert a note on pronunciation. The word "nuclear" should be pronounced "NEW-klee-ar," not "NEW-kew-lar." Language is dynamic, and word usage and pronunciation change over time according to what people do. So if people use or pronounce a word wrong long enough, the error becomes accepted. I checked a number of on-line dictionaries, and I sadly found that some of them now admit "NEW-kew-lar" as an alternate pronunciation. Others acknowledge that it is common but state explicitly that it is wrong. My reason for siding with the latter group is that the sound in a word is carried by the letters, and in "NEW-kew-lar" the sounds come in a different order from the letters. So "NEW-kew-lar" doesn't make any sense. If you say, "NEW-klee-ar," you can't go wrong, and you'll make a much better impression on some of your listeners.

It would be dishonest—it would even be ludicrous—to assert that nuclear technology doesn't present hazards to environmental quality and public health and safety. But all other energy technologies present environmental and public health and safety risks of their own, which are often not recognized by laypersons. Before embarking on the detailed discussion of nuclear energy systems that occupies much of this book, I provide my assessment of all the energy sources available to modern societies in order to put nuclear energy in context. This assessment is followed by a discussion of the highly publicized topic of global warming.

But in the first few chapters I want to provide some background information on science in general and on the modern physics underlying nuclear phenomena. Then the assessment of energy alternatives and a discussion of global warming are presented, and after that the individual branches of nuclear technology are described in chapters of their own.

PART I
BACKGROUND SCIENCE

CHAPTER 1
HOW SCIENCE WORKS

Unless you are a scientist yourself, or at least unless you have had quite a few science classes in school, you may have picked up most of your impressions of what science is, and how scientists go about their work, from the popular media. Articles in newspapers, and two-minute stories on news television, don't have space or time to present much detail, even if the reporters have any more understanding than the average person. Therefore, before I begin to discuss nuclear technology specifically, I want to provide you with some basic understanding, both philosophical and practical, of science in general.

In this book, when I speak of science, I am talking about natural science—the study of nature. It is always a good idea to define a term precisely when you begin to discuss it, so I will begin by defining science: Science is the search for natural causes of natural phenomena.

1.1 THE SCIENTIFIC METHOD

When I was in junior high school, or maybe even in elementary school, I was introduced to the "scientific method." It was claimed that science proceeds in an orderly sequence from observations and experiments to hypotheses, which progress to theories if they gain enough experimental support, and which finally become accepted as laws if they stand the test of experimental challenge for a long time. This is a misleading characterization of the workings of science. The word "law" implies a principle that is fully established and not open to question anymore; actually, everything in science is subject to correction and refinement. "Law" is an unfortunate choice of words; there is really nothing in science with higher status than a theory. Also, sometimes theory precedes experiment: Theory predicts phenomena that haven't been observed yet. Then if the phenomena are observed later, the theory is strongly supported by the observations.

I think the term "law" should apply to the actual principles that govern natural phenomena, which we seek to understand in science but which we realize that we are only discovering approximately. Our theories can then be regarded as our best approximations to the actual laws of nature. But the term "law" has long been applied not only to long-established theories, but also to some purely empirical[a] relationships that really aren't even theories, but only convenient mathematical summaries of a restricted class of observations. Don't take the term "law" too seriously!

The branches of natural science, such as physics, chemistry, and biology, are empirical. As the great physicist Richard Feynman put it, in science, "The test of all knowledge is experiment."[1]

There are two kinds of knowledge in science. First is observation. What we can see, hear, or otherwise discern by our senses may be regarded as fact, although our interpretation of what we observe may be questioned. (Just because someone thinks he saw a UFO doesn't mean he saw an alien spacecraft. But unless he is delusional, he must have seen something.) The other kind of knowledge is explanation: hypotheses or theories that make sense of our observations. A hypothesis is a tentative explanation that serves as a basis for further observations or experiments. A theory is a well tested explanation that has stood up to all experimental challenges so far. In order to qualify as a theory, a hypothesis not only must withstand all experimental tests, it must also be able to predict the outcome of further experiments, and it must be "falsifiable." That means that it must be testable by experiments of which one possible outcome is to prove the hypothesis false.[2] If a false outcome is obtained, then the theory is either discredited or shown to be limited in its validity. When a theory is confirmed by a variety of different kinds of experiments, it is more strongly supported than it would be if only one kind of experiment confirmed it. In physics, at least, theories are expressed in mathematical form (equations and formulas of algebra and calculus), which can produce numbers that can be compared to precise measurements.

Mathematics is sometimes called a "formal science." The rules of mathematics are different. In mathematics, you start with axioms, or propositions considered self-evident and accepted as true without proof. An example of an axiom: If two quantities are equal to a third quantity, they are equal to each other.[3] From a small set of axioms, other propositions are proven logically. It is a remarkable fact that the results of mathematics, a type of science that requires no confirmation from nature, are

---

[a] The word *empirical* means based on observation or experience.

indispensable to the study of natural science, which is wholly driven by confirmation from observations of nature.

A crucial feature of science is that it is never finished. As noted above, there really are no "laws," in the sense of pronouncements that are no longer open to question. Every observation and every conclusion are open to challenge. When scientists report observations of new phenomena of any significance, other scientists usually attempt to repeat the experiments or to perform other experiments that should lead to similar observations. When a group of researchers reported that they had observed nuclear fusion reactions at room temperature ("cold fusion"), there was a huge surge of excitement at the possibilities this would imply for cheap, clean energy. So a large number of independent research groups tried to duplicate the original experiments. Nobody succeeded, and the current opinion of most scientists who engaged in cold fusion research is that the original research group misinterpreted its observation. And even though we still use the terminology "Newton's laws of motion" for the principles of motion introduced by Isaac Newton, we now know that they are only valid for bodies moving slowly, compared to the speed of light, relative to the observer. For faster-moving bodies (even for modestly fast-moving objects like GPS satellites), Einstein's theory of relativity must be applied. For slow relative motion, Einstein's formulas for motion become equivalent to Newton's "laws."

That equivalence illustrates one of the most important principles in the philosophy of science, the Correspondence Principle.[4] This principle applies to situations where a well established mathematical theory on a phenomenon already exists, as in the case of Newton's "laws," which were believed for two hundred years to fully describe the motion of bodies. It states that we know in advance that any new theory, whatever its character or details, must reduce to the well established original theory to which it corresponds when it is applied to the circumstances in which the original theory is known to hold.

Besides hypotheses and theories, there is another class of explanations of phenomena that scientists find very useful: models. Convenient ways of looking at phenomena, models are not purported to be true characterizations of reality, but they enable calculations to be performed with sufficient accuracy for practical purposes. An example is the liquid-drop model of the atomic nucleus. Nobody thinks that the nucleus is a drop of liquid, but mathematical representations of the nucleus as a liquid drop, accounting for surface tension and the short-range attractive forces among neutrons and protons (analogous to the attractive forces among liquid molecules), enable accurate calculations to be performed for some nuclear phenomena, such as nuclear fission.[5]

1.2 SCIENCE IN PRACTICE

In the actual practice of science, individual scientists often have to find their own sources of funding. They write proposals to funding agencies such as the U. S. Department of Energy (DOE) or to private companies who may find the potential results of the proposed research profitable.

When a funding agency such as the DOE is interested in exploring some research topic, it usually publishes a request for proposals (RFP). Typically, more scientists submit proposals than the funding agency can support. So the process is highly competitive, and only a relative few of the proposals are selected for funding.

How does the funding agency choose? It relies on the opinions of researchers experienced in the research topic of interest. These experts review the proposals and score them according to criteria given by the funding agency, normally including technical credibility, originality, probability of success, and potential value of the results (both intrinsic value and value per dollar of funding). The nuclear science community is actually rather small, and it is often difficult to find enough reviewers who haven't submitted their own proposals. For an expert to review proposals with which his own is in competition would be a conflict of interest, so such experts are excluded *a priori* from the review process. However, by cajoling the available reviewers each to review numerous proposals, the funding agencies always seem to get the job done. The desire of the funding agencies to obtain useful results tends to eliminate some high-risk proposals (i.e., proposals with a high probability of failure) that might produce findings of exceptional importance, but otherwise this peer review process generally separates the good proposals from the bad ones and ensures that high-quality proposals are chosen for funding.

In a large establishment such as the DOE with a major mission to support scientific research, its own staff includes scientists with expertise in the areas it supports. These staff scientists monitor the progress of the research their agency supports, in order to ensure that its funding is being spent well. On one hand, this monitoring helps protect the investment the taxpayers or shareholders are making, but on the other hand, it requires the researchers to spend a lot of time writing progress reports that divert them from further progress on the research they are being supported to perform. But no system is perfect, and on the whole this one works pretty well.

When noteworthy findings are obtained in the course of a research project, and at the conclusion of the project, they are not only reported in the required progress reports or the final project report, they are also published in the open literature (unless they are "classified"—i.e., deemed by government to need access restrictions in the interest of national security—which is rare in nuclear reactor technology unrelated to weapons). However, before a paper can be published in a peer-reviewed conference proceedings or an archival journal, it must undergo a review by experts in the field, much as the original proposal did. The depth of the review depends on the requirements of the conference or journal. At the very least, the reviewers carefully judge the claims made in the paper, and they verify the mathematics to the greatest practical degree.

Usually, reviewers are not paid to perform reviews. In papers that report lengthy mathematical derivations, space limitations prohibit complete step-by-step presentations of all the equations, so only the key steps are given. Unpaid reviewers will not usually try to fill in all the omitted steps, but they will judge the plausibility of the equations. When papers report calculations by computer codes for which months of work are required to set up the input, unpaid reviewers will not try to duplicate the calculations. However, in projects of exceptional importance, reviewers may be paid to do just that.

For papers reporting experimental work, which normally requires elaborate and expensive equipment, no attempt is made to duplicate the experiments in the course of the review. However, the reviewers will judge the validity of the method, and they will perform "sanity checks" to ensure that the results do not violate accepted theories. For example, if a paper purports to unveil a perpetual motion machine, which violates the accepted principles of thermodynamics, it will not get published. (If such a thing were ever actually invented, it would turn the whole world of science on its head and the inventors could become very rich and famous. But I wouldn't buy stock in a company that claimed to have invented one.)

Because the review process is incomplete, mistakes are made and get published. However, as pointed out in the previous section, science is never finished. When new theoretical results are published, independent scientists will write proposals to test those results experimentally. When a new experimental result is published, independent groups will write proposals to perform independent experiments to obtain similar results. If a new theory is discredited by experiment, or if new experimental results cannot be duplicated, they are discarded as erroneous. So science is inherently self-correcting, and eventually the truth will emerge.

In industrial projects, such as the design of a nuclear reactor, the process is different from the kind of research project described above. The details of a design are intellectual property, and a design company is rightly unwilling to share these hard-won details with its competitors. So it holds most of its design calculations and tests proprietary. However, for something like a nuclear reactor or an airplane, a company must apply for a license to build and operate it, so it must submit its design calculations in confidence to a regulatory agency such as the U. S. Nuclear Regulatory Commission or the Federal Aviation Administration.

Before a license application is submitted, all the design calculations and test results are subjected to a review by experts within the company, much like the independent peer review performed in publicly funded research projects. Every number derived in the design will be verified at least once in this internal review process. This method ensures that the resulting design is as free of errors as humanly possible. When the product is actually built, a verification test program is conducted to confirm that the product works as intended. For example, when a newly designed airplane is built for the first time, the prototype undergoes a lengthy and highly structured flight test program before the design can be certified. New nuclear reactors undergo a start-up test program before their final operating license is granted.

In fields removed from my own experience, there may be variations from the foregoing description of how scientific research and development work is performed. However, that description summarizes my own experience over a 32-year period in nuclear research in universities and national laboratories.

CHAPTER 2
A FEW CONCEPTS FROM CLASSICAL PHYSICS

We can't get very far in discussing nuclear science without understanding some basic physics concepts like force, energy, work, acceleration, and so forth. So we shall begin by introducing these concepts briefly. For further explication, you might refer to a good high school or freshman college physics textbook, depending on your mathematical background.

This chapter is by far the most mathematical chapter in the book. That is because the mathematics is relatively simple, and you don't have to have any mathematical background to understand it. You should regard the mathematical expressions as merely a kind of shorthand to express the ideas succinctly that I explain in words when I introduce them. The verbal descriptions of the concepts are the key to understanding them; then the math just saves having to repeat the words every time.

First, we need to understand how physicists describe motion. The position of an object is followed in some kind of coordinate system, such as the two-dimensional Cartesian, or rectangular, system shown in Figure 2.1. If you specify the x- and y-coordinates, you know where the object is. Two-dimensional coordinate systems are appropriate for describing motion in a plane, such as a billiards table or a small portion of the Earth's surface where the Earth's curvature doesn't matter. For three-dimensional motion, a z-coordinate, perpendicular to both the x- and y-directions, is also defined. Other coordinate systems, such as cylindrical, spherical, and hexagonal coordinates, are also useful.

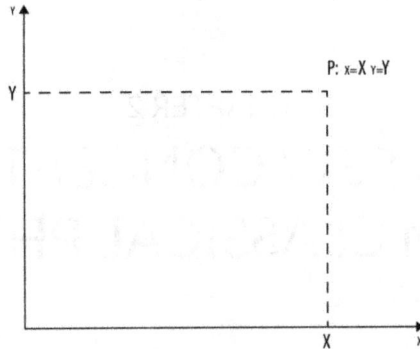

Figure 2.1—Two-dimensional Cartesian coordinate system

The velocity of the object is the rate of change of its position with respect to time at the instant of observation, or the instantaneous rate of change of position—that is, the change in its position during a very small interval of time, divided by the value of that small interval of time. Mathematicians use the concept of a "derivative" to describe such change rates; they write, for example, $v_x = dx/dt$ for the x-component of the velocity. The expression dx/dt, said as "d-x-d-t," is called the derivative of x with respect to t, and v_x is said as "v-sub-x." There are specific reasons for the use of this notation for the derivative, but if you haven't studied calculus you don't need to worry about them; just remember that it means the instantaneous rate of change of x with respect to t. Velocity is a quantity with both magnitude and direction; such a quantity is called a vector. Vectors are usually denoted either by boldface type or by placing an arrow above the symbol. The symbols *i, j,* and *k* are often used to denote "unit vectors" in the x, y, and z directions—i.e., dimensionless vectors with magnitude equal to 1 and direction along the coordinate axes. Then velocity is written

$$\boldsymbol{v} = v_x \boldsymbol{i} + v_y \boldsymbol{j} + v_z \boldsymbol{k}.$$

(When two quantities are written side by side, as in *ab* or $v_x \boldsymbol{i}$, it means they are to be multiplied. Multiplication and division are performed before addition and subtraction, so there is no ambiguity in the equation above.) The velocity vector in two dimensions is illustrated in Figure 2.2. The absolute value of velocity—the total magnitude of the velocity without regard to its direction—is called the speed, and it is equal to

$$v = \sqrt{v_x^2 + v_y^2 + v_z^2}.$$

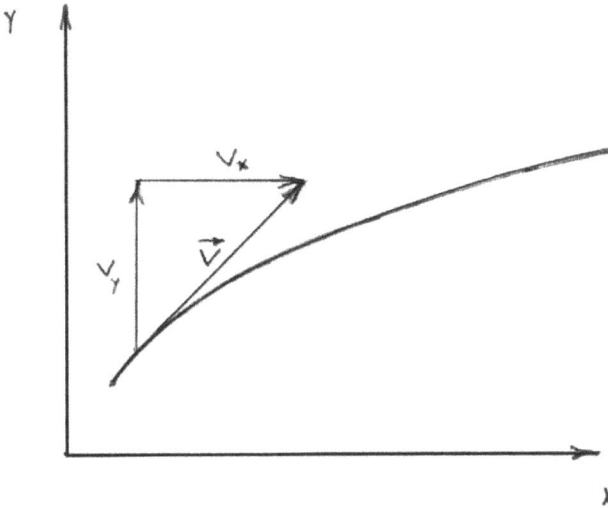

Figure 2.2—The instantaneous velocity of an object moving in two dimensions

There are many systems of units for physical quantities like velocity. We can specify velocity in miles per hour, feet per second, or furlongs per fortnight. Physicists usually use the Système Internationale (SI), an internationally adopted system based on the meter for length, the kilogram for mass, and the second for time. Velocity and speed in the SI are then expressed in meters per second (m/s).

Acceleration is the instantaneous rate of change of velocity with respect to time. Mathematicians write $a_x = dv_x/dt = d(dx/dt)/dt = d^2x/dt^2$, $\boldsymbol{a} = a_x\boldsymbol{i} + a_y\boldsymbol{j} + a_z\boldsymbol{k}$, and so forth. Acceleration has units of velocity per time, such as m/s per s (or m/s^2) in SI units.

Acceleration can either be in the direction of the velocity, perpendicular to the velocity, or a combination of the two. Acceleration in the direction of the velocity increases the speed, but lateral acceleration doesn't. Imagine an ice skater holding onto a rope as she glides along, and suppose the rope is attached at the other end to a pole fixed to the ice but free to swivel as she goes around and around. Your experience tells you that she has to pull hard on the rope, and that she will move in a circle around the pole. If the ice is perfectly slippery (no friction) and there is no air resistance to her motion, she won't have to keep thrusting with her skates; she will just glide around as long as she wants, and she won't slow down or speed up. (Real ice is never perfectly slippery, and there is air resistance, but we idealize the situation in our minds to simplify things.) Figure 2.3 shows how her velocity changes at a particular instant in time. The velocity is actually constantly changing in a direction perpendicular to her instantaneous direction of motion, and it is being deflected inwards, towards the center of the circle. This is called centripetal acceleration. As we shall see in a moment, a centripetal—inward—force is required to produce centripetal acceleration.

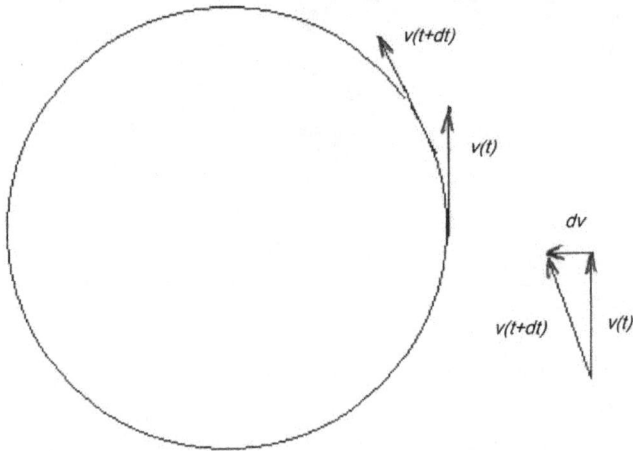

Figure 2.3—Centripetal acceleration in circular motion

In the figure, the notations $v(t)$ and $v(t+dt)$ are used. These notations mean v at time t and v at time $t+dt$. When one quantity depends on another like this, the dependent quantity is said to be a function of the independent quantity. Functional notation like this is read as "v of t" or "v of $t+dt$."

We are used to thinking of centrifugal force. When we go around a corner in a car, we feel as though we are being thrown outwards, not inwards. But that's because we experience things in our own reference frames, which accelerate with us. In our own reference frames, there is absolutely no way to distinguish between a force and an opposite acceleration (that's actually one of the principles of Einstein's General Theory of Relativity), but someone watching from an external, fixed reference frame can tell that the force exerted on us by our car seats is really deflecting us inwards, towards the center of the curve, and that the skater is being accelerated inwards, towards the pole. But we, like the skater, are moving along our curved paths at just the right rate to remain at a constant distance from the center as we constantly accelerate inwards towards it. From the fixed reference frame, one views the centrifugal force we feel as we move on a curved path as an apparent force, not a real force. There is no magic hand that suddenly pulls us outwards as soon as we deviate from a straight path.[a]

In classical physics, motion is governed by Isaac Newton's three "laws" of motion, which we have noted in Chapter 1 not to be immutable decrees from on high, but to be only descriptions that have turned out to be very accurate for objects in slow motion with respect to an observer, relative to the speed of light. The term "classical physics" refers to motion in slowly moving reference frames and to large scales relative to the atom. The "laws" of classical physics break down for motion at speeds near the speed of light or on the atomic scale, and Einstein's Theory of Relativity and/or quantum theory must be applied. (Actually, for some purposes you need the "relativistic" corrections at modest speeds. As noted in Chapter 1, if the relativistic corrections were not applied to tracking GPS satellites, we would lose track of them quickly!) In his *Philosophiæ Naturalis Principia Mathematica* (1687), Newton proposed three laws of motion;[1] these laws are formulated in a so-called "inertial reference frame"—i.e., one that is not being accelerated relative to some truly fixed reference:[b]

Newton's first law of motion: An object will stay at rest or continue at a constant velocity unless acted upon by an unbalanced net force.

Newton's second law of motion: The acceleration a of an object of mass m, measured in an inertial reference frame, is equal to the net force F acting on it divided by m; $a = F/m$, or, as it is usually stated, $F = ma$.

Newton's third law of motion: Whenever an object A exerts a force on another object B, B simultaneously exerts a force on A with the same magnitude in the opposite

---

[a] In our own accelerating reference frame, centrifugal force is as real as any other force. If we are placed in a centrifuge, as used to train astronauts and fighter pilots, the centrifugal stresses in our bodies can cause injury if the speed of the centrifuge is too great. In an accelerating reference frame, the equivalent force resulting from the acceleration is called an "inertial force"; centrifugal force is the inertial force equivalent to a centripetal acceleration. We can perform quantitative analyses in accelerating reference frames by applying the inertial forces and get correct answers (d'Alembert's Principle). But errors arise when we apply inertial forces in fixed reference frames.

[b] We probably can't really identify any such "truly fixed reference." The Universe is expanding, and all parts of it not gravitationally bound to each other are moving away from each other at accelerating rates. For practical purposes, an inertial reference frame is one whose acceleration is not relevant to the phenomena being analyzed. For many purposes, the Earth's surface can be considered an inertial reference frame, but for analyzing weather or artillery shells, to name a couple of examples, it cannot; the rotation of the Earth introduces an apparent force, called the Coriolis force, that must be taken into account.

direction. (This law is sometimes stated, "For every action there is an equal and opposite reaction," but that statement is too easily misinterpreted too broadly.) If we, instead of a pole, are holding the skater's rope at the center of her circle, we feel an outward, or centrifugal, force on the rope, and that may make us think that there is a centrifugal force acting on her. But, in our inertial reference frame, there isn't. What we feel at our end of the rope is the reaction force exerted by the skater on the rope as required by the third law, not the force on the skater by a magic hand.

In the mathematical expression of Newton's second law, $F = ma$, F is the net, or total, force—i.e., the vector sum of all forces on the object, including contact forces like friction or the pull of a rope, and forces acting within the object ("body forces") like gravity. The mass of an object can be defined either in terms of its resistance to acceleration (its inertia) or in terms of the force of gravity acting on it. It turns out that the masses defined from these two different viewpoints are the same. In classical physics, mass can be thought of as a measure of how much matter the object contains.

The SI unit of force is the newton (N), equal to one kilogram times one meter per second per second, or 1 kg-m/s^2. A newton is about 0.22481 pounds of force, or, for easy recollection, roughly a quarter of a pound. (A pound of force is the force exerted by standard gravity at the Earth's surface on a mass of one pound.)

The momentum p of an object is defined as the product of its mass and velocity: $p = mv$. Note that it is a vector quantity. Another way to write Newton's second law is $F = dp/dt$, the rate of change of momentum with respect to time. This is more general than $F=ma$, and it applies even for objects moving near the speed of light, but in that case the expression for momentum is different, including the relativistic correction discovered by Einstein.

There are analogous formulas for rotating bodies, which are consequences of Newton's second law. We define the concept of torque as a measure of the ability of a force to produce a rotation. If a force F is applied at a distance r from an axis of rotation, the torque τ (the lower-case Greek letter tau) about the axis is given by $\tau = Fr$.[c] The mass of a body can be considered a measure of its resistance to acceleration; the corresponding measure of a body's resistance to changes in its state of rotation is called the moment of inertia, usually denoted as I. It accounts both for how much

---

[c] It's actually a bit more complicated. The rotation is produced by the component of the vector force F perpendicular to the plane defined by the axis and the vector r from the axis to the point where F is applied. The torque is also a vector, and it is given by $\tau = r \times F$ ("r cross F"), where X denotes a particular type of vector operation called the cross product. See any textbook on introductory calculus for an explanation.

mass the body contains and for how it is distributed about its center of rotation. The moment of inertia of a body also depends on where the center of rotation is located. The farther from the center of rotation the mass of a body is located, the greater the body's moment of inertia about that center of rotation. A tire has a greater moment of inertia about its center of mass than a solid ball of rubber of the same mass has about its center. (The center of mass is the point around which all the mass of a body is evenly distributed, according to a specific, precise mathematical definition. It is often the center of rotation for objects in practice, from spinning tires to the rotating Earth. For application of Newton's second law for curvilinear—i.e., non-rotational—motion, all of a body's mass may be considered to be concentrated at the center of mass.) The rate of rotation of a body about a center of rotation is defined as its angular velocity, ω (Greek lower-case omega). This is the rate of change of the angle θ (Greek lower-case theta) between an arbitrary reference line perpendicular to the axis of rotation and the line to a point on the body, as shown in Figure 2.4: $\omega = d\theta/dt$. The rate of change of the angular velocity is called the angular acceleration α (Greek lower-case alpha): $\alpha = d\omega/dt$. The equation for rotational motion analogous to Newton's second law (for a rigid body in non-relativistic situations) is $\tau = I\alpha$. The product $I\omega$ of the angular velocity and the moment of inertia is called the angular momentum L, and the equation can also be written $\tau = dL/dt$. These formulations are valid only for the very simple geometry described here, and they become much more complicated when generalized.

Next, we introduce the concepts of work and energy. Work is an action of a "system" that is equivalent to the raising of a weight. A system can be any collection of matter. To illustrate this definition of work, we can consider a system that consists of an electric battery. Suppose this battery is hooked up to an incandescent light. Then the current from the battery heats up the filament in the light bulb, producing heat and light. But the same current could go into an electric motor with a shaft on which a spool is mounted. A rope could be wound around the spool, so that as the motor turned the shaft a weight on the end of the rope would be raised. As far as the battery is concerned, it doesn't matter whether it is lighting a bulb or lifting a weight; it sends out electrons in the same way. So the battery is performing work, even though the actual action on the light bulb does not raise a weight.

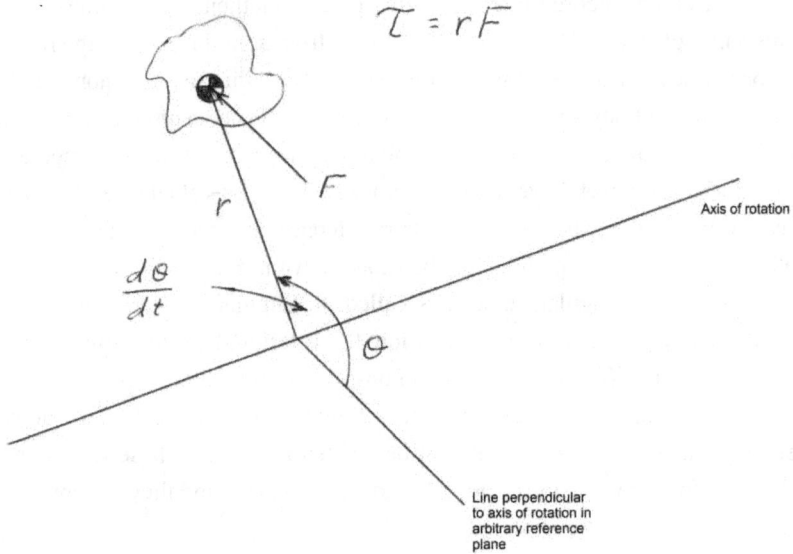

Figure 2.4—Angular velocity and torque

That is the definition of work used in the science of thermodynamics (that word literally means the dynamics of heat). It is more general than the mechanical work defined in introductory physics courses. There, work is defined for an object moved by a force. For simplicity, we consider an object being pushed along by a constant force acting in the direction in which the object is moving. If the magnitude of the force is F and the object moves a distance Δx, the work done by the force is $W = F \, \Delta x$ (Δ, the upper-case Greek letter delta, is often used in physics and mathematics to denote a change in some quantity, here the position along an x-axis). A lateral force, such as the centripetal force in circular motion, does no work because it is perpendicular to the object's motion and does not change the object's speed.

The thermodynamic definition of work enables us to give a simple definition of energy: Energy is a property of a system that enables it to perform work. We will encounter many forms of energy, and it not always obvious how they can be pressed into service to perform work, but in fact they can, as you can usually figure out for yourself with some mental struggling. We now consider some kinds of energy specifically.

Kinetic energy is the energy of motion. If a mass m is moving with speed v, its kinetic energy is $E_k = \frac{1}{2} mv^2$ (if v is small compared with the speed of light).

Potential energy is energy stored in a force field, such as a gravitational field. A force field exerts a specific force on an object at each point in space, and positive or negative work is done by the force field as the object moves along or against the direction of the force. For example, when the object moves against the force field, the field acquires the potential to do more work on the body, so the potential energy of the body is increased. (When you lift a weight in the Earth's gravitational field, and then drop it, gravity will do work on it as it falls.) In a uniform force field, as the gravitational field on Earth approximately is over a small region of the Earth's surface and in a narrow range of altitudes, the acceleration imparted by the field to an unrestrained object is the same everywhere. Note that gravity is exerted on each incremental piece of mass independently, so that all objects undergo the same acceleration when acted on only by gravity. In a vacuum, an anvil and a feather fall with the same acceleration. Our ordinary observations to the contrary result from different effects of air resistance. For gravity, the symbol g is normally used for this acceleration, and the force exerted by the gravitational field is called the weight. On the Earth's surface, g is about 9.8 m/s², or 32.2 ft/s². There aren't enough letters in English and Greek to use a unique letter for everything of interest, so W is usually used for weight as well as work unless work and weight are used in the same equations. Thus the weight of the mass m is given by $W=mg$, in accordance with Newton's second law of motion.

If an object is moving in a force field such as gravity, with no other forces acting on it, the sum of its kinetic energy and its potential energy is constant. This is the simplest form of the law of conservation of energy.

Thermal energy is the energy of molecular motion in a material. Even in a solid, atoms are not at rest, but jostle against their neighbors; we experience this jostling as temperature—the more energetic this jostling is, the hotter the material. Thermal energy is actually just molecular kinetic energy, but since the material as a whole is at rest in its own reference frame, thermal energy is accounted for separately from the kinetic energy of the bulk material.

In everyday conversation, we often refer to thermal energy as heat, but in the precise language of thermodynamics, we use heat in a different sense: Heat is energy transferred across a system boundary because of differences in temperature between the system and its surroundings. Heat is not a form of energy in itself, but takes whatever form in which energy is transferred across the system boundary. Heat is transferred by the processes of conduction and radiation. Conduction transfers energy through collisions between molecules. Radiation, which we will study in some detail in Chapter 4, is the direct transfer of energy across space. When heat is conducted from the boundary of a solid body into a fluid, flow of the fluid can increase the effectiveness of the heat transfer: The flowing fluid carries away the heat as soon as the heat is received in the fluid, and the fluid at each point is continuously replaced by cooler fluid from upstream. So a special name is given to heat transfer into a fluid: convection. (In some fields, the term advection is used for general heat transfer by a fluid, and the term convection is reserved for the type of advection in which buoyant effects caused by heating the fluid are important. In other fields, this type of advection is called natural convection. The terms are not rigorously distinguished.[2, 3])

The SI unit of energy is the joule (J), defined as one newton-meter (N-m). This definition reflects the equivalence of work and energy. A joule is also equal to 1 kg-m^2/s^2. The amount of energy delivered in a unit time is called power. The SI unit of power is the watt (W), defined as one joule per second.

The rate of heat transfer by conduction within a material is governed by Fourier's law of heat conduction, which is expressed mathematically as

$$q = -k\nabla T \, ,$$

where q is called the heat flux, which has units of energy per unit area per unit time (e.g., joules per square meter per second), k is the thermal conductivity, which is a material property in units of energy per unit length per unit time per degree, and ∇T is the temperature gradient, which is a vector in the direction of most rapid temperature increase, with magnitude equal to the increase in temperature per unit length in that direction. The gradient operator, denoted by ∇, is discussed further in Chapter 6. The symbol is an upside-down upper-case Greek letter delta, and it is pronounced "del." This equation is a vector equation, which says that heat flows in the direction of the most abrupt temperature change. The minus sign shows that heat flows from hot places to cold places.

When heat flows from a solid body with surface temperature T_w into a fluid whose temperature away from the surface is T_f, the heat flux from the surface into the fluid is given by Newton's law of cooling,

$$q = h\left(T_w - T_f\right),$$

where h is called the heat transfer coefficient, which depends on the wall and fluid materials and the flow characteristics. As you would expect, the heat flux depends on the difference between the surface temperature and the wall temperature. The heat transfer coefficient has units of energy per unit area per unit time per degree.

Radiant heat transfer occurs primarily by electromagnetic radiation. Electromagnetic radiation is explained in more detail in Chapter 4, but for now all you need to know is that an entire family of electromagnetic waves exists, from long-wavelength radio waves through medium-wavelength infrared, visible, and ultraviolet light, to short-wavelength X-rays and ultra-short-wavelength gamma rays. All these waves are fundamentally the same, and they form a continuous spectrum from the longest-wavelength waves to the shortest-wavelength waves.

Every object emits electromagnetic radiation in an amount dependent on its temperature, with a distribution over the electromagnetic spectrum that also depends on its temperature. The energy flux (energy per unit area per unit time, e.g., W/m²) from the object's surface is given by the Stefan-Boltzmann law,[4]

$$\mathcal{E}(T) = \epsilon \sigma T^4 \, ,$$

where ϵ (the lower-case Greek letter epsilon) is a dimensionless property of the material surface called the emissivity or emittance, σ (the lower-case Greek letter sigma) is a fundamental constant of nature called the Stefan-Boltzmann constant, 5.6697×10^{-8} W/m²-K⁴, and T is the absolute temperature. The symbol K denotes the kelvin, the SI unit of temperature. One kelvin is equal to one degree Celsius, but the kelvin scale is an absolute temperature scale. Zero K is absolute zero, or -273.15 °C (-459.67 °F). The Stefan-Boltzmann law is a steep relationship. For example, an increase of 10% in absolute temperature, which is about 50 °F at normal environmental temperatures, will increase the emitted heat flux by almost 50% ($1.1^4 = 1.46$). The distribution of the energy flux over the electromagnetic spectrum for a so-called "black body"—i.e., one that absorbs all the radiation incident on it and reflects none—is given by

$$\mathcal{E}_\lambda(T) = \frac{2\pi h c^2}{\lambda^5 [\exp\left(\frac{hc}{kT\lambda}\right) - 1]} \, ,$$

where h is another fundamental constant of nature called Planck's constant, 6.6256×10^{-34} J-s, c is the speed of light, λ (the lower-case Greek letter lambda) is the wavelength, and k is yet another fundamental constant of nature called Boltzmann's constant, 1.3805×10^{-23} J/K. Boltzmann's constant is named for Ludwig von Boltzmann

(1844-1906),[5] a pioneer in an approach to thermodynamics called statistical mechanics. The expression "exp" denotes the exponential function: $\exp(x) = e^x$, where e is a special number (specifically, 2.718281828...) called the base of natural logarithms. The exponential function has the very useful property that it is its own derivative: $d(e^x)/dx = e^x$.[6] In the equations here, we are using the alternate notation "exp" for the exponential function because the exponent $hc/kT\lambda$ is a bit unwieldy to write as a superscript. The symbol e is universally used for the base of natural logarithms, just as π (lower-case Greek letter pi) is universally used for the ratio of a circle's circumference to its diameter. Figure 2.5 shows the blackbody spectrum for several temperatures. The units are selected so that the numbers will not be extremely large or extremely small; a micron is 10^{-6} m—i.e., a millionth of a meter.

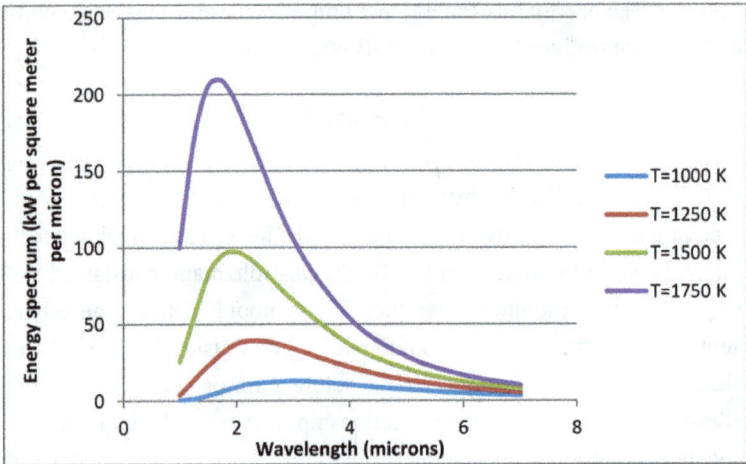

Figure 2.5— E_λ (T), the blackbody spectrum, for several temperatures

Nature is observed to follow accounting rules that we call conservation laws. We now list some that apply to classical physics.

The law of conservation of mass: The mass of an isolated system is constant. An isolated system is one whose boundaries allow nothing to pass through them. This proscription includes energy, because, as we shall see in later chapters, matter and energy are actually just different forms of the same thing. However, in classical physics, the mass associated with heat transfer and radiated energy is very small, and the mass change associated with energy transfer is negligible.

The law of conservation of momentum: When two bodies interact with each other in the absence of any external force, their total momentum remains constant; if their initial momenta are p_1 and p_2 and their final momenta are p_1' and p_2' (said as "p-one prime" and "p-two prime"), then $p_1 + p_2 = p_1' + p_2'$. This law applies even in relativistic interactions and for interactions in which mass is converted to or from energy. The additional momenta imparted to the two bodies by the release of energy are in opposite directions and cancel each other out.

The law of conservation of energy: The total energy content of an isolated system is constant. In classical physics, the energy and mass of a system are accounted separately, because conversions of mass into or from energy do not change either significantly. However, for atomic- or nuclear-scale phenomena, or for bodies moving near the speed of light, such conversions must be taken into account. A more general law combines the laws of conservation of mass and energy: The total mass-energy content of an isolated system is constant.

The zeroth law of thermodynamics: If two bodies are equal in temperature to a third body, then they are equal in temperature to each other. This is very much like the axiom of algebra quoted in Chapter 2, but it deals with a physical phenomenon rather than abstract quantities. The axioms of algebra cannot be proven, but they are a reasonable starting point for the construction of the logical structure of mathematics. The zeroth "law" of thermodynamics cannot be proven either, but it could be disproven if any experiment came up with contrary results (assuming that there was no experimental error or equipment defect). It is called the "zeroth" law because the need for it was only recognized after the first and second laws of thermodynamics were named and well established, but it is a more fundamental principle than the first and second laws.

The first law of thermodynamics: When a system undergoes a change of state, the change in the total energy of the system is equal to the heat transferred into the system minus the work performed by the system. When the conversion of mass into or from energy is included in the change of total energy, this law is completely general. The first law of thermodynamics is just a statement of the law of conservation of energy that is not restricted to an isolated system.

The second law of thermodynamics: It is impossible to construct a device that will operate in a cycle and produce no effect other than the raising of a weight and the exchange of heat with a single "reservoir" (i.e., a body whose temperature is constant and which is so large that transfers of heat into and out of it cannot change its temperature). Another way to state the second law, which can be shown to be equivalent to the preceding way, is that it is impossible to construct a device that operates in a cycle and produces no other effect than the transfer of heat from a cooler body to a hotter body. The first statement is called the Kelvin-Planck statement, and the second statement is called the

Clausius statement.[7] Like all other "laws" of physics, this law cannot be proven, it could only be disproven. But it has withstood all experimental challenges so far.

So, in practical terms, what does the second law of thermodynamics mean? In our field of nuclear technology, it means that you cannot build a "heat engine"—for example, a power plant—that takes heat from a source, such as a coal fire or a nuclear reactor core, and converts it completely into work. You always have to reject some heat to some kind of heat sink, such as a river, a lake, or the atmosphere. The efficiency of a heat engine is dependent on the temperatures of the heat source and heat sink. The hotter the source and the colder the sink, the higher the possible efficiency. It also means that the old feel-good saying that you can do anything you put your mind to is not true. There are limitations on what you can do. You can't fool Mother Nature. In fact, you won't make any money trying to defy any of the "laws" of physics, but the second law has the unique quality that it is stated in terms of impossibility.

In later chapters, these conservation laws are applied to fluids flowing in channels. Some basic equations for this kind of fluid flow are worth noting here. First is the "continuity equation," which expresses the law of conservation of mass for fluid flow in a channel. This equation is

$$\rho A v = constant,$$

where ρ (lower-case Greek rho) is the fluid density, A is the cross-sectional area of the channel, and v is the flow speed. For incompressible flow (ρ = constant), this equation says that in places where the channel is constricted, the fluid has to flow faster. The second equation of interest is called Bernoulli's equation,[8] which is derived from the law of conservation of energy. For incompressible flows, Bernoulli's equation is

$$p + \frac{1}{2}\rho v^2 + \rho g z = constant,$$

where p is the pressure, g is the acceleration caused by gravity, and z is the elevation ($\rho g z$ is the density of potential energy). This equation says that when a fluid is accelerated by a constriction in its flow channel, its pressure falls. This phenomenon is exploited in constricted channels called Venturi tubes, where the reduced pressure is called the Venturi effect.[9] The phenomenon is also the reason airplanes can fly; the fluid flow over an airplane wing is not bounded by channel walls, but the flow between fluid streamlines obeys the same principle. The airflow is accelerated over the top of the wing, where its pressure is reduced. The difference in pressure below and above the wing creates lift. Actually, the shape of a wing serves to induce a circulation around the wing that greatly enhances the acceleration of the flow over the wing, so that a great deal of lift is generated by a fairly small wing area.

Finally, I want to make a few remarks about scale and units. Physicists deal with length scales that range from nuclear particles (about 10^{-15} m) to the diameter of the observable universe (about 10^{27} m), and with time scales that range from the transit time of a gluon within a nucleon[d] (about 10^{-24} s) to the age of the universe (about 13.75 billion years, or 4×10^{17} s).[10, 11] The SI system of units is sometimes awkward to use for quantities of such large and small magnitudes, although prefixes on the unit symbol can help. For example, the prefix n- for nano, meaning 10^{-9}, and the prefix G-, for giga-, meaning 10^9, help us avoid writing the exponents all the time and simplify discussion. There is a prefix for every third order of magnitude from 10^{-24} to 10^{24} along with every order of magnitude from 10^{-3} to 10^3.[12] However, many more convenient units are in common use, such as the light year for astronomical distances. The light year is defined as the distance light travels in a year, which turns out to be 9.46×10^{15} m (9.46 petameters, or Pm). We will often encounter such practical units in the following chapters.

The foregoing discussion of classical physics doesn't even address all the basic phenomena. But now we have enough ideas from classical physics to go on to discuss atomic and nuclear structure.

---

[d] Nucleons, quarks, and gluons are discussed in the next chapter.

CHAPTER 3
ATOMIC AND NUCLEAR STRUCTURE

Ordinary matter is made of atoms and molecules. Molecules are combinations of atoms bound together; the molecular composition of a substance determines its chemical identity. Atoms are the chemical building blocks of molecules, and they come in 91 naturally occurring "elements" plus 26 man-made elements (these elements are tabulated in the Periodic Table introduced later in this chapter).[a] Atoms are the smallest units of matter that maintain their identity as chemical substances. We now know that atoms are composed of central nuclei with positive electric charge and peripheral electrons with negative electric charge. Atoms are extremely tiny. The diameter of an atom is about 10^{-10} meters, or 10^{-7} millimeters—if you line them up, you can fit about 250 million of them in an inch. Because they are so small, they cannot be seen directly with even the most powerful optical microscopes. This cannot be remedied by building better optical microscopes. It is a fundamental limitation, imposed because the wavelengths[b] of visible light (4000-7000 angstroms (Å), where an angstrom is defined as 10^{-10} meters) are much larger than the diameters of atoms (about 1 Å), so that visible light cannot focus on one atom at a time. What we know about atoms and nuclei has been inferred from a very large number of indirect observations.

I do not want to digress too far from my basic purpose of explaining nuclear

---

[a] There are 26 man-made elements as of this writing. New ones may be created in the future. Most of the man-made elements are heavier than uranium, element number 92; one lighter element, technetium, does not occur in nature (on Earth, anyway) and must be manufactured. Technetium is important in nuclear medicine, as discussed in Chapter 10.

[b] The concept of wavelength for both light (electromagnetic radiation) and matter is introduced a little further on in this chapter.

technology, but a description of a few of these observations will help clarify how inferences are made about the nature of atoms and nuclei.

3.1 SOME EXAMPLES OF EXPERIMENTS

In 1909, Ernest Rutherford did a now-famous experiment in England in which he shot alpha particles (now known to be the same as helium nuclei, with two protons and two neutrons) at thin gold foils. The amount of deflection of the alpha particles in their transit through a foil was registered by noting the location of the impact of each alpha particle on a zinc sulfide screen behind the foil; the screen would light up where it was hit by an alpha particle. Most of the alpha particles went right through the foils with very little deflection, but once in a while an alpha particle was deflected sharply. This led Rutherford to conclude that most of the mass of physical materials, even dense solids, is concentrated in nuclei that are very small compared to the total size of the atoms.

Also in 1909, the American physicists Robert Millikan and Harvey Fletcher did another now-famous experiment to determine the numerical value of the electric charge of an electron. They found that when small droplets of oil are formed in a spray nozzle, friction with the nozzle induces small electrostatic charges on some of them, which were hypothesized to be caused by the removal of a small number of electrons. The charge on a droplet can be found by suspending the droplet between horizontal metal plates connected to a voltage source like a battery. The voltage is adjusted to make the gravitational and electrostatic forces on the droplet balance, and the charge can then be calculated from the weight of the droplet and the strength of the electric field between the plates (which is determined by the voltage). The charge on a single electron is the largest number of which the charge on every oil droplet is a whole-number ("integer") multiple. Millikan and Fletcher measured the charge to be 1.592×10^{-19} coulombs, which has since been refined to 1.602×10^{-19} coulombs (the coulomb is a unit of electric charge based on electric current; it is the charge transported by one ampere of current in one second). The symbol e is sometimes used for the charge of an electron. (This is not related to the use of e for the base of natural logarithms; as noted before, we don't have enough symbols to go around.) Also, sometimes the unit is implicitly assumed: in bookkeeping the charge exchanges in chemical or nuclear reactions, we speak of charges of $+1$, -2, etc., where we really mean $+1$ e, -2 e, etc.

The wavelengths of electrons in an electron beam are much smaller than the wavelengths of visible light, and electron beams can be used to create images of objects on the atomic scale.[1] Many such images can now be found on the Internet.

Figure 3.1 shows an image of atoms on a gold surface. Seen externally in the electron microscope image, the atoms look like little spheres, as one might expect intuitively. While the images look like photographs, one must keep in mind that they were taken with electrons, and therefore they are still indirect observations— they cannot be seen by looking directly through an apparatus with the human eye.

These are very simple examples of the kinds of experiments physicists have done to determine the nature of atoms and their nuclei. As the information sought has become more difficult to obtain, the experiments have become more complicated and expensive, and drawing inferences from the experiments has required more subtle thinking. But by application of the peer review process described in Chapter 1, scientists over the course of many years have reached agreement on many features of the atom and the nucleus.

Figure 3.1—Experimental image of gold atoms. Created by Erwin Rossen, Eindhoven University of Technology, The Netherlands, and released into the public domain by the author. Accessed via Wikimedia Commons[1]

3.2 QUANTUM PHYSICS AND ATOMIC STRUCTURE

The atom consists of a very small and dense nucleus in the middle, as discovered by Rutherford, and one or more electrons around the nucleus. Nuclear structure is discussed in Section 3.4, but for now we need to know that the nucleus consists of positively charged protons and neutrally charged neutrons. (The assignment of the terms "positive" and "negative" was apparently arbitrary, but it is universally agreed on.) The term "nucleon" refers to either a proton or a neutron. The number of protons in the nucleus determines the chemical element to which the atom belongs, and the number of neutrons distinguishes the different isotopes of an element. For example, all isotopes of uranium have 92 protons, but they have different numbers of neutrons in their nuclei. Uranium-238, the most common isotope, has 238-92=146 neutrons. All isotopes of an element behave the same chemically. The electrons around the nucleus are negatively charged. The number of electrons around the nucleus is equal to the number of protons in the nucleus, so that the overall charge of an atom is neutral, unless the atom is missing one or more electrons or has picked up extra electrons, in which case it is said to be ionized.

Sometimes a single isotope of an element occurs in two different energy states. What that means is explained below. For now, we only need to know that the term "nuclide" is used to denote a unique nuclear species, with specific numbers of protons and neutrons and a specific energy state. In isotopes that occur as two different nuclides, one of them is found to be in a state that we call metastable. There is a compact nomenclature for nuclides, such as ^{99m}Tc for technetium-99m. Here, the "m" denotes a metastable nuclide.

All known nuclides are summarized in a chart called the Chart of the Nuclides, or the Segré Chart, after the Italian physicist Emilio Segré. This chart presents numerous useful data, such as mass, half-life, decay types and energies, and percent abundance, about each nuclide. A legible version of the Segré chart takes up a lot of space, so it is not reproduced in this book. However, it can be found in interactive form on the Internet,[2] or it can be ordered in wall chart or booklet form.[3] The more familiar periodic table has one box per element, so it takes up a lot less space. The periodic table of the chemical elements is shown in Figure 3.2.[4]

The distribution of the relatively light electrons in the relatively large space around the nucleus is most rigorously described by the concepts of quantum physics. The term "quantum" arises from the hypothesis by the German physicist Max Planck in 1900 that radiation is emitted or absorbed in discrete packets, or "quanta," of energy.[5] In 1905, Einstein explained the photoelectric effect[6] by theorizing that light itself occurs in such quanta. These quanta of light are called "photons." This discovery led to the resolution of seemingly contradictory observations about the nature of light: In some experiments light behaves like waves, and in other experiments light behaves like particles. The resolution of the paradox lay in the realization that light possesses a wave-particle duality wherein a quantum of light exhibits a wavelength inversely proportional to the energy of the quantum, so that it is both a particle and a wave at the same time. But this duality does not apply only to light; it applies to everything— electrons, baseballs, pickup trucks, and celestial bodies. Each of these objects has a wavelength that depends on its kinetic energy, and electrons of the proper kinetic energy have wavelengths on the atomic scale. That is why electrons in a beam, which have a much higher energy, and thus a much shorter wavelength, than quanta of visible light, can be used to take "pictures" of individual atoms.

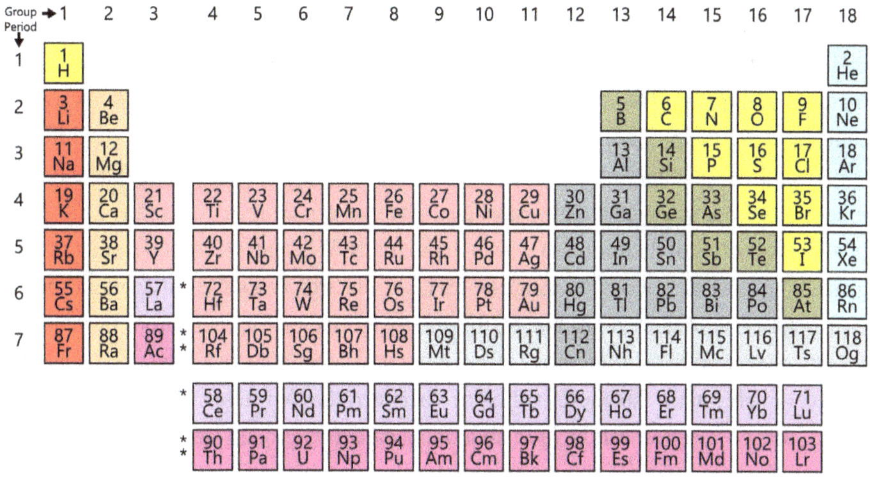

Figure 3.2—The periodic table of the elements (This figure was taken from the Wikimedia Commons and is in the public domain.)

The wavelengths of massive objects like pickup trucks are so short that there is no practical value in applying quantum physics to such objects. But for particles on the atomic or nuclear scale, quantum physics is essential to understanding. The fundamental equation governing quantum physics is called Schrödinger's Equation,[7] discovered by the Austrian physicist Erwin Schrödinger in 1927. This equation is presented below simply to illustrate the sort of mathematical expressions physicists deal with; if you have not studied mathematics, don't try to make sense of it. Schrödinger's equation describes the variation in time and space of an object's wavefunction ψ; the square of the absolute value of the wavefunction is interpreted as the probability that the object will be located in a particular place at a particular time. For an electron bound to the nucleus of an atom, the wavefunction describes a "probability cloud," which occupies the region in which the electron can be found.

$$i\frac{h}{2\pi}\frac{\partial}{\partial t}\Psi(r,t) = -\frac{h^2}{8\pi^2 m}\nabla^2\Psi(r,t) + V(r)\Psi(r,t)$$

The time-dependent three-dimensional Schrödinger Equation for a single particle

For atoms with more than one electron, another principle of quantum physics applies: the Pauli Exclusion Principle, proposed by the Austrian physicist Wolfgang Pauli in 1925. This principle states that no two identical fermions (a class of particles that includes electrons, protons, and neutrons) can occupy the same quantum state at the same time. So the probability clouds of the different electrons in an atom are different.

Also, the probability cloud of an electron depends on its energy state. The electrons in atoms are not stationary. They have energies that can be regarded as the kinetic and potential energies of their motion within their clouds. When an electron changes its energy state, its probability cloud changes with it. An electron can have only certain specific energy levels within an atom; that is, the energy of an electron in an atom is quantized. When an electron in an atom moves from a higher energy state to a lower one, it emits a photon which carries the energy difference between the two states. The lowest possible energy that an electron can have in an atom is called its ground state, and when all the electrons in an atom are in their ground states, the atom is also in its ground state.

Figure 3.3 shows the probability clouds for electrons in the first few energy levels in the hydrogen atom.[8] These images are not photographs; they are computer-generated. They are color-coded cross sections of the probability density (black=zero density, white=highest density). The letters and numbers on the rows and columns indicate "quantum numbers" that characterize the different energy states. The quantum numbers arise from the solution of the Schrödinger Equation. The image for column s and row 1 is the ground state. The probability clouds in three-dimensional space are obtained by rotating the ones shown here around their vertical axes.

Is there any hope of ever being able to track the motion of the electron within this probability cloud? No: According to Heisenberg's Uncertainty Principle (formulated by the German physicist Werner Heisenberg in 1926), which is actually equivalent to

Schrödinger's Equation, one cannot know the position and the momentum of a particle precisely at the same time. This uncertainty is expressed mathematically as $\Delta p \Delta x \geq h/4\pi$, where $h=6.626\times10^{-34}$ J-s. This number is called Planck's constant; it is one of the fundamental constants of nature, or fixed numbers that govern fundamental physical processes. The quantities Δp and Δx are the uncertainties in the momentum p and the position x, respectively. The more precisely you know an object's position, the less precisely you know its momentum, so you cannot even know which probability cloud it occupies if you know its momentum precisely. Thus, the limitation is fundamental, and the probability clouds are the most we can hope to know.[c]

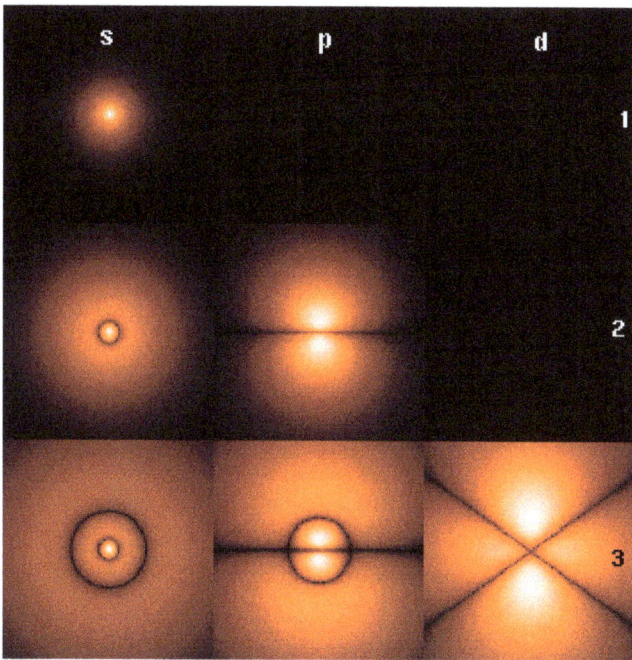

Figure 3.3—Probability clouds for electrons in the first few energy levels in the hydrogen atom. This image was obtained from the Wikimedia Commons and is reproduced under the terms of the GNU Free Documentation License, Version 1.2 or later
(http://commons.wikimedia.org/wiki/Commons:GNU_Free_Documentation_License).

---

[c] I am indebted to my friend Dr. Charles A. Wemple for the following joke: A cop stops Heisenberg and asks him, "Sir, do you know how fast you were driving?" Heisenberg says, "No, but I know exactly where I am." The cop says, "Sir, you were travelling at exactly 87 miles per hour." Heisenberg says, "Great! Now I'm lost!"

3.3 MODELS OF ATOMIC STRUCTURE

The rigorous description of atomic structure given in the previous subsection is challenging to understand and work with. Therefore, simpler ways to visualize atoms are quite useful.

The earliest model of the atom that accounted for the quantization of the energy of the electrons was the Bohr planetary model, proposed by the Danish physicist Niels Bohr in 1913. In this model, the electrons are regarded as revolving around the nucleus in circular orbits like planets revolving around the sun. This is the model implicit in commonly seen iconic drawings of atoms to symbolize nuclear energy. Figure 3.4 shows the planetary model for hydrogen.

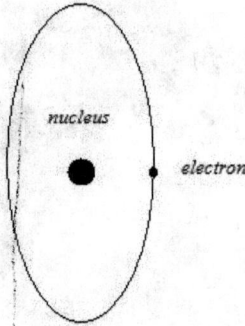

nucleus

electron

Figure 3.4—Planetary model for hydrogen

A more sophisticated model, which approximately represents the probability clouds, is the shell model. The clouds are simplified into spherical or lobed shells on which the electrons are located. The number of electrons on any shell is limited by the Pauli Exclusion Principle, but at least two electrons may be located on each shell, because they can have opposite spins, and therefore are not in exactly the same quantum state. Figure 3.5 shows a shell model representation of sodium.[9]

11: **Sodium**   **2,8,1**

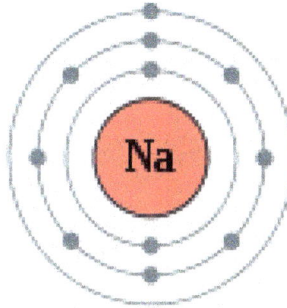

Example of a sodium electron shell model

Figure 3.5—Shell model for Sodium. This figure was taken from the Wikimedia Commons and is in the public domain subject to the Creative Commons Attribution/Share-Alike License.[10]

3.4 NUCLEAR STRUCTURE

Nuclear structure obeys the same principles of quantum physics as the electronic structure in the atom, so the nucleons occupy probability clouds just like electrons do. The planetary model of the atom is not useful for the nucleus, but the shell model is. The simplest way to visualize the nucleus is just as a clump of protons and neutrons, as shown in Figure 3.6. Another useful model of the nucleus is the liquid-drop model, mentioned in Chapter 1. A drop of liquid in a vacuum assumes a spherical shape because surface tension pulls it into the shape with minimum potential energy, which is a sphere. But this same surface tension in a large droplet that is disturbed can pinch the droplet in the middle and divide it into two droplets, as shown in Figure 3.7. The nuclei of massive atoms like uranium can undergo fission reactions, in which they divide into two smaller droplets similarly. The equations that describe the division of a liquid droplet can also be used to describe nuclear fission, so the liquid drop model has quantitative value as well as usefulness for visualization.

Figure 3.6—Model of the nucleus as a clump of nucleons

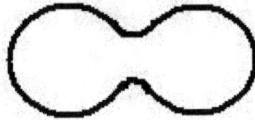

Figure 3.7—Liquid-drop model of atom undergoing fission

The neutrons and protons in the nucleus are held together by a mutual attractive force called the strong nuclear force.[11] This is one of the four known fundamental forces of nature, along with gravity, electromagnetism, and the weak nuclear force. The weak nuclear force is active in beta-decay, which we discuss at some length in Chapter 4. The protons are also subject to their mutual electrostatic repulsion. The strong nuclear force is much stronger than electrostatic repulsion for protons close to each other, but it only acts over a very short range, whereas the electrostatic force is a long-range force that acts among all the protons. Thus, the larger the nucleus, the more likely it is that the electrostatic force can cause some pieces to break off, either in nuclear fission or in alpha decay, both of which are discussed in later chapters.

Electrons are believed not to be further divisible, but protons and neutrons are believed to comprise combinations of more fundamental particles called "quarks."[12] Quarks are of several types, and the whole terminology for them, including the name "quark" itself, is an exercise in physicists' sense of humor. The name "quark" was chosen by the American physicist Murray Gell-Mann in allusion to James Joyce's *Finnegan's Wake* ("Three quarks for Muster Mark!"), and also because it sounded to Gell-Mann like the quacking of ducks. Quarks are classified according to properties dubbed color and flavor; there are six flavors, called up, down, charm, strange, top, and bottom. I wonder what a strange quark tastes like. I suppose it probably tastes like chicken!

There was some sense in Gell-Mann's choice of the word "quark," as protons and neutrons are composed of three quarks bound together. Now, this is where things get really interesting! The masses of the three quarks add up to much less than the mass of the nucleon. There are also an indeterminate number of quark-antiquark pairs that continually pop into and out of existence,[13] but the extra mass in the nucleon actually comes mostly from the zero-rest-mass particles called "gluons"[14] (so called because they "glue" the quarks together) that carry the interactions among the quarks. That is not a misprint! Particles that have no mass contain most of the mass of the nucleus!

The key to understanding this lies in the most famous equation in all of physics: Einstein's $E=mc^2$, where E is energy, m is mass, and c is the speed of light in a vacuum. This equation says that matter and energy are actually the same thing in different forms. When energy in any form is released from a source of any kind, or added to an object in any way, there is a change in the mass of the source or the object. The mass of an object is not a constant quantity as we are used to experiencing it. An object's mass as measured by an observer depends on the relative velocity of the object and the observer. The relationship is described by another of Einstein's equations,

$$m = m_o/\sqrt{1 - v^2/c^2} \, .$$

In this equation, m is the mass of the object as measured by an observer with respect to whom the object is moving with speed v, and m_o is the object's rest mass, or the mass measured by another observer moving with the object. Since m is the mass measured by an observer in motion relative to the object, m is also called the relativistic mass. This equation shows you that if v gets greater and greater, the mass gets greater and greater far out of proportion to the speed. If the speed approaches the speed of light, the mass approaches infinity. If the speed is zero, the mass is just the rest mass. Thus, if you are trying to accelerate a particle, say a proton, in an accelerator, you have to add more and more energy to increase its speed by equal amounts as its speed gets greater and greater. Nevertheless, it is still an ordinary proton, still with its original rest mass in its own reference frame. Since it is impossible to add an infinite amount of energy to an object, the only things that can travel at c, the speed at which light travels in a vacuum, are objects with zero rest mass.

Some particles, including photons and gluons, actually do have no rest mass (remember the wave-particle duality, so that even photons and gluons can be considered as particles). Photons are never at rest in any reference frame; they always travel at the speed of light in any reference frame. Let's explore that idea for a moment. If you and a friend are traveling in a railroad car and decide to play catch with a baseball, you throw the ball back and forth with some speed, say 30 miles per hour, with respect to yourselves. If the train is traveling at 60 mph, then, with respect to an observer standing on the ground, the ball is moving at either 60+30 mph = 90 mph or 60-30 mph = 30 mph, depending on which way the ball is moving. However, if you shine flashlights at each other, every-body—you, your friend, and the observer on the ground—measures the same speed for the light beams with respect to himself. This extremely counterintuitive idea is called the postulate of the constancy of the speed of light, or the second postulate of Einstein's special theory of relativity. Like any hypothesis of physics, it is subject to refutation by experiment, but all relevant experiments to date have confirmed it.

Even though they have no rest mass, photons carry energy. Similarly, gluons can-not be brought to rest and still remain gluons, but their energy is equivalent to mass and provides most of the mass of nucleons. So, when you weigh yourself on a scale, you are mostly weighing the relativistic mass of the gluons in the nucleons of the atoms in your body, even though the gluons actually have no rest mass themselves! According to the equation above for relativistic mass, if a particle has no rest mass but nonzero relativistic mass, its speed must be equal to c (zero divided by zero is math-ematically indeterminate, while zero divided by anything else is zero; the indetermi-nate value is identified by additional information). The gluons cross the very short distance between quarks at the speed of light, so the transit time across this distance is very brief. Quarks exchange gluons about 10^{24} times per second!

The exchange of gluons between quarks is what holds the quarks together in a nucleon. But the effect of this exchange is not confined within the nucleon. There is a residuum of the quark-quark gluon exchange that spills out of the nucleon. This residuum is what creates the strong nuclear force that binds nucleons together to form atomic nuclei.[11, d]

The (relativistic) mass of the quark/gluon combination in a nucleon is not the same for every state of the nucleon, and this is reflected in the rest mass of the nucle-on. The rest mass of a nucleon when it is free from other nucleons is greater than the mass of the same nucleon when it joins with other nucleons in the nucleus of an atom. The quark/gluon combinations actually give up mass when the nucleons they com-pose join with others. The lost mass is called the mass defect or the binding energy

---

[d] The terminology for these phenomena is not uniformly used among particle physicists. Some, like the author of a Lawrence Berkeley Laboratory article on the subject (http://aether.lbl.gov/elements/stellar/strong/strong.html), use the term "strong nuclear force" to denote the force be-tween nucleons. I follow that practice here. Others, like the author of the Wikipedia article cited in Reference 11, use the terms "residual strong force" and "nuclear force" for the force between nucleons, and reserve the terms "strong nuclear force," "strong interaction," and "strong force" for the force between quarks.

(remember, mass and energy are different manifestations of the same entity). When, for example, a neutron is absorbed by a nucleus, this binding energy is normally emitted in the form of a gamma ray (see Chapter 4).

A quark of one flavor can change into a quark of another flavor by emission or absorption of a particle called a W boson. This process leads to the phenomenon of beta decay, a form of radioactivity. And this observation leads us to Chapter 4, which discusses all forms of radiation.

CHAPTER 4
RADIATION

Giving a simple, comprehensive definition of radiation is a bit tricky, because the term "radiation" subsumes more than one phenomenon. But we shall use the term to mean energy transmitted across space by changes in atoms or nuclei, or the process of emitting such energy.

There are two different basic forms of such energy: energy in particles of zero rest mass, and energy in particles of nonzero rest mass. The first kind of energy is electromagnetic radiation, and the second includes alpha and beta particles and neutrons, along with some more esoteric particles that need not concern us. Most of the radiation that concerns us is emitted from nuclei that eject the radiation as they change in some way. We will consider each type of radiation specifically in its own section, but first we will discuss properties that all types of nuclear radiation have in common.

4.1 RADIOACTIVITY

Radiation emitted from nuclei was given the names alpha (the Greek letter α), beta (the Greek letter β), and gamma (the Greek letter γ) radiation, or rays, when early researchers discovered them and could tell that they were different from each other but didn't know anything about the nature of any of them. We now know what all of them are, but the original names are still used.

In each of these types of radiation, a particle is emitted from a nucleus that is for some reason predisposed to emit it. Such nuclei are called radioactive, and the process of emitting the particles is called radioactive decay. As noted in Chapter 3, the heavier the nucleus, the more likely it is to fall apart. Heavy nuclei such as uranium are predisposed to emit alpha particles, which consist of two protons and two neutrons. These are the same as ordinary helium nuclei.

Another reason for radioactivity is an unfavorable neutron-to-proton ratio. It is a fact of nuclear physics that the farther up the periodic table a nuclide lies in the progression from hydrogen to uranium and beyond, the higher its neutron-to-proton ratio will be for stable (nonradioactive) isotopes. In nuclear fission, a heavy nucleus like uranium is divided into two lighter nuclei, called fission fragments, plus two or three free neutrons. The fission fragments will have the same ratio of neutrons to protons, on average, that the uranium nucleus had. But for the fission fragments, this ratio is too high for stability, and the nuclei will reduce the ratio by converting some of its neutrons into protons and electrons. The electrons, which are ejected from the fission fragment nuclei, are called beta particles. The details of this process are discussed in Section 4.4 below.

Sometimes a nucleon in a nucleus will occupy an energy state above its ground state. This often happens when a nucleus absorbs a free neutron. At some time after it is absorbed, the newly added neutron will drop down into its ground state, and in the transition a photon will be emitted to take away the energy difference. This photon is called a gamma ray. Gamma rays are also emitted in association with alpha and beta decay when the decay event puts the daughter nucleus in an excited state.

A key feature of all these types of radiation is that you cannot predict when a particular nucleus of a radioactive substance will decay. If you watch two apparently identical nuclei of the same radioactive nuclide, one of them may decay as soon as you start to watch, and the other may never decay. Usually, the situation is somewhere in between those extremes, but you can't tell which will decay sooner and which will decay later.

However, different radioactive substances do not decay at the same rate. Some, like fission products, which are far from nuclear stability, decay very rapidly. Others, like uranium-238, decay very slowly. We can't predict the exact time of decay of any particular nucleus, but we can measure the probability that a nucleus of some particular nuclide will decay in any given time period. This probability is expressed by the "half-life"—the time interval that will pass from any arbitrary starting time until exactly one-half of the nuclei present at that starting time have decayed. So fission products tend to have very short half-lives (often less than a second), while U-238 has a half-life of more than a billion years.

If there are N_o (spoken as "N-naught") nuclei in the original sample at a time we arbitrarily denote as zero, and $N(t)$ (spoken as "N of t") at some later time t, then it can be shown that

$$N(t) = N_o \, e^{-\lambda t},$$

where λ (the lower-case Greek letter lambda) is called the decay constant. The entire expression is said as "N of t equals N-naught e to the minus lambda t." The decay constant is related to the half-life $t_{1/2}$ by

$$\lambda = 0.693/t_{1/2} \, .$$

The symbol e is the base of natural logarithms, 2.718281828... This quantity appears over and over in equations describing nature, as does the function $e^{\pm \lambda t}$, where λ and t may take on meanings other than a temporal decay constant and time, respectively. It is called the exponential function, and it describes all sorts of phenomena in which a quantity doubles or decreases by half in some specific constant increment in t beginning at any arbitrary starting time. Figure 4.1 shows exponential decay for radium-226, which has a half-life of 1600 years.

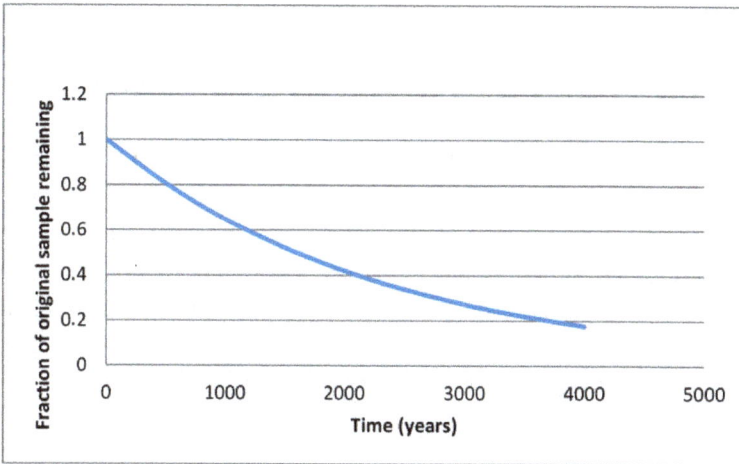

Figure 4.1—Decay of radium-226

One aspect of radioactive decay one should keep in mind is that the more radioactive a substance is (for a given number of nuclei in the original sample), the shorter its half-life and the more quickly it decays. The product of decay, called the daughter nuclide, may or may not itself be radioactive. Eventually, however, the end result of a radioactive decay chain is stable, or nonradioactive.

Also, the same nuclide may have more than one radioactive decay mode. For such nuclides, the half-life is the time it takes for the combination of decay modes to produce decay by one-half.

The SI unit for activity is the becquerel (Bq), defined as one radioactive disintegration per second. This is a very tiny amount of radioactivity, so levels of radioactivity encountered in practice give huge numbers when expressed in Bq. For this reason, the activity in a sample of a radionuclide is often measured in curies (Ci). One Ci is 3.7×10^{10} radioactive disintegrations per second, or 3.7×10^{10} Bq; it is the activity of a gram of radium-226, which the Curies studied extensively.[1] The activity of a sample depends on the half-life and the quantity of the radionuclide in the sample. As an illustration, the activity in a kilogram of U-238 is 3.36×10^{-4} Ci. Because of its long half-life (4.468 billion years), U-238 is not very radioactive.

The small size of the becquerel relative to significant levels of radioactivity, and the resulting large numbers when such levels are expressed in Bq, can lead to unwarranted alarm when those numbers are reported out of context. For example, antinuclear activists like to cite releases of radioactivity in Bq because 12.4 million Bq sounds much more frightening than 3.36×10^{-4} Ci, even though they represent the same thing.

4.2 ELECTROMAGNETIC RADIATION

Electrically charged particles attract each other if they have opposite charge, and they repel each other if they are either both positively charged or both negatively charged. For one or a collection of charged particles, one may calculate the force that another charged particle would experience if placed anywhere in the region of influence of the original particle or particles. The force that a unit charge would experience at each point in space, accounting for both magnitude and direction, is called the electric field created by the original particle or particles.

In the SI system, electric field has units of newtons per coulomb, where the newton (N), which we first met in Chapter 2, is the basic unit of force and the coulomb is the basic unit of electric charge. Recall that the newton is the force required to accelerate a mass of one kilogram by one meter per second every second ($1(m/s)/s$, or $1 m/s^2$). If a force acts on an object over a distance, it performs work. A force of one newton acting over one meter does work in the amount of one newton-meter (N-m), which is defined as one joule (J), a unit we also met in Chapter 2. We have also met the watt (W): A watt is one joule per second. One coulomb is actually quite a lot of charge: It is the total charge of $6.24150965 \times 10^{18}$ protons. The reason for this large number is that the coulomb is defined in terms of the practical unit of electric current, the ampere: One coulomb is the charge passed by one ampere of current in one second. One newton per coulomb, the unit of electric field, equals one volt per meter, where a volt (V) is defined as one joule per coulomb (1 N/coul=1 N-m/coul-m=1 J/coul-m=1V/m). The volt is a unit for a quantity called electric potential, which is just the potential energy change of a unit charge in an electric field when the charge moves over one unit of distance from one point in the field to another. (Potential energy in an electric field is exactly like potential energy in a gravitational field: If we raise a mass m by one meter on the Earth's surface, where the acceleration g imparted by gravity is 9.8 m/s^2, the potential energy of the mass has increased by the work done against the gravitational force mg; i.e., if m=1kg, the potential energy increases by 1 kg x 9.8 m/s^2 x 1 m = 9.8 kg-m^2/s^2= 9.8 J. Gravitational potential is defined as potential energy per unit mass.)

Charged particles in motion exert additional forces on each other, which we know as magnetism. An electric current in a wire (electrons moving along the wire) exerts magnetic forces on other moving charges, and the magnetic forces exerted by a bar magnet are due to the motion of the electrons in certain kinds of atoms, such as iron. The force exerted by one or a collection of moving charges on another moving charge can be calculated at any point in space from a quantity called the magnetic field, which itself is calculated from the motion of the original moving charge or charges. Like the electric field, the magnetic field has both magnitude and direction.

The electric and magnetic fields are not just convenient mathematical descriptions. They have a physical reality, as shown by the phenomenon of electromagnetic waves. Electromagnetic waves are mutually reinforcing, propagating fluctuations in the electric and magnetic fields in space; in a vacuum, these fluctuations travel at the speed of light. In fact, light is one form of electromagnetic radiation. One of the fundamental equations of electromagnetic theory says that an electric field that changes with time induces a magnetic field, and another equation states that a magnetic field that changes in time induces an electric field. (These two equations and two others are Maxwell's equations, the foundational equations of classical electromagnetic theory, after the Scottish physicist James Clerk Maxwell, ca. 1865. Maxwell's equations are discussed more in Chapter 20 and Appendix II.) So, as a photon moves along, its oscillating electric field induces an oscillating magnetic field, which induces the oscillating electric field, and so on. That's how electromagnetic waves propagate. Because photons carry energy across space, they fit our definition of radiation.

The frequency of oscillation in these fields is usually denoted by the lower-case Greek letter nu (υ), and the wavelength by the lower-case Greek letter lambda (λ). (Since there are not enough letters in the Roman and Greek alphabets to give a unique symbol to each physical entity, λ is commonly used for both wavelengths and radioactive decay constants. If both entities should appear in the same equation, one or the other would have to be given a different symbol in that case.) In a vacuum, the frequency and wavelength are related by the expression

$$\upsilon = c/\lambda \,.$$

The basic unit of frequency in the SI is the hertz (Hz), which is simply one cycle per second.

When electromagnetic waves travel through media, their speed is decreased. The frequency remains the same, so their wavelength also decreases. The energy E contained in one photon of electromagnetic radiation is related to the frequency and wavelength (in a vacuum) by the expression

$$E = h\,\upsilon = hc/\lambda,$$

where h is Planck's constant.

Figure 4.2 shows the complete electromagnetic spectrum.[2, 3] The spectrum is shown on a logarithmic scale, in which each order of magnitude (i.e., each successive factor of 10) has equal length. This depiction avoids cramming most of the information into a very small space at one end or stringing it out over vast distances at the other. Electromagnetic waves are launched by the acceleration of charged particles. At the low-energy, long-wavelength end, radio waves are created by the oscillation of electrons in antennae. Radio waves have wavelengths in a vacuum between 10^8 m (called Extremely Low Frequency waves) and 1 mm (Extremely High Frequency). Microwaves, with wavelengths around 1 cm, fall near the high-energy end of the radio waves. Closer to the middle of the electromagnetic spectrum, thermal radiation[4] is created by the transitions of electrons in atoms from excited states (higher-energy electron shells) to lower energy states, or by transitions in vibrational modes in molecules. The excitation is caused when atoms or molecules bump against each other as a result of their thermal motion. Objects at lower temperatures (up to 200 °C or so) radiate mostly in the infrared range (wavelengths of roughly 10^{-2} mm), but as temperature increases, the dominant wavelengths move into the visible range (about 0.5×10^{-3} mm, or 5000 Å). The temperature in an incandescent light bulb filament is roughly 3000 °C. At higher energy, X-rays, with wavelengths of about 1 Å, or 10^{-10} m, are normally created by deceleration of high-energy electrons (a phenomenon called "Bremsstrahlung," which is German for "braking radiation"). And near the higher-energy end of the spectrum, gamma rays, with wavelengths of about 0.01 Å, are created by the transitions of nucleons in a nucleus from excited states to lower-energy states.

Fundamentally, all electromagnetic radiation is the same, from radio waves to gamma rays; the various types of electromagnetic radiation differ only in energy.

THE ELECTROMAGNETIC SPECTRUM

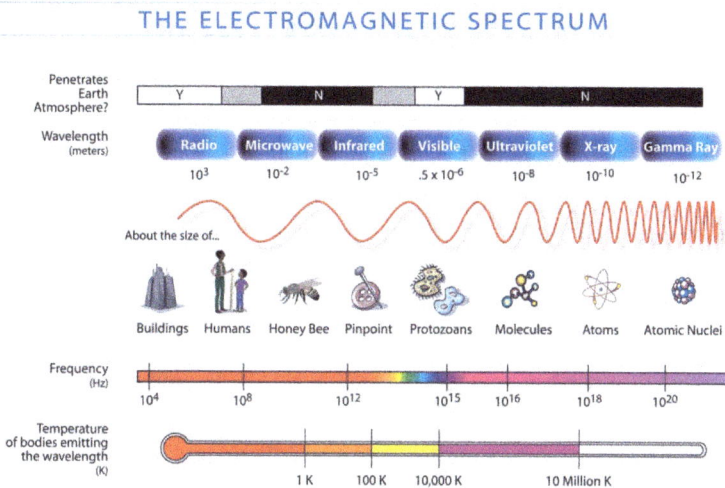

Figure 4.2—The electromagnetic spectrum from 10^4 to 10^{20} Hz. This file was taken from the Wikimedia Commons and is reproduced under the terms of the GNU Free Documentation License, Version 1.2.[5]

However, from a practical standpoint, the differences in energy have great importance. Radio waves can travel around corners by a process called diffraction. Diffraction occurs with all wavelengths (indeed, with any kind of wave, for example waves in water), but at shorter wavelengths the diffraction patterns are too small to allow bending of the waves around corners. For visible light, one example of diffraction is observed as the fuzziness at the edges of shadows.

Electromagnetic waves of different energy also differ in how they penetrate matter. Long-wavelength waves pass readily through materials that do not conduct electricity, which is why you can receive radio broadcasts inside your house with no external antenna. But the electrons in the outermost shells of conducting materials are loosely bound to their atoms, and they can be moved along in an electric current. Long-wavelength waves make these loosely bound electrons in conducting materials oscillate; this oscillation both attenuates the waves and allows a receiving antenna to pick them up. The energy of oscillation of an electron is eventually lost to the atoms by collisions, which increase the thermal energy of the materials. That is, the materials heat up.

Microwaves pass into nonconducting or weakly conducting materials, but they are strongly absorbed by certain kinds of molecules, such as water. This absorption heats materials containing these molecules; the combined effects of penetration and absorption allow us to cook water-containing foods quickly throughout in microwave ovens, rather than waiting for heat to be conducted into the interior from the surface. Besides being absorbed by materials or passing through them, electromagnetic waves can also be reflected, which is why we can see things and why radar works.

Visible light is strongly attenuated by most dense materials, but light does pass into or through some materials, such as glass and water. If you hold up your hand close to a bright light, you can see that some red light has passed through, and you can see the shadows of the bones in your fingers.

Up to this point in the spectrum, absorption of photons in materials merely causes oscillation of loosely bound electrons, if the materials are conductive, or excitation of atomic electrons or molecular vibration modes. However, at the energies of X-rays and gamma rays, photons can knock electrons from their atoms, changing the neutrally charged atoms into ions. Radiation with enough energy to do this is naturally called ionizing radiation. Ions are chemically reactive, and they may form compounds that were not present in the original material. If the material is the DNA of a living cell, the genetic information in the DNA can be altered, producing mutations or triggering cancer. A perspective on this possibility is given in Section 4.6 below.

Photons have extremely tiny energies compared to everyday objects like birds, whose kinetic energies are conveniently measured in joules. For photons and other atomic- or nuclear-scale particles, a more convenient unit called the electron volt (eV) is used. An electron volt is the energy required to increase the electric potential of one electron by one volt. The charge on an electron is -1.6×10^{-19} coulomb, so one electron volt is equal to 1.6×10^{-19} J (i.e., 1 V x 1.6×10^{-19} coul = 1 J/coul x 1.6×10^{-19} coul = 1.6×10^{-19} J). Photons of visible light have energies on the order of 1 eV, X-rays have energies on the order of a thousand eV (denoted by keV, for kilo-electron-volt), and gamma rays have energies on the order of a million eV (MeV, for mega-electron-volt).

4.3 ALPHA RADIATION

Alpha particles (or "rays") are emitted from some heavy radioactive materials such as uranium, and alpha particles were known and named before the structure of the nucleus was understood (remember the Rutherford experiment discussed in Chapter 3). We now know that alpha particles are identical to the nuclei of ordinary helium, consisting of two protons and two neutrons.

The helium nucleus is very tightly bound: It has a much higher value of binding energy per nucleon than other nuclei near it in the Chart of the Nuclides. When an alpha particle is ejected from a heavy nucleus, the binding energy its nucleons give up in becoming an alpha particle, along with the change in binding energy of the remaining nucleons as they rearrange themselves into a new nuclide, is converted to kinetic energy of the alpha particle plus the recoil energy of the remaining nucleus. Alpha particles typically have energies of a few MeV.

Because of their protons, when alpha particles hit a solid material they interact very strongly by the electrostatic force with the atomic electrons in the material. The electrostatic force between two charged objects is a long-range force. It is inversely proportional to the square of the distance between the objects, so the alpha particle interacts with many atomic electrons at the same time. Because of this simultaneous spread-out interaction, the alpha particle slows down in a very short distance. Alpha particles cannot penetrate a sheet of paper or the dead layers of tissue in human skin. (The reason Rutherford used gold foils in his experiment was that gold can be hammered into very thin sheets, which alpha particles can penetrate.) So alpha radiation impinging on you externally cannot hurt you, except at the surface of your eyes, which have no dead layers. However, if an alpha-emitting material is ingested or inhaled, it is more damaging than other kinds of radiation of equal energy. Like gamma rays, alpha particles have enough energy to ionize the atoms they interact with. Alpha particles deposit all their energy in a small volume, unlike gamma rays, which spread it out over a long track that may pass all the way through one's body. Radiation damage distributed over more tissue will heal faster than localized damage from radiation of equal energy. As mentioned above, the practical effects of radiation exposure are discussed in Section 4.6.

4.4 BETA RADIATION

Like alpha radiation, beta radiation was known before atomic structure was understood. Both types of radiation were discovered simultaneously by the French physicist Henri Becquerel in 1896 in experiments with uranium salts. It was Ernest Rutherford who figured out that beta radiation consists of electrons. (Another type of beta radiation emits positrons, which are anti-electrons—particles just like electrons but with positive charge; if an electron and a positron encounter each other, they annihilate each other and turn into a pair of gamma rays. Positrons were suggested theoretically by Paul Dirac in 1928, and not observed definitively until 1932.)[6]

These electrons are emitted from the nucleus, not from the outer electron shells of the atom. The nucleus, as we have seen in Chapter 3, does not contain electrons. In the emission of an electron from a nucleus, a neutron is converted into an electron and a proton, and the electron is ejected from the nucleus with kinetic energy of a few MeV.

From this outcome, one might be tempted to think that a neutron is a bound electron-proton pair, but that would not be correct. In the other type of beta radiation, a proton in the nucleus is converted into a neutron and a positron. Obviously, a particle cannot contain a different particle that contains a particle identical to the original particle in addition to something else.

Remember the discussion of quarks and gluons in Chapter 3. A neutron consists of two down quarks (electric charge -1/3) and an up quark (electric charge +2/3), plus the gluons constantly being exchanged among them. A proton consists of two up quarks and a down quark. In electron emission, one of the down quarks changes to an up quark and an intermediate particle called a W^- boson (said as a "W-minus boson"), which carries the negative unit charge that has to be removed for the converted nucleon to have a charge of +1. This intermediate particle immediately decays into an electron and an electron antineutrino. The process of positron emission is the reverse of electron emission: an up quark changes to a down quark (causing a proton to change into a neutron) and a W^+ boson (i.e., a "W-plus boson"), which quickly decays into a positron and a neutrino. The conversion of up and down quarks into each other is possible because the masses of the quarks and their gluons depend on the binding energy released in the formation of the nucleon, which varies for either type of nucleon depending on the type of nuclide in which it is bound. Some neutrons in nuclei are more massive than protons in the same nuclei, and some protons in nuclei are more massive than neutrons in the same nuclei.

The existence of neutrinos (and antineutrinos) was inferred in 1930 by Wolfgang Pauli from measurements of the properties of beta-decay.[7] The name was coined by the pioneering nuclear reactor physicist Enrico Fermi, an Italian who immigrated to the United States in 1938 to escape the Italian Fascist regime. Fermi quipped that the name "neutrino" means "little Italian neutron."[a] Neutrinos are like neutrons in having

---

[a] I can't find a reference for that. Everything I have found in print says he simply meant "little neutral particles." However, in a course I took in elementary nuclear engineering, the professor made the statement quoted above. I think it's cute, so I'm passing it on, even if it's apocryphal.

no electric charge, and they are little in the sense that they have a very small rest mass. Also, neutrinos have almost no interaction with matter, and their existence was not considered proven until 1956.[8]

The inference of the existence of neutrinos was based on apparent contradictions of the basic theories of physics by beta decay. When numerous nuclei of the same nuclide emit beta particles, the energies of the beta particles fill a continuous spectrum, whereas the quantum theory of the nucleus says that nuclear energy levels are quantized, so that one would expect all the beta particles to have the same energy (or, if there are several beta decay modes that lead to different excited states in the daughter nucleus, there should be a number of discrete beta-decay energies)—i.e., beta particles should be emitted only at one or more particular energies, not in a continuous spectrum. Furthermore, beta decay appears to violate the laws of conservation of momentum and mass-energy. In the reference frame of the original nucleus, the momentum of the nucleus is zero. If the nucleus kicks out a beta particle, the total momentum of the beta particle and the recoiling daughter nucleus should still add up to zero, and their two momenta should be in opposite directions, but that is not the case. Also, the total mass-energy of the beta particle and the daughter nucleus is less than the total mass-energy of the parent nucleus, which apparently violates the law of conservation of mass-energy. The apparent discrepancies can best be resolved by postulating a third particle that carries some of the energy and some of the momentum released in the decay event.

Clearly, theoretical physics had a lot at stake in the search for neutrinos. But neutrinos interact so little with other matter that it took 25 years to detect them. It is estimated that 50 trillion neutrinos from solar radiation pass through a human body every second and they do no damage to it at all!

Neutrinos are now believed to have nonzero, but very small, rest masses, on the order of 0.1 eV expressed in energy equivalent, whereas the electron, itself a small particle, has mass of 0.51 MeV, five million times greater. Neutrinos in beta decay carry energy of a few MeV; in order for a particle of such small rest mass to carry that much energy, it must move at a speed near the speed of light. Experiments appear to show that neutrinos' speeds are less than the speed of light, but so slightly less that we still can't measure the difference with confidence.[b]

Wikipedia (http://en.wikipedia.org/wiki/Neutrino, footnote 3) at least acknowledges that *neutrino* is a pun on the Italian word for neutron, *neutrone*.

[b] A widely reported observation at the Gran Sasso laboratory in Italy, which seemed to show that neutrinos were traveling faster than the speed of light, was eventually determined to be the

Beta particles, being charged, interact with the electrons in the atoms of materials like alpha particles do, but since they are only singly charged, they penetrate farther into materials than alpha particles – several millimeters in water or living tissue. This is far enough to get through the dead layers of your skin, so external beta radiation can do damage to your body when hitting it from the outside. Also, when a positron (a form of antimatter) encounters an atomic electron at close range, the two annihilate each other and turn into a pair of 0.51 MeV gamma rays. Both β^+ particles and β^- particles emit X-rays (Bremsstrahlung) as they slow down through electrostatic interactions with atomic electrons. Appropriate clothing is sufficient to protect radiation workers from direct damage from beta radiation, but the gamma rays and X-rays are more penetrating, so there is some internal damage even from beta particles that are stopped by protective clothing. Of course, beta radiation emitted from ingested or inhaled materials is even more damaging than externally impinging beta radiation.

4.5 NEUTRONS

Free neutrons are not usually found in nature. They have no electric charge, so they are free to collide with atomic nuclei without having to overcome the force of electrostatic repulsion as protons must. And when they get close enough to other nucleons in a nucleus, they are subject to the short-range but powerful strong nuclear force, which tends to pull them in and hold them to the nucleus. So free neutrons are soon absorbed by the materials they encounter.

However, nuclear fission reactors run on a neutron chain reaction, as we will explore in detail in Chapter 6. In and near a nuclear reactor core, there are great numbers of free neutrons, which are released by the fission reactions in the reactor core. Also, there are some other sources of free neutrons, such as some heavy nuclei that decay by spontaneous fission.

As we have seen, the mass of the absorbed neutron in the new nucleus is different from the mass of the free neutron. The mass difference is released as a gamma ray. So free neutrons are another penetrating radiation that causes damage to tissue, and nuclear reactor cores and other sources of free neutrons must be shielded to protect people working near them.

---

result of a computer programming error in the data reduction software.

4.6 BACKGROUND RADIATION, HEALTH EFFECTS, AND RADIATION PROTECTION STANDARDS

Many people think that radiation is an unusual phenomenon to which they would only be exposed in unusual circumstances, but that is not at all true. Numerous natural sources of radiation bombard us all the time, and almost everyone in industrialized countries routinely submits to artificial sources of radiation voluntarily. We now look at these natural sources of radiation, which we call background radiation, and the artificial sources of radiation we choose to accept.

About half of our exposure to natural radiation comes from cosmic rays,[9] which are particles raining down on the Earth from space. This rain of particles consists of about 90% protons, 9% alpha particles, and 1% electrons. Some come from the Sun, but most come from outside the solar system. They range in energy from about 100 MeV to over 10^{14} MeV. This latter energy is equal to 16 J, which is a good fraction of the kinetic energy in a tennis ball served by Serena Williams (80 J)—in a single proton! The cosmic rays themselves don't usually make it to the Earth's surface. Instead, they collide with nuclei in the upper atmosphere, releasing cascades of other particles, including beta particles and some exotic particles called mesons.

Most of the remainder of our exposure to natural radiation comes directly or indirectly from uranium and thorium in soils. These radioactive elements are very common in soils, although more so in some places than in others. Uranium-234, a minor constituent of natural uranium, decays into thorium-230, which decays into radium-226, which decays into radon-222, all by alpha-particle emission. Radon is a gas, and it percolates though the soil and out into the atmosphere, where it is breathed in by you and me. When concentrations of radon in the air build up, as in basements, the exposure is increased. Radon is believed to be the second most frequent cause of lung cancer, after smoking, and to account for 21,000 cancer deaths per year in the United States alone. Radon-222 is an alpha emitter, so inhalation of radon exposes the sensitive tissue of the lungs to alpha particles.

A small amount of our exposure to natural radiation comes from sources inside our bodies. Carbon-14 is produced in the upper atmosphere by a reaction between cosmic rays and atmospheric nitrogen. This carbon-14 eventually makes its way down to the Earth's surface in the form of carbon dioxide, where it is taken up by plants that are eaten by animals. One part per trillion of the carbon in every living organism is carbon-14, which beta-decays with a half-life of 5730 years. When the organism dies, it ceases to take up more carbon dioxide, so the concentration of carbon-14 in its remains decreases predictably. This phenomenon is the basis for radiocarbon dating.

Also, 0.0117% of natural potassium is potassium-40, which beta-decays with a half-life of 1.277×10^9 years. Carbon and potassium are required constituents of the human body, so you are irradiating yourself all the time with carbon-14 and potassium-40.

Exposure to artificial sources of radiation is primarily for medical purposes. Most people in industrial nations receive dental X-rays about once a year. Most women over 40 years old in advanced nations also receive X-rays for mammograms. Some radioisotopes (radioactive isotopes) are used for medical diagnostic purposes, and some people receive radiation for treatment of diseases like cancer.

Radiation energy absorbed by matter is measured in several different units. The SI unit of radiation dose is the gray (Gy), defined as one joule of radiation energy absorbed per kilogram of material. Since different kinds of radiation do more damage to tissue than others (because, as noted above, their energy is deposited over different volumes of tissue), a different unit is defined to express the dose of gamma radiation to which a dose of another type of radiation is equivalent. The SI unit of dose equivalent in tissue is called the sievert (Sv), also in units of J/kg. However, doses and dose equivalents are often expressed in rad or rem, respectively, which were in use long before the gray and the sievert were defined. One rad = 10^{-2} Gy, and one rem = 10^{-2} Sv. Millirad (mrad) and millirem (mrem) are also used, which are equal to 10^{-3} rad and 10^{-3} rem, respectively.

The relationship between dose equivalent (DE) and dose (D) is

$$DE = D \times QF \times MF,$$

where QF is the "quality factor" for the particular type of radiation, and MF is a further "modifying factor" that accounts for additional hazards from radionuclides that become concentrated in specific types of tissue. Table 4.1 gives approximate values of QF. MF can be as large as five.[10]

Henceforth in this text, the terms "dose" and "dose equivalent" are usually not distinguished: The term "dose" is used for both. Usually, "dose equivalent" is the intended meaning. The distinction can be inferred from the units given. If they are in rad, they express dose, and if they are in rem, they express dose equivalent.

Table 4.1—Approximate values of quality factor (QF) for different types of radiation (adapted from Reference 10)

Radiation type	QF
Gamma rays, X-rays, and beta rays	1
Neutrons with energies less than 10 keV	3
Neutrons with energies greater than 10 keV	10
Alpha particles emitted within tissue	10

Table 4.2 shows the average annual natural whole-body dose received by people in the United States as presented by Glasstone and Sesonske.[11] This average value is 102 mrem, but individual exposures vary quite a bit. Some locations have more uranium or thorium in the soil than others, and the cosmic ray dose varies considerably with altitude. In Denver, the cosmic ray dose is about twice that at sea level. According to Glasstone and Sesonske, natural background dose rates in the United States vary from about 80 to 200 mrem per year.

Table 4.2—Average annual whole-body radiation dose from natural sources in the United States (adapted from Reference 11)

Source Dose	(mrem)
Cosmic rays	44
Uranium and thorium and their decay products and external sources of potassium-40	40
Internal potassium-40	14
Carbon-14	4
Total	102

Glasstone and Sesonske say that the figures in Table 4.2 are based on one of several sets of estimates and are lower than most. Alan Waltar, in a more recent publication, gives higher numbers.[12]

A source of radiation exposure most people don't think about is increased cosmic radiation received in air travel. Because of the reduced atmospheric shielding, one receives about 0.7 extra mrem per hour at an altitude of 10,000 m (about 33,000 feet). So a single 4-hour coast-to-coast flight gives you about 3 mrem above your exposure at sea level; a professional airline pilot making, say, three such flights a week can increase his annual dose by 468 mrem, or about 459% of the average natural background!

For artificial sources, Waltar gives average doses for Americans of 53 mrem/yr from medical procedures, 10 mrem from consumer products such as lantern mantles and smoke detectors, and less than 1 mrem/yr each from nuclear weapons fallout and the nuclear industry.

The effects of radiation on living organisms can be classified into early effects, delayed effects, and genetic effects. Early effects require very high doses of radiation, which are only experienced in very unusual circumstances. For example, in the Chernobyl nuclear power plant accident (which is discussed in detail in Chapter 13), about 50 firefighters received lethal doses. Table 4.3 shows probable early effects of acute whole-body doses of radiation.[13] The qualifier "probable" reflects the fact that some individuals are more resilient than others, and the term "acute" means that the entire dose is received essentially at once.

Table 4.3—Probable early effects of acute whole-body
doses of radiation (adapted from Reference 13)

Acute dose (rem)	Probable somatic effect
5-75	Temporary reduction of white blood cell level in some people.
75-200	Vomiting, fatigue, and loss of appetite in 5-50% of exposed persons within a few hours, and moderate blood changes. Recovery is achieved within a few weeks from most symptoms.
200-600	Vomiting within 2 hours, severe blood changes, hair loss within 2 weeks. At lower end of range, most individuals recover within 1-12 months. At upper end, only 20% survive.
600 or more	Vomiting within 1 hour, severe blood changes; death within 2 months for at least 80% of individuals exposed to 600-1000 rem, 100% for more than 1000 rem. Survivors require long convalescence, may suffer permanent impairment.

Doses that are spread out over time are less damaging than doses received all at once. Consider the effects of sunshine, in which the ultraviolet rays cause sunburn if too much is received at once. Four straight hours in direct sunlight without sunscreen will burn most Caucasian people badly, but sixteen periods of fifteen minutes on successive days will only give most Caucasians a tan. The body has healing mechanisms that repair radiation damage, if the dose rate is slow enough that these mechanisms can keep up with the damage rate.

Much higher acute doses to limited regions of the body can be used in radiation therapy. In 2019, I received radiation therapy for prostate cancer. Over a four-week period, my prostate was exposed to 6000 rem of high-energy X-rays (up to 10 MeV, which were only considered X-rays and not gamma rays because they were generated by Bremsstrahlung and not radioactive decay). Some surrounding tissues received as much as 3000 rem. The cancer cells were killed, but the healthy tissues recovered.

Delayed effects include eye cataracts and cancers. These illnesses do not appear immediately after exposure, but require latency periods that usually extend over many years. Delayed effects can appear after the latency period following an acute dose, or they may appear after the accumulation of many years of a low dose rate. For example, in the early 1900s, some women were employed by watchmakers to paint the numbers and hands of watches with paint containing radium that glowed in the dark. They shaped the points of their brushes with their lips, unavoidably ingesting some of it. The radium accumulated in their bones, where it continuously irradiated them, and about 100 of them eventually died of bone cancer.[14]

Genetic effects are changes in genes or chromosomes in egg or sperm cells. These changes produce mutations in offspring, and these mutations are usually detrimental, sometimes even lethal. But not always. Beneficial mutations are the mechanism of evolution.

All of the delayed illnesses and genetic effects are observed in populations that have no unusual exposure to radiation. Some of them are known to be caused by factors other than radiation; for example, some chemicals are known to cause mutations and cancers. (Chemicals produced by charring meat, as on a charcoal grill, are implicated in some cancers.[15] Tobacco smoke is a witches' brew of carcinogens.) Mutations seem to occur spontaneously as a result of errors in the complicated process of DNA replication, and some cancers, such as prostate cancer, seem to be a probable consequence of aging. Which of these cancers are caused by the natural background radiation to which everyone is exposed, and which are caused by other factors, is difficult to determine. Since most people would apparently develop some form of cancer if they lived long enough, the incidence of cancer in human populations is very high, and it is very difficult to distinguish cancers caused by low doses of artificially imposed radiation from cancers that would have occurred anyway. A proposal has been made to perform experiments on mice in salt beds deep in the Earth, in which the uranium concentration (hence the radon concentration) is extremely low, and which are well shielded from cosmic radiation.[16] Then the background dose rate would be as low as possible, and effects of very low doses of imposed radiation might be discernible. But these experiments have not yet been performed.

It is reasonable to suspect that there may be some level of radiation exposure—a threshold—below which no delayed effects are experienced. The same mechanisms that repair damage suffered in acute exposure work on damage suffered in accumulated exposure. There is even a hypothesis—called the "hormesis hypothesis"—which suggests that, since human beings evolved with constant exposure to natural background radiation, they are not only adapted to it but also need it. Thus, if this hypothesis is true, exposure to radiation at natural background rates is good for you![17] This hypothesis is far from being accepted by the health physics community, but there is some empirical evidence for it.

Since the specific effects of very low levels of radiation are so difficult to establish, it has been assumed for setting radiation protection standards that no amount of radiation exposure is safe, and the health effects of low doses can be extrapolated from measurable effects of large acute doses. This approach is called the Linear No-Threshold (LNT) hypothesis. This hypothesis, the threshold hypothesis, and the hormesis hypothesis are illustrated in Figure 4.3.

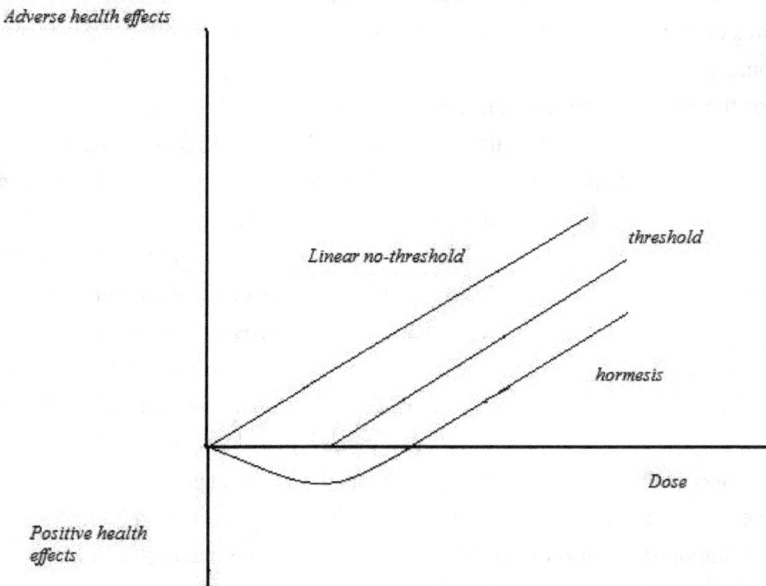

Figure 4.3—Three hypotheses on health effects of low-level radiation

Radiation protection standards have been established for people who work with or near sources of radiation, and also for the general population that might be exposed to artificial radiation accidentally.

Because protection standards are based on the LNT hypothesis, it is common for anti-nuclear activists to claim that any amount of radiation is harmful. However, there is no scientific basis for making such a claim. No one has done the required experiments to assess the effects of increased doses in amounts comparable to or less than natural background radiation doses.

However, research has been reported very recently[18] that shows the risk of cancer from low-level radiation to be extremely low, and far below the risk predicted by the LNT hypothesis. Radiation damages DNA by double-strand breaks in the double helix of DNA. When the double helix is damaged by one or a few double-strand breaks, the damage is readily repaired. Only when a double helix is damaged by multiple double-strand breaks, as happens in high-dose-rate radiation, is the repair prone to the kind of errors that can give rise to cancer. These findings provide a strong empirical basis for rejecting the LNT hypothesis and revising radiation protection standards based on a realistic model of the risk from radiation. Reference 18 also reports a study begun in 1975 at the Tohoku Radiological Sciences Center in Japan (ironically, near the site of the damage caused by the earthquake of 11 March 2011, discussed further in Chapter 14), "showing low doses to have promoted immunological response, rather than suppressing it"—tangible evidence for the hormesis hypothesis.

Table 4.4 gives the latest limits established by the U. S. Nuclear Regulatory Commission (NRC) for radiation exposure to individuals working in an environment of elevated radiation intensity.[19] Table 4.5 gives the NRC's limits for exposure of members of the general public to radiation emitted by industrial facilities such as nuclear reactors. The limits are quite complicated, depending on how the dose is distributed in a person's body. However, the basic whole-body dose limits can be succinctly expressed simply as 5 rem per year for occupational exposure and 0.1 rem per year for the general public.

Table 4.4—Occupational Radiation Exposure Limits

Source: 56 FR 23396, May 21, 1991, unless otherwise noted.

§ 20.1201 Occupational dose limits for adults.

(a) The licensee shall control the occupational dose to individual adults, except for planned special exposures under § 20.1206, to the following dose limits.

(1) An annual limit, which is the more limiting of--

(i) The total effective dose equivalent being equal to 5 rems (0.05 Sv); or

(ii) The sum of the deep-dose equivalent and the committed dose equivalent to any individual organ or tissue other than the lens of the eye being equal to 50 rems (0.5 Sv).

(2) The annual limits to the lens of the eye, to the skin of the whole body, and to the skin of the extremities, which are:

(i) A lens dose equivalent of 15 rems (0.15 Sv), and

(ii) A shallow-dose equivalent of 50 rem (0.5 Sv) to the skin of the whole body or to the skin of any extremity.

(b) Doses received in excess of the annual limits, including doses received during accidents, emergencies, and planned special exposures, must be subtracted from the limits for planned special exposures that the individual may receive during the current year (see § 20.1206(e)(1)) and during the individual's lifetime (see § 20.1206(e)(2)).

(c) When the external exposure is determined by measurement with an external personal monitoring device, the deep-dose equivalent must be used in place of the effective dose equivalent, unless the effective dose equivalent is determined by a dosimetry method approved by the NRC. The assigned deep-dose equivalent must be for the part of the body receiving the highest exposure. The assigned shallow-dose equivalent must be the dose averaged over the contiguous 10 square centimeters of skin receiving the highest exposure. The deep-dose equivalent, lens-dose equivalent, and shallow-dose equivalent may be assessed from surveys or other radiation measurements for the purpose of demonstrating compliance with the occupational dose limits, if the individual monitoring device was not in the region of highest potential exposure, or the results of individual monitoring are unavailable.

(d) Derived air concentration (DAC) and annual limit on intake (ALI) values are presented in table 1 of appendix B to part 20 and may be used to determine the individual's dose (see § 20.2106) and to demonstrate compliance with the occupational dose limits.

(e) In addition to the annual dose limits, the licensee shall limit the soluble uranium intake by an individual to 10 milligrams in a week in consideration of chemical toxicity (see footnote 3 of appendix B to part 20).

(f) The licensee shall reduce the dose that an individual may be allowed to receive in the current year by the amount of occupational dose received while employed by any other person (see § 20.2104(e)).

[56 FR 23396, May 21, 1991, as amended at 60 FR 20185, Apr. 25, 1995; 63 FR 39482, July 23, 1998; 67 FR 16304, Apr. 5, 2002; 72 FR 68059, Dec. 4, 2007]

Workers in radiation environments wear dosimeters to track their cumulative exposure. When I worked for the Idaho National Laboratory, I would occasionally need to visit the Advanced Test Reactor, a small high-power-density light-water reactor used for research, and I was given a dosimeter to wear each time. My exposure (which was never high enough to measure) was recorded, and it remains on file at the Laboratory. In Table 4.4, the annual whole-body limit to dose is 5 rem. An acute whole-body dose of 5 rem is enough to cause temporary depression of white blood cell levels in some people, as shown in Table 4.3, but if the same dose is spread out over a year, its somatic effects (i.e., observable symptoms) are nil.

Table 4.5—Radiation dose limits for the general public

Source: 56 FR 23398, May 21, 1991, unless otherwise noted.

§ 20.1301 Dose limits for individual members of the public.

(a) Each licensee shall conduct operations so that —

(1) The total effective dose equivalent to individual members of the public from the licensed operation does not exceed 0.1 rem (1 mSv) in a year, exclusive of the dose contributions from background radiation, from any administration the individual has received, from exposure to individuals administered radioactive material and released under § 35.75, from voluntary participation in medical research programs, and from the licensee's disposal of radioactive material into sanitary sewerage in accordance with § 20.2003, and

(2) The dose in any unrestricted area from external sources, exclusive of the dose contributions from patients administered radioactive material and released in accordance with § 35.75, does not exceed 0.002 rem (0.02 millisievert) in any one hour.

(b) If the licensee permits members of the public to have access to controlled areas, the limits for members of the public continue to apply to those individuals.

(c) Notwithstanding paragraph (a)(1) of this section, a licensee may permit visitors to an individual who cannot be released, under § 35.75, to receive a radiation dose greater than 0.1 rem (1 mSv) if—

(1) The radiation dose received does not exceed 0.5 rem (5 mSv); and

(2) The authorized user, as defined in 10 CFR Part 35, has determined before the visit that it is appropriate.

(d) A licensee or license applicant may apply for prior NRC authorization to operate up to an annual dose limit for an individual member of the public of 0.5 rem (5 mSv). The licensee or license applicant shall include the following information in this application:

(1) Demonstration of the need for and the expected duration of operations in excess of the limit in paragraph (a) of this section;

(2) The licensee's program to assess and control dose within the 0.5 rem (5 mSv) annual limit; and

(3) The procedures to be followed to maintain the dose as low as is reasonably achievable.

(e) In addition to the requirements of this part, a licensee subject to the provisions of EPA's generally applicable environmental radiation standards in 40 CFR part 190 shall comply with those standards.

(f) The Commission may impose additional restrictions on radiation levels in unrestricted areas and on the total quantity of radionuclides that a licensee may release in effluents in order to restrict the collective dose.

[56 FR 23398, May 21, 1991, as amended at 60 FR 48625, Sept. 20, 1995; 62 FR 4133, Jan. 29, 1997; 67 FR 20370, Apr. 24, 2002; 67 FR 62872, Oct. 9, 2002]

As shown in Table 4.5, the NRC regulations require any entity operating any facility in which radioactive materials are used, as a condition of their operating license, to limit exposure to any member of the public to a whole-body dose of 0.1 rem per year, but not more than 0.002 rem in any hour. (There are exceptions for such members of the public as patients receiving irradiation for medical diagnostics or treatment.) Such facilities include nuclear power plants. A dose rate of 0.1 rem per year is about the same as the average natural background radiation dose rate.

However, there is another principle that licensees must follow, called the ALARA principle. ALARA stands for "as low as reasonably achievable." The figures in Tables 4.4 and 4.5 are limits, but when a licensee can take "reasonable" measures to reduce exposure further, it is required to do so. Actual dose rates to members of the public from nuclear reactor operations[11] are usually less than 1 mrem per year, or about 1% of natural background. People routinely accept greater doses than that voluntarily (if not consciously) by living in the Rocky Mountains or by taking commercial airline flights.

In order to ensure that public exposure to radiation is limited to the regulatory limits, the NRC has also established limits on concentrations of specific radionuclides in air and water at the boundary of a licensee's restricted area. Examples of such limits are shown in Table 4.6.[20] These limits are in addition to any naturally occurring concentration of these radionuclides.

Table 4.6—US NRC CONCENTRATION LIMITS FOR AIR AND WATER

Radionuclide	Solubility	Air Conc. Limits (pCi/L)	Water Conc. Limits (pCi/L)	Health effects
Hydrogen-3 (Tritium)	Insoluble	100	1,000,000*	Low-energy beta-emitter. When in the form of water, it can become organically bound by replacing hydrogen atoms in a body's cells. As water, it crosses the placenta and irradiates fetuses when pregnant women are exposed.
	Soluble	100	1,000,000	
Strontium-90	Insoluble	0.006	-	Beta-emitter. Behaves like calcium and concentrates in bones.
	Soluble	0.03	500	
Iodine-131	Insoluble	0.2	1000	Beta-emitter. Concentrates in the thyroid, esp. via the milk pathway.
	Soluble	0.2	1000	
Cesium-137	Insoluble	0.2	1000	Beta- and gamma-emitter. Resembles potassium and collects in muscles.
	Soluble	0.2	1000	
Radon-222	no decay products	10	-	Damage mostly from short-lived alpha-emitting decay products deposited in bronchial walls and can cause lung cancer.
	with decay products	0.1	-	

Radium-226	Insoluble	0.0009	60	Alpha-emitter. Similar to calcium, concentrates in bones. Primary route of exposure is ingestion.
	Soluble	0.0009	60	
Natural uranium	Insoluble	0.00009	-	Primarily alpha-emitter, also chemically toxic, esp. to kidney. Inhalation or ingestion increases chances of lung and bone cancer.
	Somewhat soluble	0.0009	-	
	Soluble	0.003	300	
Plutonium-239	Insoluble	0.00002	-	Alpha-emitter. Main health danger comes from inhalation of fine particles or incorporation in cuts.
	Somewhat soluble	0.00002	20	
	Soluble	-	-	
Americium-241	Insoluble	0.00002	20	Alpha- and gamma-emitter. Decay product of Pu-241; of special concern for workers handling reactor-grade Pu.
	Soluble	0.00002	20	

Source: Nuclear Regulatory Commission, 10 CFR Part 20, Appendix B (Washington, D.C.: US Government Printing Office, 1994).

* Drinking water standards set by the Environmental Protection Agency (EPA) are based on a limit of 4 millirem per year via the drinking-water pathway only. So allowable concentrations under the EPA would be generally less than 1/10 of those given in the table. In the case of a few radionuclides, like tritium, the allowable limit is even lower (20,000 pCi/L).

4.7 PRACTICAL APPLICATIONS OF RADIATION

From the discussions in Section 4.6, one might infer that radiation is at best an unwanted but unavoidable presence in our environment or an undesirable but necessary side effect of technologies like nuclear power. However, that conclusion would be completely mistaken. Radiation has many practical uses that make all our lives better and some of our lives possible.

Such uses are examined extensively in the book by Alan Waltar,[12] which I recommend. In this section, I briefly summarize a few applications. Radiation for medical purposes is important enough to warrant a chapter of its own, Chapter 10, so no further mention of it is made in this chapter.

4.7.1 Smoke detectors

Smoke detectors are present in almost every household in the United States, and all hotels and motels. They use an alpha-emitter such as americium-241 to ionize the air in a chamber that has openings to allow the entry of air from outside the chamber (collisions between the alpha particles and the air molecules strip electrons off the molecules). A voltage applied across the chamber by a battery or household wiring causes a small electric current to flow through the air in the chamber. This current is sensed by an appropriate metering circuit. If smoke from a fire enters the chamber, the current is reduced, and the metering circuit senses the reduction and triggers an alarm.

4.7.2 Food irradiation

Some 76 million Americans become ill each year from food-borne illnesses, 325,000 of them are hospitalized, and 5,000 die.[21] Furthermore, a significant fraction of harvested food spoils before it is eaten, especially berries and seafood. (I get very annoyed when raspberries that looked okay when I bought them get moldy before I eat them the next day.) Much of this misery and waste could be prevented by food irradiation. Indeed, food irradiation is already being applied on a limited basis in the United States and abroad. However, vociferous opponents have successfully blocked broader application.

Why? Because they claim that irradiation causes chemical changes which may be incompatible with human metabolic processes, perhaps causing cancer or other diseases. By that logic, we should never have begun to cook food, as heat surely does the same thing. In fact, as noted above,[15] it is known that some carcinogens are produced by broiling meat. I don't know about you, but every time I cook anything I am performing a unique chemical experiment!

Food irradiation bombards foods either with electrons or X-rays produced by electron accelerators or with gamma rays emitted by cobalt-60 or cesium-137. The energy of the incident radiation is enough to kill pathogens, but not enough to cause nuclear reactions. (Nuclear reactions are discussed in the next chapter.) Therefore, no changes in the nuclear structure of the target materials are possible, and the irradiated food does not become radioactive.

4.7.3 Mutant food

Biological evolution occurs when segments of a population with traits better suited to survival leave more surviving offspring than other segments of the population. In nature, this process is called natural selection. If the gene pool were static, less competitive genes would be winnowed out, and the gene pool would become less diverse. However, genetic mutations spontaneously appear as the result of natural radiation or other environmental influences. Usually, mutations are maladaptive and lead nowhere, but occasionally mutations are adaptive and lead to improved genotypes.

Using artificial radiation to accelerate the mutation process, geneticists can increase the frequency of adaptive mutations. Numerous crop varieties have been developed that have specific desired traits (resistance to disease or drought, higher yields, or better flavor, for example) by bombarding seeds with radiation.

4.7.4 Industrial processes

Radiation is used for measurement, manufacture of materials, sterilization, inspection, and many other industrial processes. Attenuation of beta and gamma rays is a very accurate means of measuring thicknesses of sheet materials; beta rays are used for thin, penetrable materials like paper and plastic, and gamma rays are used for thicker, denser materials like sheet metal. The rubber in tires is commonly vulcanized by gamma radiation, leading to longer-wearing tires than previously attainable. Radiation is used to sterilize surgical instruments, contact lens solutions, cosmetics, bandages, and many other personal products. Radiographs made by gamma rays or neutrons are used to verify the integrity of aircraft welds in metal components or bonds in composite materials.

The foregoing applications are only a few of the beneficial uses of radiation discussed by Waltar. Far from being an unequivocally bad by-product of nuclear technology, radiation is like fire: dangerous when out of control, but enormously useful and beneficial when properly applied.

4.8 PLUTONIUM

I devote a special section to plutonium, because so much misinformation has been circulated in the popular media about the hazards of plutonium that I feel compelled to single it out for separate discussion.

Plutonium is produced in nuclear reactors, either as a by-product or intentionally. It comes in several isotopes, most notably Pu-239, which can be used as reactor fuel or in nuclear weapons. Plutonium-238 can be produced for radioisotopic generators in spacecraft; the process required for this is different from the process that produces Pu-239. Plutonium-240, Pu-241, and Pu-242 are created by the successive absorption of neutrons by Pu-239.

Plutonium has often been proclaimed to be the most toxic substance known to man.[22] While I was working at the Idaho National Laboratory, a writer to the opinion page of the Idaho Falls Post Register, protesting the use of the laboratory's Advanced Test Reactor to produce plutonium-238 for radioisotopic power supplies in unmanned spacecraft, claimed that one atom of Pu-238 can cause cancer.

These concerns are not valid, and they show an amazing lack of perspective. That writer lives in Jackson, Wyoming, where the radon concentration is exceptionally high—an average of more than 4 picocuries (one picocurie, or pCi, is 10^{-12} Ci) per liter of air indoors.[23] The alpha particles emitted by radon-222 have almost exactly the same energy as those emitted by Pu-238 (5.4897 MeV and 5.4992 MeV, respectively). Once emitted, an alpha particle is distinguished only by its energy; it doesn't matter where it came from. A concentration of 4 pCi per liter works out to about 70,000 atoms per liter for Rn-222, and your lungs contain a liter or two of air after exhalation. So that writer has between 70,000 and 140,000 atoms of radon in her lungs all the time when she is indoors in Jackson. The half-life of Rn-222 is only 3.8 days, compared to 88 years for plutonium-238, so the radioactivity of a given number of Rn-222 atoms is much more intense than that of the same number of Pu-238 atoms. Furthermore, Rn-222 decays to polonium-218, which decays by emitting another alpha particle, of 6.1 MeV, with a half-life of 3 minutes. Pu-238 decays to U-234, which has a half-life of 245,500 years. It all works out that the radon that is in her lungs while she is indoors has the same effect as about 600 million atoms of Pu-238. Yet she continues to live in Jackson.

The toxicity of plutonium is of two types: chemical and radioactive. The chemical toxicity is comparable to that of caffeine.[24] The radioactive toxicity depends on how it is taken in by the body and in what chemical form it is contained. Metallic plutonium, especially in the small particles in which it is most likely to be taken in, is pyrophoric—i.e., it burns spontaneously in air. Thus, plutonium oxide is the most likely form to be taken in. Only 0.04% of ingested plutonium oxide is absorbed; the rest is excreted. Inhalation is more dangerous. Reference 25 cites U.S. Department of Energy data that show a 1% increase in lifetime cancer risk with the inhalation of 5000 plutonium particles 3 microns in width (i.e., 3×10^{-6} m). The size of an atom is less than 1 nanometer (10^{-9} m),[26] so there are on the order of 27 billion plutonium atoms in each of the 3-micron particles (this number is obtained by dividing the volume of a sphere 3 microns in diameter by the volume of a sphere 1 nanometer in diameter).

The idea that only one atom of Pu-238, or any other radioactive substance, can cause cancer is correct in a very perverted sense. Cancer is believed to be caused by the proliferation of mutant cells. Mutations can be caused by radiation. A single alpha particle, beta particle, or gamma ray can cause a mutation. So a cancer can result from one such particle. However, the vast majority of damage caused by radiation does not lead to cancer. Either the cell is repaired or it dies. The body sustains radiation damage all the time, and only rarely do cancer-causing mutations result. To say that you are at a significantly increased risk of cancer because you absorb one and only one atom of plutonium is simply not true. It is equivalent to saying that one photon of ultraviolet radiation from the sun can cause skin cancer.

The fact is that plutonium is toxic, but not exceptionally so. Plutonium is chemically less toxic than many common substances. The radioactive toxicity of Pu-238, per atom, is much less than that of radon, a hazard that we all live with. Plutonium-239, with a half-life of 24,000 years, is much less toxic per atom than Pu-238. Don't lose any sleep over "the most toxic substance known to man."

CHAPTER 5
NUCLEAR REACTIONS

When another particle of any kind impinges on a nucleus, some sort of interaction occurs. Such interactions are classified as nuclear reactions. The simplest kind of nuclear reaction is just the deflection of the incoming particle. This deflection is called scattering. But other reactions are also possible, including absorption, expulsion of nucleons, nuclear fission, and nuclear fusion, depending on the type of nucleus and the type of incoming particle, and the energy of their relative motion.

If the incoming particle is another nucleus, it must overcome electrostatic repulsion before it can do anything other than scatter, because the nuclei are both positively charged. To do this, it must have at least a certain minimum kinetic energy, which depends on the identity of both particles. This minimum kinetic energy is called the threshold energy for the reaction.

But if the incoming particle is a neutron, it can wander into the nucleus with essentially no kinetic energy, and a reaction will occur as soon as the neutron comes close enough to a nucleon in the nucleus for the strong nuclear force to come into play. Neutrons of very low energy, and of certain higher energies, are likely to be absorbed, but most higher-energy neutrons are more likely to scatter.

High-energy photons can also interact with nuclei, either by exciting nucleons to higher energy states or by actually knocking nucleons out of the nucleus. All such reactions have energy thresholds, because the nucleons can only occupy discrete energy levels in the nucleus, so if the incoming photon doesn't have enough energy to raise the easiest-to-excite nucleon to its next level, the photon will just be scattered.

There is a common notation for nuclear reactions. This notation is best explained by an illustrative example. Suppose the incoming particle is a neutron, and it is absorbed by a nucleus of nuclide A. The absorption process releases the binding energy of the additional neutron, and this binding energy is emitted as a gamma ray. Because it has a new neutron, the nucleus has changed identity, and it is now nuclide B. The notation for the reaction is

$$A(n,\ \gamma)B\ .$$

This simply means that a nucleus of nuclide A absorbs a neutron and emits a gamma ray and in the process becomes a nucleus of nuclide B. This type of reaction is called an "n-gamma" reaction, abbreviated as an (n,γ) reaction.

The probability that a reaction of a specific type will occur is measured by an effective cross-sectional area presented by the nucleus to the incoming particle, which is called the cross section. The cross section for a nuclear reaction isn't really equal to the cross-sectional area of the nucleus, because the nature of nuclear reactions is more complicated than a simple collision of two rigid bodies. But the magnitude of the cross section is often roughly the same as the actual cross-sectional area for such a collision, so the term is reasonable. Cross sections are measured in units called "barns." The name was chosen because a cross section of a barn is fairly large, and for a neutron to hit a nucleus when the cross section is about a barn is as easy as hitting a barn door. One barn is equal to 10^{-24} square centimeters, which is actually not as big as a barn door at all. Cross sections are usually denoted by the symbol σ (lower-case Greek sigma). This type of cross section is more precisely called the microscopic cross section, to distinguish it from a related quantity that depends additionally on the density of the material as a whole. If there are N nuclei of the nuclide of interest in a unit volume (say, a cubic centimeter), and the microscopic cross section of the nuclide for a specific reaction is σ, then the macroscopic cross section for the reaction is defined as $\Sigma=N\sigma$, where Σ is the upper-case Greek letter sigma. The macroscopic cross section is useful in calculating reaction rates.

Next, we consider specific types of reactions. Most of the information on these reactions given below is taken from the introductory nuclear engineering textbook by Glasstone and Sesonske.[1]

5.1 ELASTIC SCATTERING

Elastic scattering is analogous to the collision of solid objects like billiard balls. An elastic scattering event conserves kinetic energy—i.e., no energy of the incoming particle is lost in excitation of nucleons in the target nucleus. As with all nuclear reactions, elastic scattering also conserves momentum.

Elastic scattering events occur between all sorts of particle pairs, but the most important kinds of elastic scattering in our areas of interest are collisions between the atoms in a material and collisions between free neutrons and nuclei. Collisions between atoms are not nuclear reactions (they involve electrostatic repulsion between the electrons of the colliding atoms), but we discuss them as needed. Collisions between neutrons and nuclei are at the heart of how nuclear reactors work.

Elastic scattering of neutrons can occur in two ways. The first way is called potential scattering. This type of scattering is a simple deflection of the incoming neutron by the action of the strong nuclear force as the neutron gets close to nucleons in the target nucleus. Because the strong nuclear force is attractive, the neutron is pulled in towards the nucleus rather than being pushed away, but it doesn't get close enough to be absorbed. This deflection is shown in Figure 5.1. The other way is called resonance scattering or compound nucleus scattering. In this type of scattering, the incoming neutron is actually absorbed by the target nucleus, the neutron and the target nucleus briefly forming a compound nucleus in an excited state, but then either the same neutron or a different one at the same energy level is expelled, leaving the target nucleus in its ground state again. The term "compound nucleus" simply means the original nucleus plus the extra neutron. The term "resonance" means that at certain energy levels the neutron is particularly attracted to the target nucleus to form a compound nucleus.

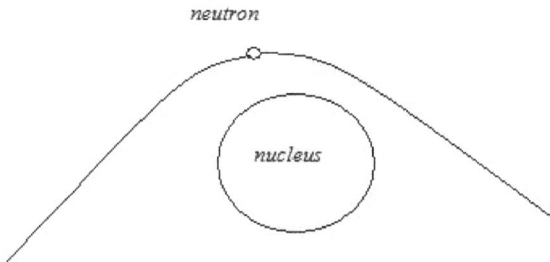

Figure 5.1—Deflection of neutron by nucleus in potential elastic scattering

If you have ever played pool, you may have discovered that when the cue ball strikes another ball exactly dead on, the cue ball stops completely and the target ball zings off with the same velocity the cue ball had coming into the collision. Even if the cue ball hits the target ball off-center, the target ball usually takes off with considerable velocity and the cue ball exits the collision with a slower speed than it had coming in. But if you shoot a billiard ball into a bowling ball, the billiard ball just bounces off the bowling ball with almost its original speed, while the bowling ball hardly moves. (If the balls are supported on a very rigid, smooth surface, the bowling ball will move off with a very small velocity, but on a felt table it probably won't even budge.)

These behaviors of balls are dictated by the laws of conservation of momentum and energy. Neutrons scattering elastically off nuclei behave the same way. If a neutron strikes a light nucleus, it can lose a lot of its initial energy in one collision. If it strikes a hydrogen nucleus, which consists of only one proton, it can lose all of its energy in one collision, because the proton and neutron have almost identical masses. However, if a neutron strikes a heavy nucleus like uranium and scatters elastically, it doesn't slow down very much.

Neutrons just released from fission reactions have energies of a few MeV. As discussed in Section 5.4 below, in some fuel materials, such as uranium-235, fission is induced much more easily by neutrons of very low energy than by neutrons with the high energy they have just after release by fission. Neutrons stop losing energy, on average, once their energy falls to the energy of thermal motion of the materials in which they are bouncing around. They are then in thermal equilibrium with the materials. Such neutrons are called thermal neutrons. The slowing down of neutrons to lower energy is called moderation, and materials added to reactors to promote moderation are called moderators. In many reactors, water is used both as a moderator, because it contains hydrogen, and as a coolant, because of its excellent heat transfer properties. In some reactors, a gas (such as helium) is used as the coolant because it can operate at higher temperatures than water can. Helium would be a good moderator if it were dense enough, but as a gas it doesn't moderate neutrons significantly. Therefore, another material is needed to act as a moderator, and graphite has excellent moderating properties. Different types of reactors are discussed in detail in later chapters.

5.2 INELASTIC SCATTERING

In reality, there are no perfectly elastic materials—i.e., materials that bounce back to their exact original shape without absorbing any energy. Thus, when billiard balls collide, some of the energy of the incoming ball is converted to thermal energy inside the balls, and the balls exit the collision with just a little less kinetic energy than the incoming ball brought to the collision.

However, in nuclear collisions, if the energy of the incoming neutron is less than the threshold energy for excitation of a nucleon in the target nucleus, the collision will be perfectly elastic. Also, if the neutron expelled from the compound nucleus in resonance scattering leaves the target nucleus in its ground state, that collision also will be perfectly elastic. These perfectly elastic collisions are possible because the energy levels of nucleons in a nucleus are discrete.

But if the ejected neutron in resonance scattering leaves the target nucleus in an excited state, the combined kinetic energy of the ejected neutron and the recoiling target nucleus is reduced by the amount of excitation energy left in the target nucleus. In that case, the collision is said to be inelastic.

The threshold energy for inelastic scattering varies from about 0.1 MeV to several MeV. It is generally higher for lighter nuclei. For example, in oxygen it is about 6 MeV.

When a nucleus scatters a neutron inelastically, the excitation energy is usually ejected as a gamma ray after a short time. Gamma emission also completes the (n,γ) absorption reaction noted above. The (n,γ) reaction is also called radiative capture. The new nucleus produced by radiative capture has one more neutron than the parent nucleus had, so the neutron-proton ratio has increased. The higher neutron-proton ratio often causes instability—i.e., the new nucleus is radioactive. This process is called neutron activation, and the new nucleus is called an activation product. Normally, an activated nucleus decays by beta-emission.

Neutrons can also lose kinetic energy inelastically if they scatter off nuclei bound in solids or in molecules. Then the neutrons can leave energy behind by exciting the molecular or crystalline bonds among the atoms, or by breaking those bonds altogether.

Figure 5.2 shows the scattering cross sections for carbon-12, the most common isotope of carbon.[2] Figure 5.2 (a) shows the elastic scattering cross section and Figure 5.2 (b) shows the inelastic scattering cross section. The nearly vertical portion of Figure 5.2 (a) for neutron energies near 0 MeV compresses a region between 0 and 0.001 eV in which the cross section drops rapidly (the so-called $1/v$ region, where v is the neutron speed—see below) and a region from 0.001 eV and 0.1 MeV in which the cross section is nearly constant. Figure 5.2 (b) shows the threshold energy of 4.8 MeV for inelastic scattering.

5.3 OTHER NEUTRON ABSORPTION REACTIONS

In addition to resonance elastic scattering, inelastic scattering, and (n,γ) reactions, which begin by the absorption of a neutron by a nucleus to form a compound nucleus, other outcomes are possible from such absorption. One is the fission reaction, discussed in the next subsection. Others are the emission of an alpha particle (the (n,α) reaction) or a proton (n,p), and reactions in which more than one nucleon depart from the compound nucleus $(n,2n)$, (n,np), $(n,3n)$, etc. The higher the energy of the incoming neutron, the more likely it is that reactions other than radiative capture or elastic scattering will occur.

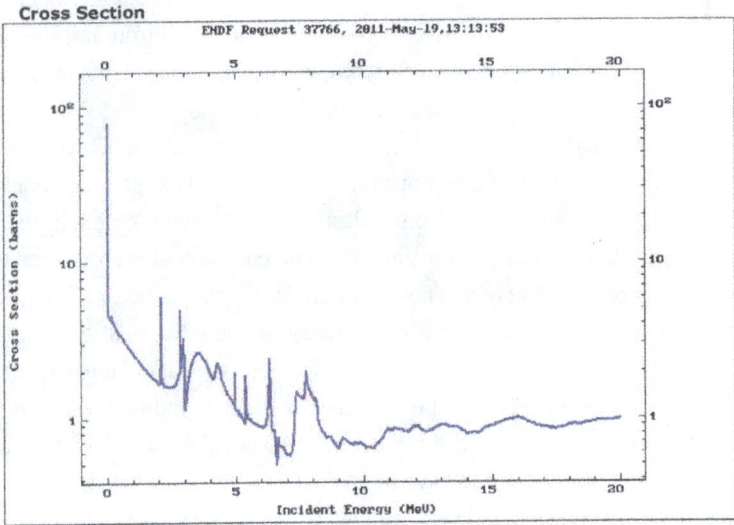

(a) **Elastic scattering cross section**

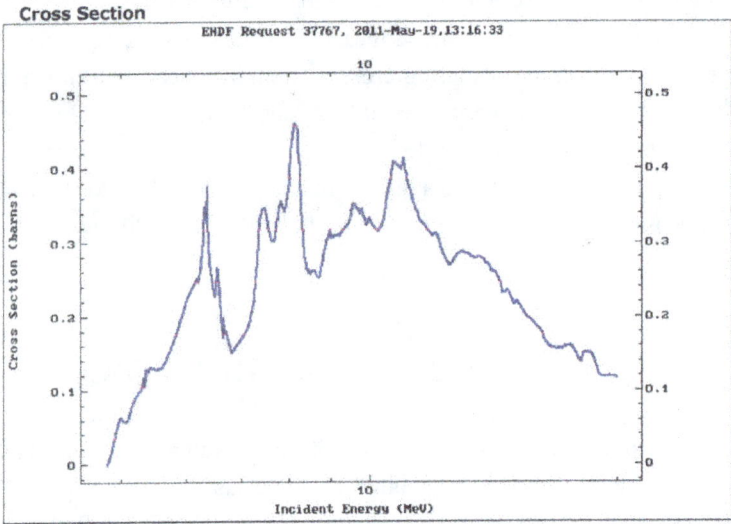

(b) **Inelastic scattering cross section**

Figure 5.2—Elastic and inelastic scattering cross sections for C-12. These plots were generated by Reference 2 and are in the public domain.

Figure 5.3 shows the microscopic absorption cross section as a function of energy for a generic nuclide.[3] Although cross sections for individual nuclides can show large variations in general character from the behavior shown in Figure 5.3, this generic nuclide illustrates some principles that apply to many nuclides of interest.

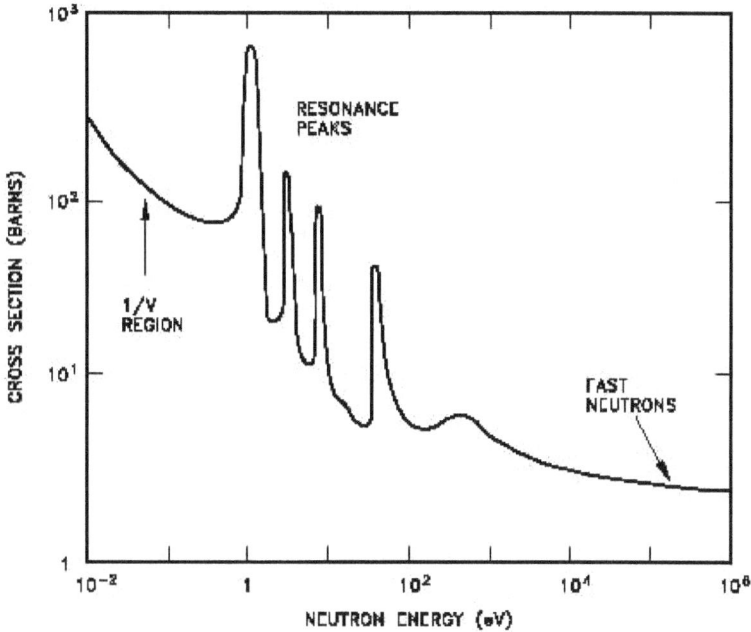

Figure 5.3—Generic neutron absorption cross section (from Reference 3)

First, at low energies, the cross section is nearly inversely proportional to the relative velocity of the neutron and the target nucleus. This downward-sloping section is called the $1/v$ ("one-over-v") region of the energy range. Then there is a region of sharp peaks in the cross section, called resonances. Resonances occur when the combined neutron kinetic energy and binding energy precisely matches the difference between an excited state of the compound nucleus and the ground state. The compound nucleus will initially be in the excited state, but the excitation energy will be ejected in one or more gamma rays, leaving the resulting nucleus securely in its ground state. Resonance peaks are often several orders of magnitude greater than the adjacent values of the cross section. Beyond the resonance region, the cross section falls off gradually.

5.4 NUCLEAR FISSION

Figure 5.4 shows the binding energy per nucleon as a function of atomic mass number (i.e., the number of nucleons in the nucleus).[4] Individual data points are shown by the dots, and a smooth curve is constructed as a best fit to the data points. The curve shows values for stable isotopes, except for the heavy elements like uranium in which all isotopes are radioactive. You can see that the point for ^{4}He is well above its neighbors, as noted in Chapter 4, and you can also see that the curve has a maximum value at about A=60. Thus, heavy elements like uranium (A mostly between 232 and 238) can release energy if they divide into two lighter elements. This division process is called nuclear fission. The lighter elements are called fission fragments.

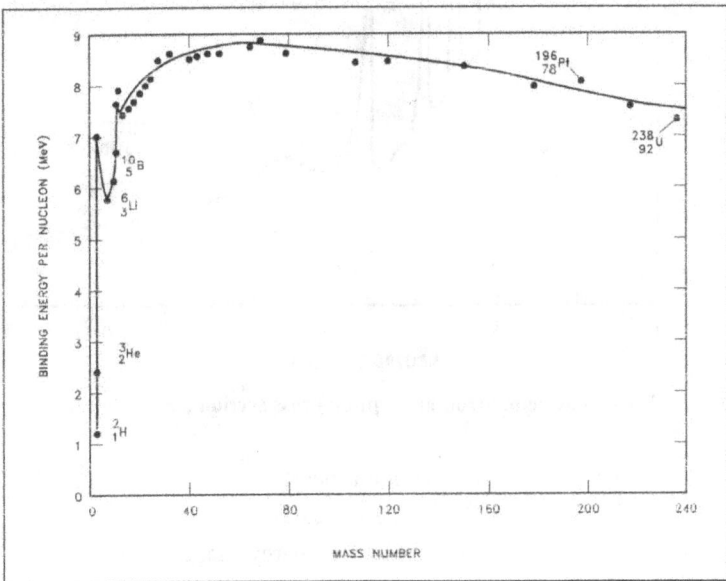

Figure 5.4—Binding energy per nucleon versus mass number (from Reference 4)

Nuclei that undergo fission don't always divide in the same way. There is a whole assortment of fission fragments into which a given nucleus can divide, as shown in Figure 5.5 for U-235.[5] As shown in the figure, the fission fragment distribution is dependent on neutron energy. In addition to the fission fragments, the fission reaction emits two or three free neutrons.

As noted in Section 4.1, stable radionuclides of greater mass number have a higher ratio of neutrons to protons than stable radionuclides of lesser mass number. When a nucleus divides in fission, the fission fragments have, on average, the same ratio of neutrons to protons as the parent nucleus had, but the fission fragments have much lower mass number. Therefore, they are highly unstable—i.e., they are radioactive with very short half-lives. The fission fragments generally decay by a series of beta decay events, which convert neutrons in their nuclei to protons, thereby reducing the neutron-to-proton ratio until eventually they become stable, or nonradioactive, nuclides. The assorted nuclides developed in these decay chains are called fission products.

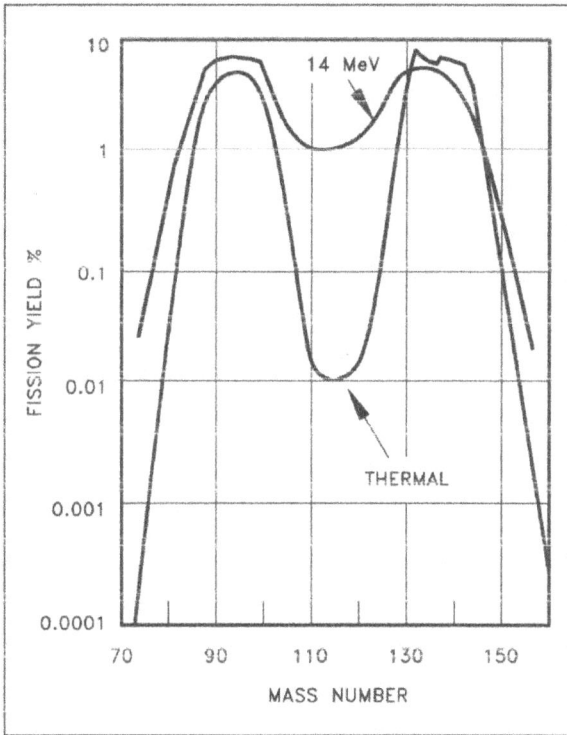

Figure 5.5—Fission fragment distribution by mass number for U-235 (from Reference 5)

In many heavy nuclides such as some isotopes of uranium and plutonium (especially Pu-240), fission occurs spontaneously, but most isotopes of uranium and thorium, which are the naturally occurring elements suitable for use as fuel in nuclear fission reactors, normally need to absorb a neutron to become unstable enough to undergo fission (even if spontaneous fission occurs, it occurs so slowly that it does not affect the fission rate significantly). Such neutrons are provided by nuclei that have already undergone fission. Neutrons emitted from fission reactions can go on to induce subsequent fission reactions, which emit more neutrons, which cause more fissions, and so on. This is the basis of the neutron chain reaction, illustrated in Figure 5.6, which permits nuclear reactors to operate. The fission chain reaction is discussed in much more detail in Chapter 6.

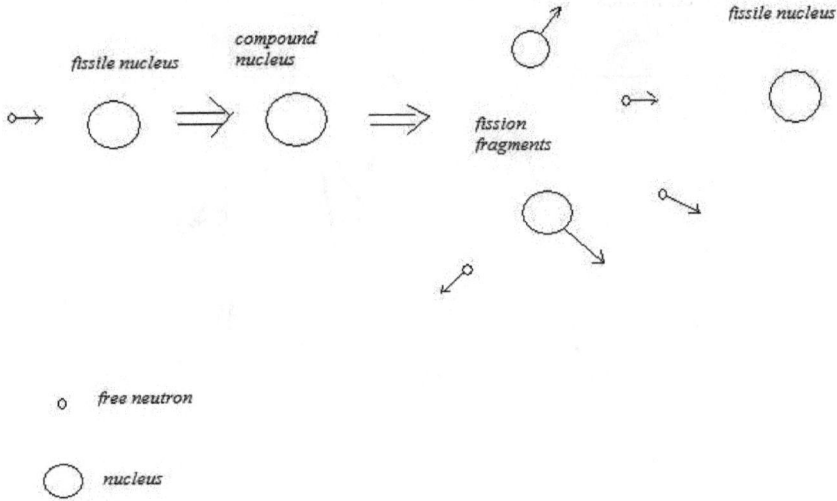

Figure 5.6—The fission chain reaction

Table 5.1 shows the distribution of the energy from one fission reaction of uranium-235. The total energy is 203 MeV, of which 10 MeV is carried by neutrinos and lost. Thus, an energy of 193 MeV is available for heating a coolant to generate electricity or to provide process heat. Contrast this to the energy available from the combustion of one atom of carbon to carbon dioxide: 4.1 eV. Thus, one atom of uranium-235 can provide 47 million times as much useful energy as one atom of carbon. This is the key to the enormous potential of nuclear energy to provide environmentally benign electricity or transportation fuel. The fuel cycle, like any other fuel cycle, has adverse environmental impacts, as we shall study in detail in later chapters. But there just isn't much fuel, or much waste, to deal with, compared to the fuel requirements or the waste burden of coal. So the overall environmental impact of nuclear power is relatively small compared to the impacts of coal and other fossil fuels. The latter environmental impacts are discussed in Chapter 8.

Table 5.1—Distribution of energy in fission of U-235[6]

Energy carrier	Energy (MeV)
Fission products	168
Instantaneously emitted gamma rays	7
Fission neutrons	5
Beta particles emitted by fission products	7
Gamma rays emitted by fission products	6
Neutrinos	10
Total	203

Figure 5.7 (a) shows the fission cross section for U-235, and Figure 5.7 (b) shows the fission cross section for U-238. The cross section for U-235 fission is similar to the generic absorption cross section shown in Figure 5.3. Fission occurs for neutrons of all energies, and in fact it is most likely for neutrons of the lowest energy, the thermal neutrons. On the energy scale, the thermal energy depends on the temperature of the material, but thermal neutron cross sections are usually tabulated at 20 °C, at which the neutron energy is 0.025 eV, corresponding to a speed of 2200 m/s. Nuclides that can undergo fission by absorbing low-energy neutrons are called fissile. In contrast, the fission cross section for U-238 is negligible below about 1 MeV. Nuclides that can undergo fission, but in useful quantities only with high-energy neutrons, are called fissionable.

Figure 5.8 shows the total cross section of U-238.[2] The cross section has a $1/v$ absorption tail at very low energies, then it is dominated by scattering up to the resonance region, and it falls off slowly above that, with a couple of very broad resonances at high energies. The resonance peaks are due to (n,γ) reactions called "parasitic absorption," or absorption in fuel that does not induce fission. The presence of these parasitic resonances gives rise to a very important inherent safety property in reactors fueled by "low-enriched uranium"—i.e., uranium in which U-238 dominates the isotopic mix in the fuel. When the temperature of the fuel increases, as in an upward power fluctuation, the resonances broaden because of the Doppler effect, which is much like the Doppler effect that enables police officers to measure the speed of your car with radar, except that it arises from the relative speeds of neutrons and nuclei. By a consequence of the Doppler effect that is best shown by mathematics beyond the scope of this book,[7] this "Doppler broadening" increases the parasitic absorption rate in U-238, which reduces the neutron population available to induce fission in U-235, thereby reducing the power. The opposite effect occurs when the temperature of the fuel decreases, as in a downward power fluctuation. Therefore, Doppler broadening provides an inherent and instantaneous natural feedback mechanism that stabilizes the reactor power level.

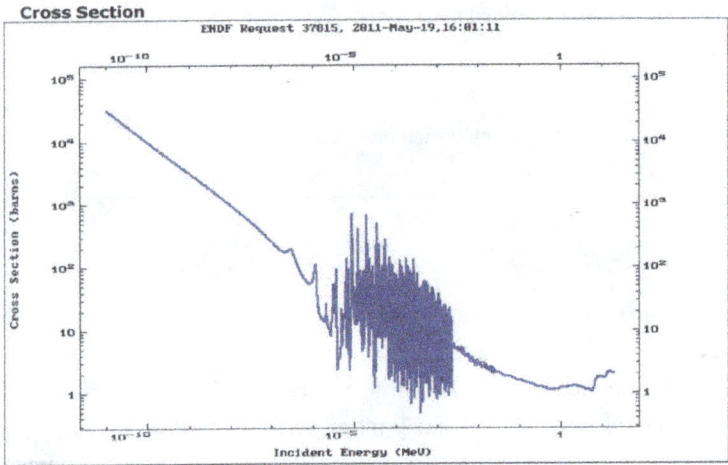

(a) U-235

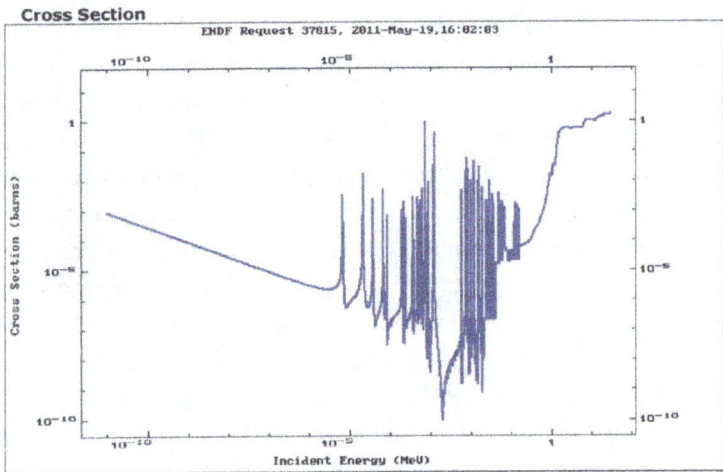

(b) U-238

Figure 5.7—Microscopic fission cross sections of U-235 and U-238 (this figure was generated by Reference 2 and is in the public domain)

Cross Section

EMDF Request 37828, 2011-May-19,16:23:55

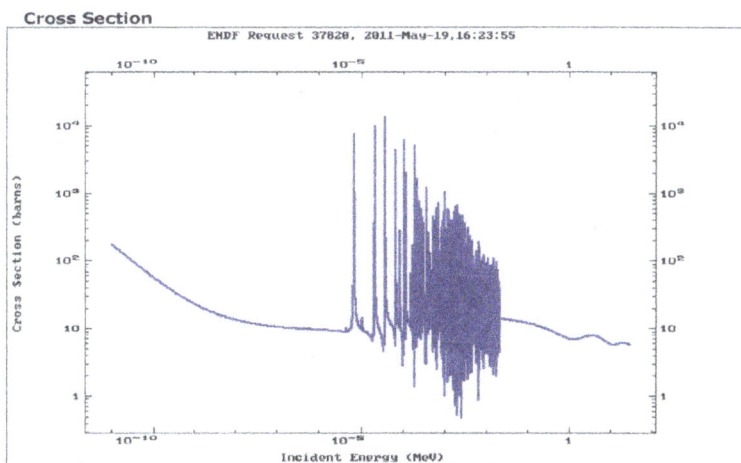

Figure 5.8—Total cross section of U-238 (this figure was generated by Reference 2 and is in the public domain)

5.5 NUCLEAR FUSION

Referring to Figure 5.4, you can see that light elements can release energy if they combine into heavier elements, whereas heavy elements can release energy by dividing into lighter elements, because in each case the resulting nuclei move towards the peak in binding energy per nucleon. The combination of two light nuclei into one heavier nucleus is called nuclear fusion.[8]

Heavy elements like uranium can undergo fission simply by absorbing a neutron, which can enter the nucleus without resistance because it has no electric charge. However, light nuclei cannot fuse unless they first overcome their mutual electrostatic repulsion, which is also called the Coulomb barrier. The easiest fusion reaction to bring about is the deuterium-tritium reaction, or the D-T reaction. The D-T reaction requires a total incoming kinetic energy of about 100 keV in the D and T particles to overcome the Coulomb barrier. Deuterium[9] is the isotope of hydrogen that has a neutron in the nucleus in addition to the single proton of ordinary hydrogen. About one in every 6500 atoms of hydrogen in nature is deuterium. Deuterium is nonradioactive. Tritium[10] is the isotope of hydrogen that contains two neutrons in addition to the proton. Tritium occurs naturally in the atmosphere in minute amounts because it is produced by cosmic rays acting on nitrogen, but it does not accumulate because it is radioactive, with a half-life of only 12.32 years. For industrial purposes, it is produced from lithium by the reactions $^6Li(n,\alpha)T$ and $^7Li(n,\alpha)T+n$. Tritium is used to make luminescent watch dial numerals and hands. It emits beta particles, which are shielded by the glass or sapphire crystal and the watch casing.

The D-T reaction can be written as $D(T,\alpha)n$, or in more detail,

$$D + T \rightarrow \alpha \ (3.5 \ \text{MeV}) + n \ (14.1 \ \text{MeV}).$$

This depiction shows that each D-T reaction produces 17.6 MeV of energy, which is initially carried as kinetic energy of the alpha particle and the neutron.

The products of the fusion reaction are not radioactive, but the neutrons will eventually be absorbed by the reactor vessel to produce activation products. Activation products are generally not intensely radioactive, and activated structure can be treated as low-level waste. Thus, the radioactive waste burden of a fusion reactor using the D-T reaction would be much less than that from a fission reactor of equal power output. (I say "would" because fusion reactors are still in the experimental stage, and no power-producing fusion reactor has yet been built.)

However, more difficult fusion reactions to produce would reduce the waste burden even more. The D-D reaction, which has two about equally probable outcomes, would produce a neutron in only one of the branches, and tritium in the other. Tritium is radioactive, but it has a short half-life, so it is not a long-term waste disposal problem. Of course, some of the tritium would react with the deuterium to produce 14.1 MeV neutrons. The D-D reactions are

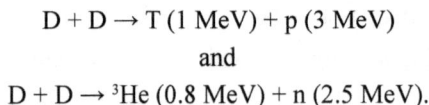

$$D + D \rightarrow T \ (1 \ \text{MeV}) + p \ (3 \ \text{MeV})$$
$$\text{and}$$
$$D + D \rightarrow {}^3He \ (0.8 \ \text{MeV}) + n \ (2.5 \ \text{MeV}).$$

Even more difficult reactions to produce have only stable charged particles as their products. Two of these reactions would have no radioactive waste burden at all, and the charged particles could be used to generate electricity directly, without the use of an inefficient thermal cycle:

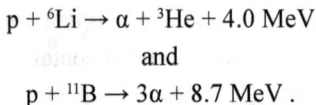

$$p + {}^6Li \rightarrow \alpha + {}^3He + 4.0 \ \text{MeV}$$
$$\text{and}$$
$$p + {}^{11}B \rightarrow 3\alpha + 8.7 \ \text{MeV} .$$

Fusion reactor technology is discussed in some detail in Chapter 20.

5.6 PHOTONEUTRON REACTIONS

High-energy photons can dislodge nucleons from a nucleus. The photoneutron reaction, in which the dislodged nucleon is a neutron, is mentioned here because it has an application in the nuclear medicine technology discussed in Chapter 10. Photoneutron reactions are written as $A(\gamma,n)B$.

CHAPTER 6
THE NEUTRON CHAIN REACTION

The neutron chain reaction, introduced in Section 5.4, is illustrated in more detail in Figure 6.1 below.

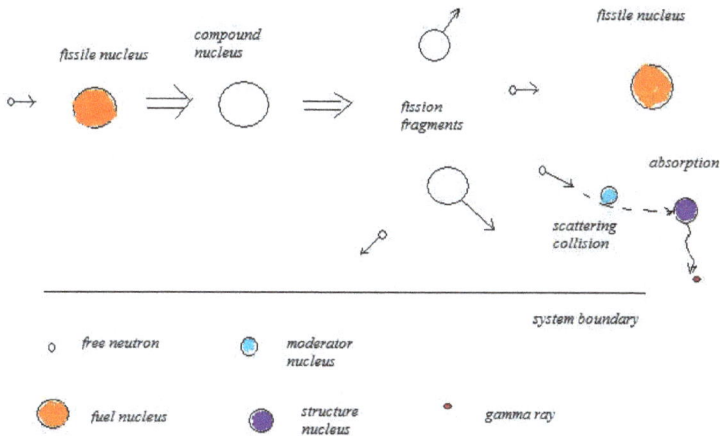

Figure 6.1—The neutron chain reaction, showing fission, scattering, leakage, and parasitic absorption

When a neutron is emitted from a nucleus in fission, it will have one of several different fates, as shown in the figure. It may be absorbed by structural or shielding material or coolant, it may be absorbed by a fuel nucleus without inducing fission, or it may cause fission in another nucleus. In order to design nuclear reactors and manage their operation, nuclear engineers must be able to keep track of these different outcomes mathematically. How they do it is discussed in Section 6.2, but first some terminology and general principles must be introduced.

6.1 THE BASICS

If one and only one neutron from a fission reaction, on average, induces a subsequent fission, the chain reaction will simmer along at a constant rate, or a steady state. This condition is called criticality, or a critical condition. If less than one of these neutrons, on average, induces a subsequent fission, the chain reaction will die out. This condition is called subcriticality, or a subcritical condition. And if more than one of these neutrons, on average, induces a subsequent fission, the chain reaction will grow. This condition is called supercriticality, or a supercritical condition.

Most of the neutrons emitted in fission are emitted during the fission reaction, but some of them are emitted later during the decay of fission products. The neutrons emitted immediately are called prompt neutrons, and the neutrons emitted later are called delayed neutrons. The fission products that emit delayed neutrons are called delayed neutron precursors. In thermal fission of uranium-235, the principal fuel of most nuclear reactors, an average of 2.42 neutrons is emitted per fission, and of these an average of 0.0158 is a delayed neutron, so the delayed neutron fraction is 0.0065. Since the delayed neutron fraction is not the same for all fissionable nuclides or for all neutron energies, for generality the delayed neutron fraction is usually denoted by the symbol β (Greek lower-case beta) in equations. The half-lives of the delayed neutron precursors for delayed neutron emission are generally between a quarter of a second and 30 seconds.

In order for the neutron chain reaction to grow from zero to the desired level for power production, the reactor must be supercritical for some period of time. It is only because some of the neutrons are delayed that the reactor can be kept under control during periods of supercriticality. If all the neutrons were prompt, the time between successive generations in the chain reaction would only be about a millisecond, and control mechanisms would not be able to operate quickly enough to keep up with the growth or decay of the chain reaction. When the chain reaction is able to sustain itself on prompt neutrons alone, the condition is called prompt supercritical. In a water-cooled reactor, prompt supercriticality causes steam explosions. In a steam explosion, the water flashes abruptly into steam, possibly with enough force to damage or destroy core or pressure vessel components, as happened at Chernobyl in 1986 and a small experimental reactor in Idaho in 1961.[1] Modern reactor design is strictly constrained to prevent prompt supercriticality. If nuclear fuel is highly enriched (i.e., if its percentage of U-235 or Pu-239 is artificially raised to more than about 90%), and if the configuration of the mass of fissile material is suitably arranged, prompt supercriticality may lead to a nuclear explosion as in a fission bomb. It is actually not easy to create the arrangement necessary for a nuclear explosion, and it is inherently impossible for such an arrangement to develop in a nuclear reactor in any accident scenario.

Individual neutron chains—the sequence of fission reactions, neutron emissions, subsequent reactions, and so on, which follow from an initial fission event—are like families. Some families die out, and others proliferate. But one can imagine an idealized average family that represents the population of a society as a whole. As the population of this average family goes, so goes the population of the society. Likewise, one can imagine a conceptual average neutron chain that collectively represents the diverse neutron chains in the reactor. In neutron chain reactions, the multiplication factor k is defined as the ratio of the number of neutrons in each generation in the conceptual average chain to the number of neutrons in the generation before it. That is, if N_i is the number of neutrons in the ith generation and N_{i+1} is the number of neutrons in the next generation, $k=N_{i+1}/N_i$. For a critical reactor, $k=1$, for a subcritical reactor, $k<1$, and for a supercritical reactor, $k>1$.

A useful parameter related to the multiplication factor is the reactivity, designated by the lower-case Greek letter rho (ρ). Reactivity is defined as $\rho=(k-1)/k$, which is the fractional deviation of k from criticality. For a critical reactor, $\rho=0$, for a subcritical reactor, $\rho<0$, and for a supercritical reactor, $\rho>0$. Prompt supercriticality can be shown to occur when $\rho \geq \beta$, the delayed neutron fraction. From the definition of reactivity, we see that $k\rho+1=k$; then the multiplication factor at the onset of prompt supercriticality is $k_{ps} \approx 1+\beta$, since $k \approx 1$ (the symbol $\approx$ means "approximately equals").

The region in a nuclear reactor where the fuel is contained and the chain reaction occurs is called the core. Thermal energy is generated in the core, so coolant passages are needed in the core to provide for removal of this energy as heat, and the core contains coolant. The core also contains structural materials to support the fuel. Neutrons released in the core by fission may be absorbed by any of the materials in the core, or they may leak out of the core altogether into the surrounding materials. Usually, the core is surrounded by a material with a low neutron absorption cross section to encourage the neutrons to bounce back into the core. The surrounding region is called a reflector. The core and reflector are surrounded by shielding and enclosed within a pressure vessel or tank (most types of coolant are under pressure, but liquid metal coolant is not, and it only needs a tank to contain it). This typical arrangement in a generic nuclear reactor is shown in Figure 6.2.

In the figure, the core is shown to comprise fuel rods and coolant channels. This is a typical arrangement, but it is not the only possible one. Reactors have been proposed with molten-salt cores that serve both as the fuel and the coolant. Reactors with cores consisting of a bed of billiard-ball-size spheres, or "pebbles," have been tested and are the subject of intense current research and development. Then the coolant, which is a gas, flows through the interstices between adjacent pebbles. The figure shows the coolant entering the upper coolant plenum and flowing downwards through the core to exit from the lower coolant plenum, but in some designs the coolant flows upwards through the core.

Figure 6.2 also shows control rods, which are tubes containing neutron-absorbing materials that are inserted into or withdrawn from the core to adjust reactivity. Reactor control is much different from control of more familiar systems such as automobiles. In a car, if you want to go faster you push the accelerator pedal down more. When you reach the desired higher speed, you leave the accelerator pedal pushed down farther than its original position. In a nuclear reactor, when you want to increase power, you withdraw control rods. But when you reach the desired higher power level, you put the control rods back in to their original positions. Neglecting second-order effects, the critical condition is the same for any reactor power. [Second-order effects are phenomena such as temperature-related changes in nuclear reaction cross sections by the Doppler effect (see Section 5.4), if the coolant flow rate is not adjusted to maintain constant temperature.]

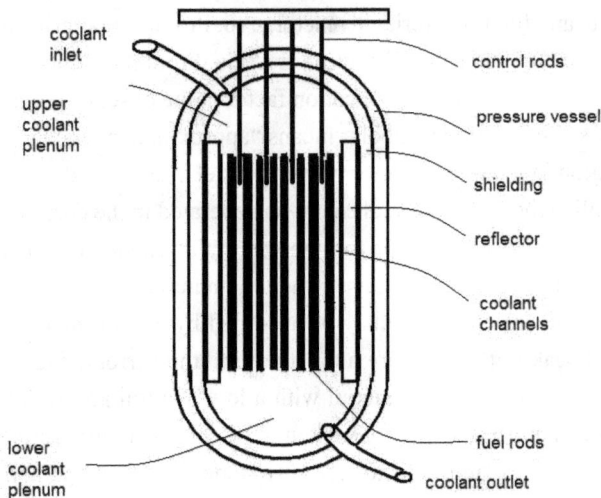

Figure 6.2—Typical layout of core, reflector, coolant plena, and reactor vessel

The process of keeping track of what happens to the neutrons is fundamentally an accounting procedure that uses a balance equation. For the core as a whole, the balance equation may be expressed conceptually as follows:

The rate of increase of the neutron population in the core	=	The rate at which neutrons are produced by fission	-	The rate at which neutrons are absorbed within the core	-	The net rate at which neutrons leak from the core

The net leakage rate is the difference between the rate at which neutrons leak out into the reflector and the rate at which they bounce back into the core from the reflector. The difference is due to absorption in the reflector and net leakage from the reflector into the surrounding shielding materials, where stray neutrons are mopped up by absorption.

Each term in this equation for the core as a whole is the sum over the entire core of the same process in each tiny piece of the core. Because the processes vary substantially within the core, it is necessary to calculate these processes locally, for each small region within the core, and then sum them up for any desired information about the core as a whole. How to do this is discussed in the next section.

6.2 HOW NUCLEAR ENGINEERS CALCULATE THE NEUTRON BALANCE

The discipline of calculating the neutron balance is called nuclear reactor physics, which is a subfield of nuclear engineering. Nuclear reactor physics is not the same as nuclear physics. The nuclear physicist studies what happens inside the nucleus. The reactor physicist analyzes the balance of free neutrons in reactors. One kind of product of nuclear physics research is microscopic reaction cross section data, which reactor physicists use in their accounting calculations.

Reactor physicists use two fundamentally different approaches to calculate the neutron balance. One, the deterministic approach, uses mathematical methods to solve the balance equation for each point in the reactor. The other, the statistical approach, samples the fates of individual neutrons introduced into the reactor. These two approaches are discussed in the following subsections.

6.2.1 Deterministic methods

Deterministic methods find a quantity called the neutron flux at each point in the reactor. Free neutrons in the reactor are flying around in all directions, with energies that vary from their initial energy after the fission event (a few MeV) to thermal energy (about 0.0253 eV). Imagine a unit cross-sectional area, say one square centimeter, at some point in the reactor, as shown in Figure 6.3. Now imagine an arrow perpendicular to the cross-sectional area, as also shown. The number of neutrons that cross this area per unit time (say, per second) in the direction of the arrow (not including neutrons going in the opposite direction), with a particular kinetic energy, is called the energy-dependent angular neutron flux, written as $\varphi(r, \Omega, E, t)$, where r is the position vector (relative to some arbitrary origin of a coordinate system), Ω (upper-case Greek letter omega) is a unit vector in the direction of the arrow, E is energy, and t is time (φ is the lower-case Greek letter phi). The parenthetical format indicates that φ is a function of, or depends on, the variables enclosed in the parentheses. The expression is spoken as "φ of r, Ω, E, and t." If you sum up the energy-dependent angular neutron fluxes for all possible directions of the arrow, the result is called the energy-dependent neutron flux, written as $\varphi(r, E, t)$. If you sum up the energy-dependent neutron fluxes for all possible energies, the result is called the one-speed neutron flux or the total neutron flux, $\varphi(r, t)$. The total neutron flux is the number of neutrons per unit time crossing perpendicularly to unit areas in all possible orientations centered at a specific point and at a particular time. One can also divide the whole energy range into "energy groups" and add up the fluxes for all energies within the separate groups, in the so-called "multigroup" method. The neutron flux at any level of this hierarchy is equal to the neutron density (number of neutrons per unit volume) at that hierarchy multiplied by the speed characterizing that hierarchy (e.g., for the one-speed neutron flux, $\varphi = nv$, where n is the total neutron density and v is the average neutron speed).

Figure 6.3—Unit area and perpendicular vector

Typical values for the total neutron flux in an ordinary nuclear power reactor are on the order of 10^{13} neutrons per square centimeter per second. If most of these are thermal neutrons, with speeds of about 2200 m/s, or 220,000 cm/s (which corresponds to a kinetic energy of 0.0253 eV), then the approximate neutron density is $n=\varphi/v\approx10^{13}/2.2\times10^5$ neutrons per cubic centimeter $= 4.5\times10^7$ n/cm^3. This is in contrast to the number densities in solids and liquids (about 10^{22} atoms or molecules per cubic centimeter) or in air (about 2.7×10^{19} molecules per cubic centimeter).

The neutron fluxes are the key to calculating most of what you need to know about the behavior of a reactor. For example, a reaction rate R (reactions per unit volume per unit time, e.g., reactions per cubic centimeter per second) for a particular reaction is given by $R(r,t)= \varphi(r,t)\Sigma(r)$, where $\Sigma(r)$ is the macroscopic cross section at r for the type of reaction in question. Then, if the cross section of interest is the fission cross section, the total fission rate for the whole reactor may be found by summing the fission rates throughout the reactor. Summing that rate over time tells you how much fuel is consumed, how much energy is produced, etc.

Equations can be written for the energy-dependent angular neutron flux, the energy-dependent neutron flux, the multigroup neutron fluxes, and the one-speed neutron flux. The simplest equation, that for the one-speed neutron flux, is presented below for illustration. It is called the one-speed diffusion equation. This is an example of a partial differential equation, which means that it involves partial derivatives. Recall the idea of a derivative introduced in Chapter 2. The x-component of the velocity was written as $v_x=dx/dt$, the derivative, or rate of change, of x with respect to t. When a quantity is a function of more than one variable, the rate of change of the quantity with respect to only one of the variables is called the partial derivative of the quantity with respect to that variable, and it is written $\partial f/\partial x$ if f (the function) is the quantity of interest and x is the variable with respect to which the derivative is being taken. The one-speed diffusion equation is

$$\frac{1}{v}\frac{\partial\varphi}{\partial t} = \nabla \cdot [D(r)\nabla\varphi] - \Sigma_a(r)\varphi + v\Sigma_f(r)\varphi + S_{ext}(r,t),$$

where it is implicit that $\varphi=\varphi(r,t)$. In this equation, D is called the diffusion coefficient, Σ_a is the total macroscopic absorption cross section, Σ_f is the total macroscopic fission cross section, v is the average neutron speed, υ (the lower-case Greek letter nu – note the subtle difference from v) is the average number of neutrons released in fission reactions, and S_{ext} is any neutron source other than neutron-induced fission (e.g., it may include spontaneous fission). The symbol ∇ (read as "del") is shorthand for the "gradient operator," which has different forms in different coordinate systems. In Cartesian (rectangular) geometry, it is

$$\nabla = \frac{\partial}{\partial x} i + \frac{\partial}{\partial y} j + \frac{\partial}{\partial z} k,$$

where i, j, and k are the unit vectors introduced in Chapter 2. This operator takes the indicated derivatives of the quantities immediately following it.

The one-speed diffusion equation is an equation for $\varphi(r,t)$. That is, using appropriate methods, one may find, or solve for, φ at every position r and every time t of interest. There are standard formal methods that enable some partial differential equations to be solved with pencil and paper, or "analytically," but even the one-speed diffusion equation is too complicated for such methods when the spatial variations of the cross sections and diffusion coefficient are taken into account. Therefore, computer programs, or codes, using numerical methods for the solution of the equations, are written for practical application. (For pedagogical purposes, the one-speed or two-group equations have been solved analytically in very simple geometries with uniform materials.[2]) Furthermore, the macroscopic cross sections and diffusion coefficient are actually energy-dependent, and the values used in the one-speed or multigroup equations must be properly calculated using the energy-dependent neutron flux, which is not known at the outset. Usually, these values (the "group constants") are evaluated in separate calculations with simplified geometry and a finely divided, or many-group, energy spectrum.[3] These group constants are then used in the one-group or few-group diffusion equations, which are solved for the whole reactor.

The results of such calculations may look like Figure 6.4, which shows the spatial variation of the thermal neutron flux and fission power in a proposed pebble-bed reactor.

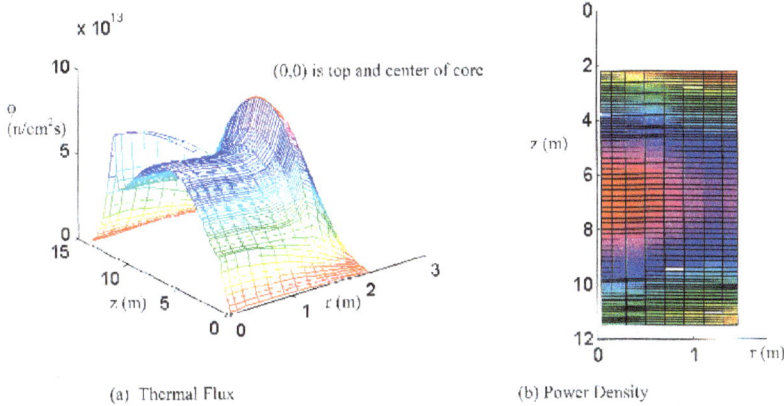

(a) Thermal Flux

(b) Power Density

Figure 6.4—Thermal flux and power density in proposed HTR-Modul 200 pebble-bed reactor.[4] This figure was taken from an externally released report of an agency of the U.S. government and is in the public domain.

6.2.2 Statistical methods

In an alternative to deterministic methods, one may track individual neutrons from fission source points in the core to their ultimate fates. Since there are about 10^7 free neutrons in every cubic centimeter of the core, and the core comprises millions of cubic centimeters, it is impractical to track every neutron in the core. Statistical methods, also called Monte Carlo methods, construct a statistical sample of the different fates of neutrons in a reactor.

Since the fission source distribution is not known at the outset, an initial distribution is assumed, and sample neutrons are chosen by a random number generator that picks the origin point and direction for each source neutron arbitrarily, but with a "weighting" that obeys the assumed source distribution. Based on the reaction cross sections, which can be shown to be equivalent to reaction probabilities along the neutron's path, the neutron is allowed to go in its initial direction until it undergoes a reaction. How far it goes, what the reaction is, and the direction and energy of any emergent neutrons are chosen similarly—by random numbers weighted according to the probabilities of the possible outcomes.

After a large number of neutrons have been tracked—typically about a million—a new source distribution is calculated from the distribution of fission reactions observed in following those neutrons. The whole process is repeated, or cycled, until the neutron source distribution is essentially the same for successive cycles. The source distribution is then said to have converged. Except for the refined source distribution, the results of these first cycles are discarded, and then, with the converged source, many new cycles are followed. The neutron source distribution continues to be updated, but it no longer changes very much.

Because of the probabilistic selection of the outcomes of all the neutron reactions, the results for such quantities as reaction rates will not be the same for two arbitrarily chosen cycles. As many cycles are performed, the results acquire statistical distributions, which can be characterized in terms of average values and variances. The variance in a statistical distribution is a standardized measure of how widely or narrowly scattered the individual values in the distribution are. Many natural phenomena follow a "normal," or "Gaussian," distribution; Figure 6.5 shows Gaussian distributions with large and small variances around the same average value (also called the mean value), arbitrarily chosen as 1. The probability that the event in question happens in a small region dx around x is $f(x)dx$. The total probability, equal to the area under the curve from $x = -\infty$ to $x = +\infty$, is 1 for both curves, meaning that if the event is going to happen, it will surely happen somewhere; the probability that the event will happen at all is a separate issue.

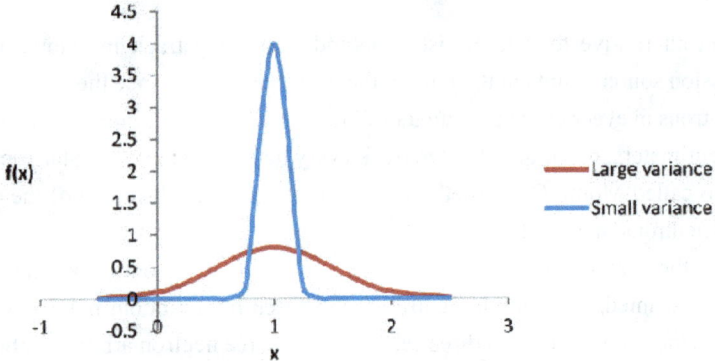

Figure 6.5—Two Gaussian distributions. The mean value is the same for both distributions, but the variance is smaller for the more highly peaked distribution.

The variance is also a measure of how much the average value of a sample of some quantity, such as the fission rate in a small region around a point in the reactor from a sample of source neutrons, differs from the average of all the values of that quantity, i.e., the fission rate near that point from all the source neutrons. (Of course, most source neutrons won't induce fissions in any particular small region of the reactor, but the more source neutrons you track the more fission reactions you are likely to have in that small region.) The more neutron histories you sample, the nearer the sample comes to the real source, and the smaller the variance becomes. Likewise, the more cycles you perform, the more closely the average over many cycles comes to the actual value of the quantity being sought. Therefore, enough cycles are performed, with enough neutrons per cycle, to reduce the variances to desired levels. Typically, 100 cycles of a million neutrons each may be run; this can take a week or more even on a powerful desktop scientific workstation.

In deterministic methods, the equations are written in a coordinate system appropriate for the geometry of the reactor. For example, if the core can be idealized as an assembly of rectangular parallelepipeds, a Cartesian geometry can be used, as in the gradient operator as written above. If the core is basically cylindrical, the equations can be written in a different form, in which the spatial variables are radial position and angular position around the axis of symmetry (these variables replace x and y in the equation), along with height z. But for complicated geometries, such as one in which cylindrical rods are placed in rectangular fuel assemblies, approximations must be made in the representation of at least some components; these approximations reduce the accuracy of the calculations. In statistical methods, the equations are not solved directly, so it is possible to represent very complicated geometries very accurately. For this reason, statistical methods are considered to have the greatest accuracy of available techniques.

However, to obtain good statistics in small regions, one must track many neutrons into those regions. To keep the calculation from becoming too time-consuming, one employs variance reduction techniques such as "particle splitting" to send more particles into regions of greater importance or interest, and fewer into unimportant regions. There is a great deal of art in the use of statistical methods in reactor physics, and those who are good at it are in high demand.

6.2.3 Computer code quality control

Reactor physics computer codes typically contain tens of thousands of lines of computer program instructions. If you have never seen programming language, you may find Figure 6.6 instructive; this figure presents instructions in the FORTRAN programming language for adding the integers 1 through 10. It is very easy to introduce errors, or "bugs," into long, complicated programs. Usually, a bug prevents a code from running at all, and often a bug will cause the computer to produce such unphysical results that it is immediately obvious that something is wrong. However, the risk that bugs will produce small but important errors in computer code calculations is great enough that practical computer codes are subjected to rigorous programs of quality control.

```
i=0
do j=1,10
i=i+j
end do
```

Figure 6.6—Instructions in FORTRAN for summing the integers 1 through 10

There are two principal aspects of computer code quality control: verification and validation.[5, 6] Verification addresses the correctness of the coding and the accuracy of the code's calculations. Validation addresses the ability of the code to produce results in agreement with physical experiments.

Verification is accomplished by meticulous checking of the coding and by confirming the ability of the code to produce the correct answers to problems for which there are known exact analytical solutions, or analytical "benchmarks" (see Reference 7 for an example).

Validation is accomplished by comparing the predictions of the code to relevant experimental measurements, or experimental benchmarks. However, before one can judge a code confidently by such comparisons, one must have a sound basis for trusting the accuracy of the experimental measurements. In reactor physics and criticality safety (a related field in which the goal is to ensure that such assemblies as spent fuel in shipping casks cannot become critical in accident scenarios), there are major international programs, led by the Idaho National Laboratory, for evaluating as many potential experimental benchmarks as possible.[8, 9]

A computer code by itself cannot tell you anything. To apply a code to a specific problem, one must provide the code with information about the problem in the form of input, which consists of properly formatted data about the problem geometry, physical properties, and all other relevant information. The input data as a whole are called a computational model. For complicated reactor problems, computational models can contain thousands of lines of data. Therefore, the models as well as the codes must be verified and validated.

The accuracy requirements for nuclear reactor physics are very high. For example, for predictions of the multiplication factor k, the difference between criticality ($k=1$) and prompt supercriticality ($k \approx 1+\beta \approx 1.0065$) is less than 1%. For predictions of

k to be meaningful, they must be accurate to about 1% of this difference; i.e., k needs to be predicted with an accuracy of 10^{-4} or better.

Fortunately, modern reactor physics computer codes and computational models are capable of such accuracy. They are so good that when the predictions of a well validated code/model combination differ significantly from experimental measurements, the error is as likely to be in the experimental procedure as in the calculation!

6.3 SOME IMPORTANT REACTOR PHYSICS PHENOMENA

This section introduces some aspects of the chain reaction that apply to most or all nuclear reactors.

6.3.1 Reactivity feedback

The idea of reactivity feedback phenomena was introduced in Section 5.4 in the discussion of Doppler broadening of the resonances in the absorption cross section of uranium-238. In general, a desirable feedback phenomenon in a nuclear reactor responds to a fluctuation in reactor power by a natural and immediate change that tends to restore the power to its level before the fluctuation. Such feedback phenomena provide operational stability—i.e., in the absence of control inputs, the reactor will run at essentially constant power.

It is possible for adverse feedback phenomena to occur. In such feedback processes, a fluctuation in reactor power would produce physical changes that would cause the fluctuation to grow. If the fluctuation were a power increase, the power would continue to increase, probably at an increasing rate. This type of behavior is unstable. Instability must be avoided unconditionally. An unstable operating regime was discovered in the startup testing of the MAPLE reactor in Canada that was intended to replace the aging NRU reactor, as discussed in Section 10.1.2. The problem was determined to be insurmountable, and the program was cancelled. The reactors will never be permitted to operate.[10] An instability caused the Chernobyl accident, which is discussed in detail in Chapter 13.

Feedback in nuclear reactors arises from changes in a number of different physical properties of reactor materials, such as fuel temperature, moderator temperature, moderator density, and coolant density. The effects of these changes are characterized mathematically by parameters called reactivity coefficients. The interested reader can find the mathematical description of nuclear reactor feedback in Appendix I. However, the following text explains the concepts verbally.

Recall the definition of reactivity ρ given in Section 6.1.—i.e., the fractional deviation of the multiplication constant from (usually) k=1. Positive reactivity means that the reactor power is increasing, and negative reactivity means that it is decreasing. The power coefficient of reactivity α_p (alpha-sub-P) is defined as the change in reactivity per unit change in power, for very small changes in power. For stability, α_p must be negative: When power increases, reactivity must decrease, and vice versa. The power coefficient of reactivity depends on the changes in the physical properties mentioned above:

Total change in reactivity = change in reactivity caused by changes in fuel temperature
+ change in reactivity caused by changes in moderator temperature
+ change in reactivity caused by changes in moderator density
+ change in reactivity caused by changes in coolant density.

For each term in this equation, the changes in reactivity are those that occur for that small unit change in reactor power. The first term on the right-hand side of the equation then becomes

change in reactivity per unit change in fuel temperature times change in fuel temperature per unit change in reactor power,

and the other terms can be expanded similarly. We can write

α_{T_f} = *change in reactivity per unit change in fuel temperature,*
α_{T_m} = *change in reactivity per unit change in moderator temperature,*
α_{D_m} = *change in reactivity per unit change in moderator density,*
and α_{D_c} = *change in reactivity per unit change in coolant density.*

Then the equation for α_p becomes

$\alpha_P = \alpha_{T_f}$ x *change in fuel temperature per unit change in reactor power*
$+ \alpha_{T_m}$ x *change in moderator temperature per unit change in reactor power*
$+ \alpha_{D_m}$ x *change in moderator density per unit change in reactor power*
$+ \alpha_{D_c}$ x *change in coolant density per unit change in reactor power.*

An increase of power heats things up and makes them expand, so inherently the rates of change of the temperatures with respect to power are positive and the rates of change of the densities with respect to power are negative. Therefore, in order for α_p to be negative as required, it is sufficient that $\alpha_{T_f} < 0$, $\alpha_{T_m} < 0$, $\alpha_{D_m} > 0$, and $\alpha_{D_c} > 0$. These conditions are designed into the reactor configuration and verified during reactor startup testing.

For boiling-water reactors (BWRs), the coolant void fraction f_v can be used in place of the coolant density. The coolant void fraction is the fraction of the volume occupied by a fluid that is empty space (note that for neutron absorption, vapor is essentially empty space). The void fraction increases dramatically when the coolant boils, but even when the coolant remains liquid its expansion effectively creates empty space between coolant molecules. Then α_{D_c} is replaced by the void coefficient of reactivity,

$$\alpha_{f_v} = \text{\textit{change in reactivity per unit change in void fraction,}}$$

and the corresponding term in the equation for α_p is replaced by

$$\alpha_{f_v} \times \text{\textit{change in void fraction per unit change in reactor power.}}$$

An increase in power causes an increase in void fraction, so for stability α_{f_v} must be negative.

As an illustration of how physical changes in a reactor determine whether the reactor is stable, we will examine how the void coefficient accounts for a change in void fraction.

In an ordinary water-cooled reactor (called a light-water reactor, or LWR—see Chapter 11), the hydrogen nuclei in the water molecules behave towards neutrons in two conflicting ways. For one thing, they act as a moderator—i.e., they slow the neutrons down from fission energy to thermal energy, at which they are much more readily absorbed by fuel nuclei to induce fission than they are at high energy. An increase in moderation increases the fission rate, and vice versa. But the hydrogen nuclei also have a significant absorption cross section for thermal neutrons. This "parasitic" absorption decreases the thermal neutron population and reduces the fission rate. When the void fraction increases, both the parasitic absorption rate and the moderation rate are reduced. When the void fraction decreases, both rates are increased. For either increases or decreases in void fraction, moderation and absorption are in competition to increase or decrease the thermal neutron population and consequently the fission power production rate. Which one wins determines whether the reactor is stable or not.

If a particular critical state of an LWR is characterized by a large thermal neutron population, a change in the void fraction may not cause a proportional change in the population of moderated neutrons. Then the change in neutron absorption is more important, and the reactor is unstable. Such a critical state is called overmoderated. On the other hand, if the critical state is characterized by a relatively small thermal neutron population, a change in the void fraction may cause a disproportionally large change in the population of moderated neutrons, dominating over the change in absorption. Then the reactor is called undermoderated, and it is stable. Details of the reactor design determine whether the reactor is overmoderated or undermoderated, and careful design analysis is devoted to ensuring undermoderation—i.e., a negative void reactivity coefficient—in all operating conditions. However, even though modern design methods are extremely accurate, it is necessary to verify the negative void coefficient, along with the proper signs of all other reactivity coefficients, through rigorous testing before the reactor can be permitted to operate. The MAPLE reactor is a case in point.

6.3.2 Reactivity control

A reactor running in a steady state must be critical (reactivity $\rho=0$), and a reactor must be supercritical ($\rho>0$) or subcritical ($\rho<0$) to change in power. Fine-tuning of the reactivity is accomplished by inserting or withdrawing control rods, which are made of materials such as boron, which strongly absorb thermal neutrons. Nuclides that significantly absorb neutrons without inducing fission are called neutron "poisons." It is actually only the ^{10}B isotope, which comprises only 20% of natural boron, that has a high thermal neutron absorption cross section (3840 barns), but it is not necessary to separate this isotope from the more common ^{11}B, which has a thermal neutron cross section of only 5 barns.

Reactor startup and power increases are possible because reactors are provided with a margin of excess reactivity: If the control rods are withdrawn past the positions at which the reactor is critical, it will become supercritical and the power can be increased. As fuel is consumed, the excess reactivity decreases. For the purpose of maintaining sufficient excess reactivity, reactors may be provided with neutron poisons that are depleted along with the fuel so that the excess reactivity remains constant until they are used up. Poisons supplied in the correct amount and distribution for this purpose are called burnable poisons.

Often, boron is the burnable poison of choice. It may be provided in rods distributed through the fuel rod lattice, it may be dissolved in the coolant, or it may be a combination. Poisons dissolved in the coolant are called soluble poisons. Another burnable poison, gadolinium oxide, can be mixed with the fuel.

6.3.3 Fission-product poisoning

Some fission products are poisonous to thermal neutrons. Two especially important fission product poisons are xenon-135 and samarium-149.

Xenon-135 is both a fission fragment and a product of the beta-decay chain that begins with the fission fragment tellurium-135:

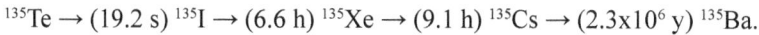

$$^{135}\text{Te} \rightarrow (19.2 \text{ s}) \ ^{135}\text{I} \rightarrow (6.6 \text{ h}) \ ^{135}\text{Xe} \rightarrow (9.1 \text{ h}) \ ^{135}\text{Cs} \rightarrow (2.3 \times 10^6 \text{ y}) \ ^{135}\text{Ba}.$$

(I is iodine, Cs is cesium, and Ba is barium.) Xenon-135 absorbs thermal neutrons by the reaction $^{135}\text{Xe}(n,\gamma)^{136}\text{Xe}$, which has a huge cross section of 2.6×10^6 barns (typical cross sections have magnitudes closer to 1 barn). Fission fragments are the immediate results of the fission reaction, which may be very short-lived; fission products are longer-lasting, and may be fission fragments themselves or the results of decay chains as shown above.

In steady reactor operation, the ^{135}Xe concentration reaches an equilibrium level, where production of ^{135}Xe from fission and the decay chain shown above is balanced by its consumption from neutron absorption and its depletion by its own decay to ^{135}Cs. However, when the reactor is shut down, the fission and neutron absorption processes no longer occur, but the production from the decay chain continues. No more ^{135}Te is produced once the fission reactions cease, but it takes time for all of the ^{135}Te present at the time of shutdown to decay into ^{135}Cs. The subsequent decay of ^{135}Cs into the stable nuclide ^{135}Ba is irrelevant, because ^{135}Cs has such a long half-life. The supply of ^{135}Te is quickly lost to decay, but the resulting ^{135}I lingers longer because of its longer half-life, as does the ^{135}Xe itself. During the first few hours after shutdown, in a typical LWR the concentration of ^{135}Xe builds to about twice the equilibrium value it had before reactor shutdown, then decays away over the next 20 hours or so.[11]

The huge thermal neutron absorption cross section of ^{135}Xe adds so much negative reactivity during the first 20 or 30 hours after shutdown that even if all the control rods are fully withdrawn, there is not enough excess reactivity to permit the reactor to be restarted. The equilibrium concentration of ^{135}Xe depends on the equilibrium neutron flux, as does the peak concentration after shutdown. The higher the equilibrium neutron flux, the higher the peak concentration and the longer it takes for the ^{135}Xe to decay away enough for restart to be possible. Unnecessary down time is very costly: Each day of shutdown at a typical power reactor costs about $1M in lost revenue. Therefore, thermal reactors operate at neutron fluxes less than about 5×10^{13} neutrons per square centimeter per second so that they can be restarted within a reasonable time after shutdown.

Even at such restricted neutron flux levels, xenon poisoning prevents thermal nuclear reactors from being stopped and restarted at will. Therefore, thermal reactors are not suitable for on-demand load-following electricity production. Instead, they are appropriate for base-load electricity supply.

Samarium-149 is produced by successive beta-decay events from the fission fragment neodymium-149 via promethium-149:

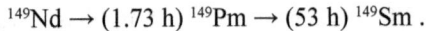

$$^{149}\text{Nd} \rightarrow (1.73 \text{ h}) \; ^{149}\text{Pm} \rightarrow (53 \text{ h}) \; ^{149}\text{Sm} .$$

Samarium-149 is stable. It absorbs thermal neutrons by the reaction $^{149}\text{Sm}(n,\gamma)$ ^{150}Sm. This reaction has a cross section of 4100 barns, which is a large cross section, although not so great as that of ^{135}Xe.

Like ^{135}Xe, ^{149}Sm accumulates to an equilibrium level in steady reactor operation. Production of ^{149}Sm from the decay chain shown above is balanced by its consumption from neutron absorption. When the reactor is shut down, the neutron absorption process no longer occurs, but the production from the decay chain continues. No more ^{149}Nd is produced once the fission reactions cease, but the ^{149}Nd and ^{149}Pm present at shutdown decay into ^{149}Sm. Furthermore, since ^{149}Sm is stable, it does not decay away but reaches a new, higher concentration in a few half-lives of ^{149}Pm. Sufficient excess reactivity must be provided to overcome the negative reactivity imposed by the built-up concentration of ^{149}Sm and restart the reactor. Once the reactor power is restored to its pre-shutdown level, the excess ^{149}Sm burns out by the (n,γ) reaction and the ^{149}Sm concentration returns to its pre-shutdown equilibrium value.

6.3.4 Reactivity response in accident conditions

The worst accident that a nuclear reactor is designed to sustain without loss to the systems, structures, and components necessary to ensure public health and safety is called the design-basis accident (DBA).[12] Usually, this is a complete loss-of-coolant accident (LOCA). For all reactor types, there are certain general principles that govern the reactivity response in a LOCA.

When the reactor loses coolant, the temperature in its core will increase. This temperature increase shuts the neutron chain reaction down by one or more of the built-in negative feedback responses. These responses are immediate and not dependent on any actuated engineered mechanical or electrical systems. They are simply the natural physical changes that result from temperature change.

After the chain reaction stops, the core continues to generate heat by the radioactive decay of the fission products, but at a much reduced rate. However, since the primary coolant has been lost, the core may still overheat unless this decay heat is somehow removed. Most currently operating reactors depend on engineered systems called emergency core cooling systems (ECCSs) to remove the decay heat. Examples of ECCSs are shown in the discussions of specific reactor types in subsequent chapters. Future reactors will feature passive safety, in which natural heat-removal processes will take the heat away from the core without operator intervention or mechanical or electrical actuation.

When the ECCS or passive cooling reduces the core temperature, the chain reaction may be able to start up again. Then negative feedback responses come into play again to stabilize the reactor, but at a much lower power level than that at which the reactor operates when coolant is flowing normally. To prevent this recriticality, some kind of system is present to provide a large negative reactivity insertion. Typically, this system comprises special control rods called "SCRAM" rods,[a] but it may involve boron dissolved in emergency cooling water or neutron poisons contained in small spheres dropped into appropriate locations in or near the core. Insertion of SCRAM rods or other shutdown reactivity control systems is said to "SCRAM" the reactor. In reactors with true passive safety, the SCRAM system has to operate without operator intervention.[13, 14]

---

[a] Dr. Charles A. Wemple provides the following explanation of the acronym SCRAM: The world's first nuclear reactor was the CP-1 ("Chicago Pile 1") reactor, designed by Enrico Fermi and built in a squash court at the University of Chicago under his supervision (http://en.wikipedia.org/wiki/Chicago_Pile-1). The shutdown rod for CP-1 was suspended by a rope hanging from a pulley over the squash court and tied to a railing near the instrument panel. A technician with an axe stood at the ready to cut the rope, if given such instructions. Hence, the safety control rod axe man, or SCRAM.

CHAPTER 7

HEAT REMOVAL
AND POWER GENERATION

In order for the heat generated by a nuclear reactor core to be used, it must be carried out of the core by a flowing coolant, which is normally water. Usually, the heat is converted to electricity by the same sort of thermodynamic cycles that are used by fossil-fuel-burning power plants. However, the heat may be used directly. As discussed in Chapter 16, high-temperature gas-cooled reactors may be used in the future to produce hydrogen for transportation fuel by electrolysis of water, which takes place much more efficiently at the high temperatures achieved by the helium coolant emerging from the cores of such reactors than at the lower temperatures achieved by other kinds of thermal plants.

In this chapter, we first discuss the basic principles of thermodynamics that govern the conversion of heat into electricity. Then we look at how these principles are applied in a generic nuclear reactor.

7.1 BASIC THERMODYNAMICS

The basic conservation laws applicable to the thermodynamic cycles used in nuclear reactors are the purely classical principles discussed in Chapter 2: the law of conservation of mass, the law of conservation of momentum, and the first and second laws of thermodynamics. (The zeroth law of thermodynamics also applies, but it is taken for granted and doesn't need to be expressed in separate equations.)

A thermodynamic cycle is a process in which a thermodynamic system undergoes a sequence of changes that eventually returns it to its initial state. A thermodynamic system is any specific collection of matter. A particular thermodynamic state is a condition in which all physical properties have specific values. In a nuclear reactor, the coolant flows in a closed loop, and one may follow a small "fluid element"—a specific packet of the fluid—around the loop from any arbitrary starting point. This fluid element, which is identical to all the other fluid elements one may choose, may be regarded as the thermodynamic system to be analyzed. The idea of a heat engine—a system that converts heat to work—was introduced in Chapter 2; here, this circulating fluid element is the heat engine of interest. In a cooling fluid, pressure, temperature, and density are sufficient to define the state of the system. In an ideal gas, any two of those variables are sufficient, but in a fluid that undergoes phase changes in the cycle (i.e., from liquid to vapor and back), the density undergoes changes at constant temperature and pressure as the fluid vaporizes or condenses, so the value of the density is needed along with temperature and pressure to define the state completely. This last point is worth emphasizing because of its significance in thermodynamic cycles: If a fluid is boiled at constant pressure, its temperature remains the same throughout the boiling process, from the point where the liquid just begins to boil until it has all been transformed into vapor.

It is noted in Chapter 2 that the second law of thermodynamics says that if a thermodynamic system operating in a cycle converts heat to work, it cannot do so with perfect efficiency—i.e., it cannot convert all of the heat it receives into work. It must reject some of the heat into a heat sink. It can be shown that a consequence of the second law is that the most efficient possible thermodynamic cycle is one in which the system receives all of its heat reversibly at a constant temperature (which we denote as T_1) and rejects all of the waste heat reversibly at a lower constant temperature (which we denote as T_2). This idealized cycle is called the Carnot cycle, proposed by the French scientist Nicolas Léonard Sadi Carnot in 1824. If the heat added at T_1 is Q_1, the heat rejected at T_2 is Q_2, and the work done in the cycle is W, then the efficiency of this cycle (the fraction of the heat input that is converted into work) is

$$\eta = \frac{W}{Q_1} = \frac{Q_1 - Q_2}{Q_1} = 1 - \frac{Q_2}{Q_1} = 1 - \frac{T_2}{T_1}$$

where η is the lower-case Greek letter eta.

In this equation, the temperatures must be expressed in absolute temperature, in which zero is absolute zero, the temperature at which molecular motions reach the lower limit permitted by quantum mechanics. For our purposes, we can consider molecular motions to cease at absolute zero. Obviously, we can't use temperature scales like the Fahrenheit and Celsius scales in this equation, because we would predict different values of efficiency depending upon which scale we used! Actually, the relationship in the equation is used to define absolute temperature scales. In different absolute temperature scales, the size of a degree is arbitrary, but the zero points are the same and the numerical values of the temperatures are proportional. For example, a degree in the Rankine scale, which is the absolute version of the Fahrenheit scale, is equal to 5/9 of a degree in the Kelvin scale, which is the absolute version of the Celsius scale. Absolute zero is equal to -459.67 °F and -273.15 °C. Water freezes at 32 °F, or 491.67 °R, which is 0 °C or 273.15 K. (In the SI system of units, the degrees symbol is not used for degrees on the Kelvin scale. The temperature unit in the SI system is just called the kelvin, denoted by K.) Then 491.67/273.15=9/5, in accordance with the definitions of the Rankine and Kelvin scales.

In the Carnot cycle, the heat transfer processes must be reversible. However, heat transfer across a non-infinitesimal temperature difference is not reversible. But heat flows from warmer bodies to cooler bodies at a rate proportional to their temperature difference, so heat transfer across an infinitesimal temperature difference would take place at an infinitesimal rate—i.e., it would take forever. So the Carnot cycle is not really achievable. It is only an idealized cycle that shows an upper limit to the efficiency of a heat engine. In the Carnot cycle, all other processes must also be performed reversibly. Again, this requirement is an idealization that cannot be achieved in practice. Any mechanical equipment for performing work suffers from friction, which is an irreversible process.

With these practical limitations set aside, we can imagine how a Carnot cycle could work. We could take a fluid such as water, which boils at 212 °F at standard atmospheric pressure (14.7 pounds per square inch, or psi) but boils at much higher temperatures at higher pressures. For example, it boils at about 545 °F at 1000 psi. We could start with a liquid-vapor mixture of water (steam) at 212 °F and 14.7 psi, and compress it to 1000 psi. The compression pump would be insulated so that no heat would be transferred into or from the steam, and we are imagining that there would be no friction or other irreversibility in the pump. If the initial mixture were properly selected (point 1 in Figure 7.1), there still would be a little vapor left at 1000 psi (point 2 in the figure), so the mixture would remain "saturated" (mixed liquid and vapor) throughout the process, and as the temperature rose as a result of the compression, at all times it would follow the boiling point of water at the increasing pressures. (Note that the temperature rises or falls during compression or expansion, even without any heat transfer into or out of the fluid. The thermal energy of the fluid increases or decreases because of the work done on or by the fluid.) Thus, at the end of the compression process, the temperature would be 545 °F. Then we could boil it at constant pressure and temperature until only a tiny amount of liquid remained (point 3 in the figure). We could then pass it through a turbine in which it would do work by turning the turbine shaft. The turbine would be insulated, so that no heat would be lost during this step, and it would again be frictionless. As the steam did work in the turbine, its temperature and pressure would fall. We would duct the steam out of the turbine when its pressure and temperature once again became 14.7 psi and 212 °F (point 4). Next, we would send the steam through a condenser, where the necessary waste heat would be rejected at constant temperature and pressure, and more of the vapor would be condensed into liquid. Finally, the liquid-vapor mixture would pass again into the compression pump, and the cycle would be completed.

This process is shown on a diagram of pressure versus specific volume (the reciprocal of density) in Figure 7.1. The bell-shaped curve is called the vapor dome, which on the left of the peak shows the relationship between pressure and specific volume when liquid just begins to boil, and which on the right of the peak shows the relationship when the liquid has all just turned to vapor. The peak itself is called the critical point, at and above which there is actually no difference between liquid and vapor. The process is illustrated schematically in Figure 7.2.

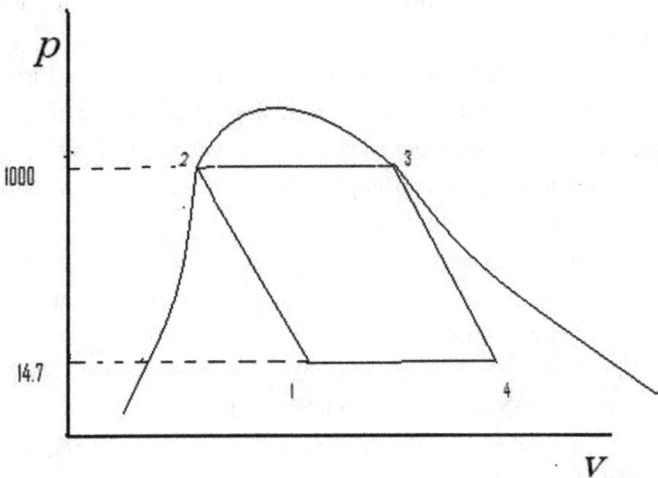

Figure 7.1—The Carnot steam cycle on a *p-v* diagram

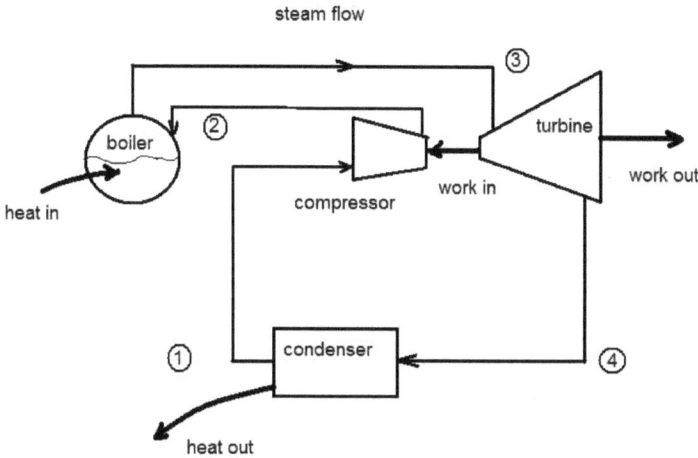

Figure 7.2—Processes in Carnot steam cycle

In real cycles, heat is not transferred across infinitesimal temperature differences, and the machinery is not frictionless. However, one could construct machinery that would operate approximately in a Carnot cycle. This is not done in practice because the compression process in step 1-2 requires a lot of work, and even if the irreversibilities in this process were small, they would reduce the efficiency of the cycle more than using an alternate cycle in which the steam in the condenser is allowed to become all liquid. Because liquids are essentially incompressible, it takes very little work to increase the pressure of a liquid. Figure 7.3 illustrates the simplest such cycle, called the Rankine cycle. The pump pressurizes the liquid in step 1-2, in which the pressure rises a great deal while the specific volume remains almost constant. The temperature, which is not shown in the p-v diagram, remains almost constant at 212 °F during this step. Then, in step 2-3, heat is added to bring the temperature to the boiling point at 1000 psi. This violates the requirement in the Carnot cycle that all the heat has to be added at a single temperature, and therefore introduces inefficiency. In step 3-4, additional heat is added at constant pressure and temperature to boil the water completely to vapor. The turbine process in step 4-5 is the same as the turbine process in the Carnot cycle, but in the condenser enough heat is removed to reduce all the steam to liquid.

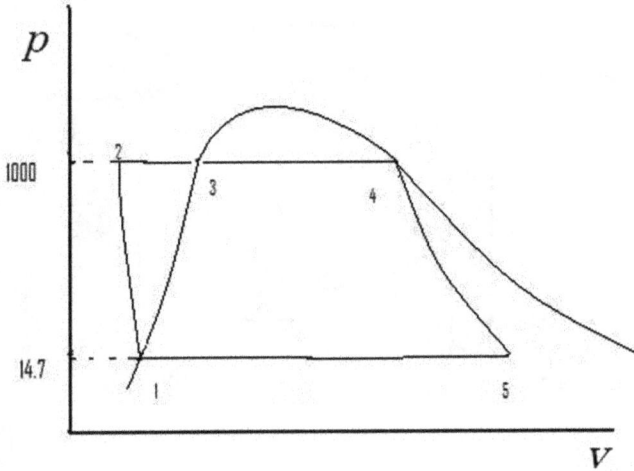

Figure 7.3—The Rankine cycle on a *p-v* diagram

The Rankine cycle is more practical than the Carnot cycle, and it is more efficient than real-world approximations to the Carnot cycle, but it does suffer from the inefficiency introduced by adding heat at lower temperature than the boiling point at the higher pressure. In practice, the inefficency is avoided by transferring heat *within* the system to raise the temperature in step 2-3 of Figure 7.3. This transfer still must take place reversibly, so a means must be found to expose the flow of step 2-3 to an increasing temperature along the flow path. Ideally, this increasing-temperature flow path would be found inside the turbine (step 4-5) in the opposite direction from the flow in the turbine that performs work. Such an idealized arrangement, called a regenerative cycle, is depicted in Figure 7.4.

It would actually be impractical to insert counterflow piping inside the turbine as shown in Figure 7.4, so in practice steam is bled from the turbine at several places to heat the flow to the boiler. Cycles applying this approach are called extraction cycles, and they are the standard method for converting heat to electricity in modern power plants. For more details, you can consult most textbooks on thermodynamics. A more focused treatment can be found in Reference 1.

By the use of extraction cycles, usually with even more sophisticated features, practical power plants can come quite close to the Carnot efficiency of a cycle operating between the maximum and minimum temperatures of the system.

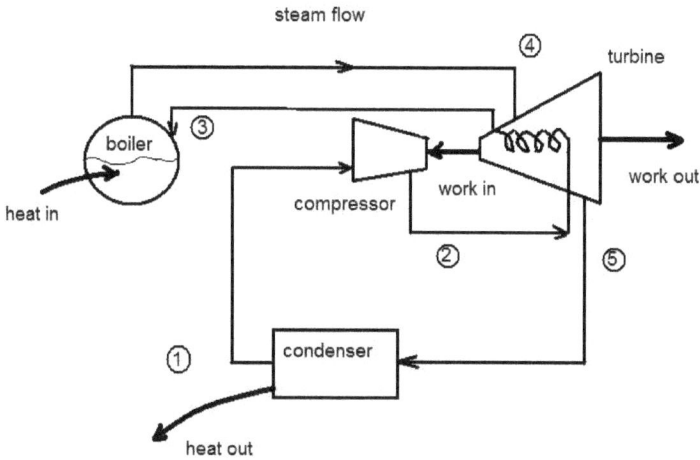

Figure 7.4—A regenerative cycle shown schematically

One can also build heat engines in which the working fluid does not change phase during the cycle—i.e., it does not change from liquid to vapor and back. The Otto cycle, illustrated in Figure 7.5, is an example of a gas cycle. The gas in the cylinders of an internal combustion engine does not go through a cycle, because it is drawn into the cylinder in step 0-1 and exhausted to the atmosphere in step 4-1, but the Otto cycle is a good approximation to the internal combustion engine for such purposes as calculating efficiency.

The reason that steam cycles are so commonly used in power plants is that a great deal of energy can be transferred into the working fluid as it boils without raising its temperature. The heat required to change a liquid to vapor at its boiling point at constant pressure is called the heat of vaporization. For water at 212 °F (100 °C), the heat of vaporization is 970.3 Btu/lbm,[a, 2] whereas the specific heat of the liquid phase at the same temperature (the heat required to change the temperature by 1 °F) is 1.01 Btu/lbm-°F.[3] So to absorb the same amount of heat required to boil liquid to vapor at 212 °F, the liquid would have to rise in temperature by 960.7 °F (970.3/1.01) if it didn't boil.

---

[a] The Btu is the British thermal unit, a common unit of energy in the U.S. and Great Britain. It is the amount of energy needed to raise the temperature of one pound (lbm, for pound of mass) of water by 1 °F at standard atmospheric pressure. There are several exact definitions of the Btu, which vary according to the temperature at which the definition is made. They are all approximately 1055 J.

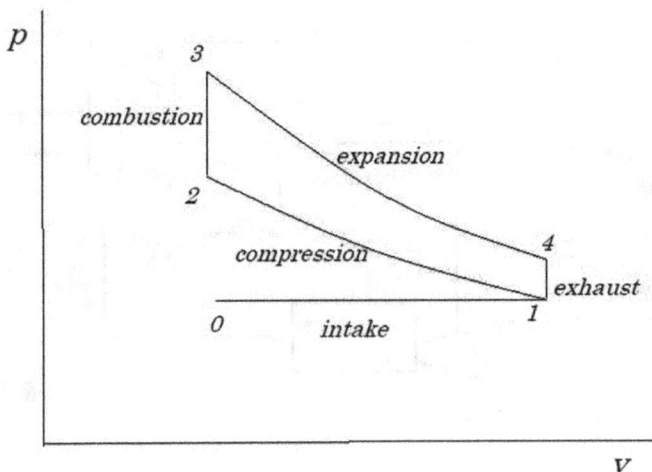

Figure 7.5—The Otto cycle

The advantage of not raising the temperature is that materials that would not be usable at high temperatures can be used in turbines, piping, and other components.

Next, we examine the layout of typical nuclear power plants and see what kinds of equipment are used to produce nuclear heat and ultimately convert it to electricity.

7.2 THE GENERIC NUCLEAR REACTOR

There are several types of nuclear reactors, which differ considerably in some fundamental aspects of their designs. However, there are some features common to most reactor types. This section briefly lays out these features as a foundation for the detailed discussion of specific reactor types in the following chapters.

Figure 7.6 shows our generic nuclear reactor and power plant. The heart of the system is the reactor core, which is contained within a reactor vessel. The reactor vessel and its contents are illustrated in Figure 6.2, which is repeated below for convenience as Figure 7.7. This vessel is usually a pressure vessel, but for cores cooled by liquid metal or a molten salt, and for some research reactors, it is simply a tank. Some kind of coolant passes through the core to remove the heat. Because the coolant can pick up some radioactivity, either by neutron activation of the coolant itself or by contamination from failed fuel elements, it typically circulates in a closed path called

the primary coolant loop. (It is abnormal for fuel elements to release radioactive contaminants into the coolant, but the possibility must be accounted for.) In boiling-water
reactors and some gas-cooled reactor designs, the primary coolant passes directly into
the turbine, but in most reactor types, heat from the primary coolant is transferred into
a secondary coolant loop in components called heat exchangers or steam generators,
which are not shown in Figure 7.6. Then the secondary coolant goes through the turbine. In the secondary loop and beyond, the nuclear power plant is not much different
from conventional thermal power plants such as coal plants.

One type of steam generator is shown schematically in Figure 7.8. The primary
coolant enters at the top, separates into numerous tubes (of which only three are
shown), and recombines at the bottom where it flows out. The secondary coolant
enters at the bottom, flows upwards in the spaces around the tubes, reverses direction
at the top to flow downwards in an annulus outside of the space containing the tubes,
and leaves the steam generator vessel near the bottom. The countercurrent flow of
primary and secondary coolants enables heat to be transferred at small temperature
differences: Where the cold secondary coolant enters the heat exchange space, it encounters primary coolant from which most of the heat has already been removed;
where it exits the heat exchange space, it has been heated all along its flow path, and
the primary coolant from which it receives heat in the upper ends of the tubes is hot.
The temperatures nearly match along the entire flow paths of each coolant.

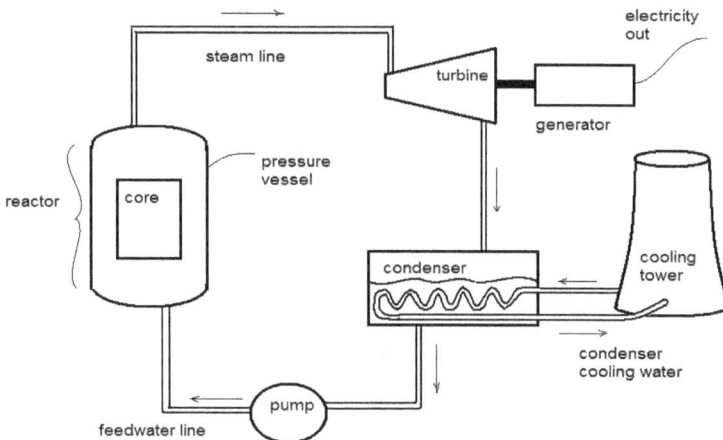

Figure 7.6—Layout of typical reactor power plant

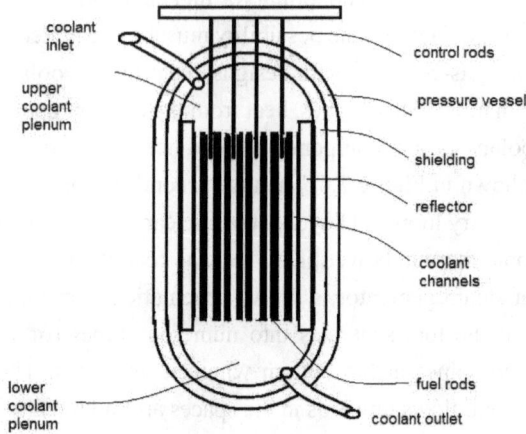

Figure 7.7—Generic nuclear reactor.

Where the secondary loop passes through the condenser, it exchanges heat with another stream of water. This water may be taken directly from a river or lake in a once-through flow path, or it may be a loop that passes into a cooling tower, where it rejects heat into the atmosphere.

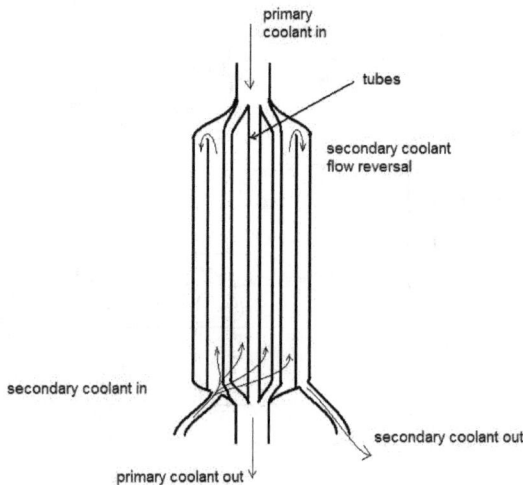

Figure 7.8—Once-through steam generator for pressurized-water reactors

The cooling tower is the iconic symbol of nuclear reactors in the public mind. Actually, cooling towers are also used in coal plants, but the public seems to associate them with nuclear plants. Cooling towers come in two types—wet and dry. In a dry cooling tower, all the waste heat is exchanged with the atmosphere, so there is no requirement for a stream of cooling water to be continuously supplied by a river or lake at all. Unfortunately, dry cooling towers are limited in their capacity, so they are only suitable for power plants up to about 300 MWe (i.e., megawatts of electric power). In wet cooling towers, a stream of water from a river or lake is circulated (i.e., it is reused repeatedly) over the outside of the flow channels containing the reactor coolant, and evaporative cooling (by evaporation of the river or lake water) is used to reduce the required water flow rate from the river or lake by about 97%, as compared to the "once-through" use of the river or lake water without cooling towers. With wet cooling towers, a typical 1000 MWe light-water reactor requires about 16,000 gallons per minute of makeup water from a river or lake.[4] That's about 30 cubic feet per second (cfs), which isn't much from a large river.

PART II

NUCLEAR POWER IN CONTEXT: ENERGY ALTERNATIVES AND ENVIRONMENTAL EFFECTS

CHAPTER 8
A COMPARISON OF ENERGY RESOURCES

No energy technology, including nuclear energy, is free of risks to human safety or of adverse environmental effects. The use of energy creates prosperity and enables modern health care to extend the average person's life far beyond what was normal even a century ago. But these benefits come at a cost. The purpose of this chapter is to examine the risks to human health and safety and the environmental consequences of all of the major sources of energy available to human society. Awareness of these risks and consequences enables us to see the adverse impacts of nuclear energy in a more realistic context.

8.1 COAL

Coal is the largest source of energy for electric power worldwide.[1] Combustion of coal is also used directly for heating. Worldwide consumption of coal in 2014 was 8.9 billion tonnes,[a] and consumption is expected to increase by 0.6% per year through at least 2040. Globally, about 40% of electricity production comes from coal; as of 2016 the fraction is 30% in the United States and 69% in China. In the U.S., over 90% of the coal consumed is used for electricity. As a fraction of total energy consumption worldwide, coal accounted for 30% in 2015.[2]

Coal power plants produce electricity by burning coal to boil water, and the steam

---

[a] A tonne, or metric ton (t), is 1000 kg. Our familiar 2000-pound ton is the short ton. Another unit called ton is the long ton, which is 2240 pounds. The long ton and the metric ton are almost identical.

thus generated passes through a turbine, spinning a shaft that turns the rotors in an electric generator.

Coal is the product of the transformation of organic matter covered by sediment and pressed by the weight of the overlying sediment. There are several grades of coal and its precursors, which include peat, lignite, sub-bituminous coal, bituminous coal, steam coal, anthracite coal, and graphite, listed in the order of increasing purity of carbon. The higher grades burn more cleanly, but graphite does not burn well and is not commonly used as a fuel. Anthracite coal is most often used for space heating, as it produces less particulate pollution than lower grades. Most of the coal used to generate electric power is bituminous coal, sub-bituminous coal, and lignite.

The environmental and public-health consequences of coal use include the impacts of mining, air pollution from coal combustion, and the problem of disposing of the ash remaining after the coal is burned.

Coal is mined by either underground mining or removal of surface layers. The latter method, also called surface mining or strip mining,[3] is used when the coal deposits are close to the surface of the ground. Strip mining is extremely destructive, destroying the surface ecosystem directly. Reclamation is required, but the damage done by strip mining cannot be completely undone. In particular, in the Appalachian Mountains the large amounts of waste rock left after strip mining generally contain materials that release acids into runoff water and cause stream acidification, degrading aquatic ecosystems.[4] A particularly egregious form of surface mining is mountaintop removal, in which as much as 400 feet of soil and rock lying above the coal deposits ("overburden") are removed and dumped into adjacent hollows and valleys in order to get to the coal.[5] After the mining operations cease, some of the overburden is returned to the mountaintop to create a flattened landscape, but much of the overburden has to be left in the hollows and valleys. Since the 1960s, mountaintop removal in Appalachia has flattened at least 470 mountains, buried more than 700 miles of streams, and destroyed 380,000 acres of forest, according to the Natural Resources Defense Council.[6] Mitigation requirements are weak and mitigation efforts cannot reverse the environmental destruction caused by mountaintop removal. Figure 8.1 shows the devastation caused by mountaintop removal.[7]

Underground coal mining is like any other subsurface hard-rock mining process. To get to the coal deposits, miners must remove rock lying between the surface and the deposits and between deposits, and this rock is a waste material containing toxic metals that leach out into rainwater runoff. However, the volumes of this waste rock are much smaller than the volumes of overburden removed in surface mining. A poignant immediate impact of underground mining is the danger to miners.

Figure 8.1—A mountaintop removal operation. This picture was taken from Reference 7 and is in the public domain.

In the U.S., mining safety has been greatly improved over earlier times, to the extent that coal mining in the U.S. is only slightly more dangerous than driving. The fatality rate in the U.S. from coal mining is on the order of 50 per year. But in some countries, the safety record is much worse. In China, the annual death toll is in the thousands (6027 in 2004). Another hazard of mining is black lung disease (pneumoconiosis), a life-shortening disease caused by inhalation of coal dust. About 4000 new cases of black lung disease occur annually in the U.S., and about 10,000 occur annually in China.[8]

Public health is affected by the air pollution released by burning coal and by the toxic materials contained in the fly ash residue. A key point in understanding the impact of burning coal is the huge amount of coal involved. A typical 500-MW coal-fired power plant will consume roughly 3,000 tons of coal every day.[9] A typical modern railroad hopper car for coal holds 100 tons;[10] then that coal plant needs 30 of those hopper cars every day. The output of coal mines in the U.S. is about a billion tons per year,[8] most of which is burned for energy.

Burning this enormous amount of coal produces commensurate amounts of air pollution. In the U.S., "scrubbers" remove most of the particulate matter from the exhaust, but in many other countries, including China, such equipment is rarely used and everything goes up the stack. Coal contains many impurities, such as mercury and about 40 different radionuclides including uranium and thorium.[11] According to the World Health Organisation of the United Nations, as of 2014 approximately 7 million people die prematurely each year worldwide from the adverse effects of air pollution.[12] About 4 million of these deaths are due to indoor air pollution from "inefficient cooking practices using polluting stoves paired with solid fuels and kerosene."[13] Additional lethal indoor air pollution comes from radon. (Radon is a natural decay product of uranium and thorium in soil and rock; it is a gas that diffuses out of the ground and enters homes from basements and crawl spaces.) In the U.S., radon is second to smoking as a cause of lung cancer, and about 20,000 deaths per year are attributed to indoor radon in the U.S.[14] Worldwide, about 8 million people die each year of lung cancer,[15] and between 3% and 20% of these deaths are due to radon.[16] Using the upper limit of this range, radon kills 1.6 million people per year. That leaves at least 1.4 million deaths per year attributable to outdoor air pollution, principally motor vehicle exhaust and power plant emissions. If only a third of these deaths are caused by coal, we can still blame coal plant emissions for over 450,000 deaths per year. The discussion of the Chernobyl accident in Chapter 13 states that the eventual total death toll from the Chernobyl accident is expected to be around 4000. I don't want to trivialize the Chernobyl accident—it was a disaster any way you look at it—but a sobering perspective can be gained by comparing these two numbers. Four hundred fifty thousand deaths per year for coal divided by 4000 deaths per Chernobyl-type accident equals 117 accidents per year. That is, it would take at least 117 Chernobyl-type accidents every year to kill as many people as our dependence on coal is killing right now. And the chance is extremely remote that a nuclear power plant accident as bad as the Chernobyl accident will ever happen again. (The Fukushima Daiichi accidents have released a lot of radioactivity, but most of this has been in the form of iodine-131, which is very short-lived; the Chernobyl accident blew out the entire radioactive inventory of the core. Half the core fell in chunks onto the surrounding grounds, and the rest went up in a smoke plume that drifted over much of Europe. The public health consequences of the Fukushima Daiichi accidents are not likely to approach those of Chernobyl within an order of magnitude. See Chapter 14.) Even if scrubbers prevent particulates from going up the smokestacks of a coal-fired power plant, there is a huge amount of ash that must be disposed of. This ash is collected in mounds at the power plant sites, and it is not isolated from the biosphere. In Kingston, Tennessee, on 22 December 2008, a dike retaining such a mound, which occupied 80 acres, failed and released a billion gallons of toxic sludge into the surrounding environment, including waterways. The contamination will take years to clean up, and the long-term health effects of the spill are unknown.[17]

Eventually, coal resources will be depleted and other energy sources will be required. Although data on coal reserves are poor, some attempts have been made to estimate future production trends. Recent estimates of the time at which coal production will peak have varied from 20 years to more than 200 years into the future. However, as high-quality coal resources are used up, lower-quality coal will have to be used in their place, which will make air pollution worse, require more coal per kilowatt, and increase the cost of coal power. This shift to lower-quality coal is already taking place.[18]

A well publicized proposal to reduce carbon dioxide emissions from coal plants (by as much as 90%) is called carbon sequestration. Essentially, instead of releasing exhaust fumes into the air, a coal plant would force its exhaust into the ground. But forcing carbon dioxide into the ground would take energy, increasing the coal requirement for a power plant by as much as 40% (exacerbating the environmental impacts of mining commensurately) and increasing the cost of electricity from coal plants by up to 90%. Coal industry spokesmen say that by 2025 these problems will be solved and electricity from coal plants with carbon sequestration will be cheaper than it is from today's coal plants.[19] Personally, I remain skeptical.

8.2 PETROLEUM

The word *petroleum*[20] means "rock oil." The term is sometimes used to include natural gas, but here we will use it only for liquid hydrocarbon compounds found in underground formations. The form of petroleum found in the ground is broadly called crude oil, which consists of a mixture of different hydrocarbons along with some other organic compounds and various impurities such as iron. Crude oil is graded according to the dominant molecular weights of the hydrocarbons in the mixture. Mixtures consisting mainly of lighter hydrocarbons such as octane (C_8H_{18}) are called light crude oil.[b] The precise definition of light crude oil varies from country to country, but light crude oils are generally considered to be those with a density less than about 900 kg/m^3.[21] Any crude oil that can be extracted by the usual method of drilling is called conventional crude oil. Conventional crude oils have densities up to about 933 kg/m^3. Heavy crude oil and extra-heavy crude oil are grades of crude oil that do not flow easily; some of these heavy grades are found in oil sands such as those in

---

[b] In chemistry, the composition of molecules is written in the form $X_xY_y...$, where X and Y denote the symbols for their constituent elements, and x and y denote the number of atoms of each in the molecule.

the Orinoco region of Venezuela.[22] The heaviest grade of crude oil is bitumen, which is so thick that it does not flow at all. This form of crude oil occurs in the tar sands of Alberta, Canada. Some of the non-hydrocarbon organic compounds found in crude oil contain sulfur, which is undesirable. Crude oils containing less than 0.5% sulfur are called sweet crude oil, while oils containing more sulfur are called sour.[23] Oil shale,[24] a rock containing kerogen,[25] a complicated mixture of organic compounds, can be used to produce liquid hydrocarbons.

Like coal, petroleum was formed from layers of organic materials covered and compressed by sediment. But whereas coal was mainly formed from terrestrial deposits, petroleum was mainly formed from deposition in oceans.[20] Much of our petroleum resource lies under oceans still, but a lot of it was formed in shallow seabeds that have been uplifted and are now high and dry.

Petroleum does not occur in cavities in rock like fuel in a tank; instead it occupies pores in sedimentary rock. It is under pressure from the overlying rock, and when a petroleum field is first penetrated the oil is squeezed out vigorously by that pressure, sometimes emerging from the well in a "gusher." But as the oil is depleted the pressure falls, and the remaining oil has to be forced out. Eventually, this doesn't work any longer, and some of the oil remains in the ground unextracted.[26]

Petroleum is used mostly for transportation fuels, but a good deal of it is used to make chemical products, especially plastics and asphalt[27] (which is used mainly for road paving), and some is used for residential and commercial heating and even for generating electricity. In the United States, the proportions of use are 70% for transportation, 24% for chemical products, 5% for residential and commercial heating, and 2% for electricity; in other countries the proportions differ.

The attractions for transportation of petroleum fuels, especially gasoline and diesel fuel, are that they have greater energy content per unit volume than any other fuels,[28] and that they occur in liquid form at ordinary temperatures and pressures, so that they are easy to carry around in thin-walled tanks. The energy content of gasoline and diesel fuels is in the range of about 33-39 MJ/L, while that of ethanol is 23.5 MJ/L.[c] That is why your gas mileage goes down when you use gasoline containing ethanol.

Petroleum resources and consumption rates are huge. In 2015, petroleum accounted for about 33% of worldwide total energy consumption,[2] which made it the world's largest single energy source. Proven reserves, which are resources that can be recovered commercially under current economic conditions, were estimated in 2017

---

[c] The joule (J) is the unit of energy in the standard metric system, and the liter (L) is the standard unit of volume. This system was introduced in Chapter 2. The joule is equal to 2.8×10^{-7}

by the U.S. Energy Information Administration at 1.727×10^{12} barrels.[29] Additional resources, which may become commercially recoverable in the future, add considerably more; oil shales were estimated in 2016 as 6.05×10^{12} barrels.[30] As of 2015, annual worldwide consumption is about 3.5×10^{10} barrels.[31] Hydrocarbons like those found in petroleum can also be made from coal by the Fischer-Tropsch process.[32] At current rates, total proven resources could last about 50 years; as proven reserves are depleted, prices will rise, and more resources will become commercially recoverable. Total known reserves would last much longer than that at current consumption rates. However, even now the easily extracted conventional oil resource is being depleted, and producers are turning increasingly to heavier grades of oil, which are harder to extract and require more energy to refine or break down into transportation fuels, and to hydraulic fracturing ("fracking"), which has both great potential and serious drawbacks.[33] Furthermore, demand will inexorably increase as more and more people in developing countries acquire enough money to buy cars.

Such enormous use of petroleum comes at a high price. Environmental and health effects are of several kinds, some of which are better known than others. Among the best publicized are oil spills, which, although they are preventable, occur with depressing regularity. One of the most serious spills was the disastrous "BP oil spill" in the Gulf of Mexico.[34] This spill flowed out of a broken pipe in a deepwater drilling operation and lasted for three months. Most large oil spills are the result of accidents of oil tanker ships. Spills in marine environments are more serious than spills on land because even small amounts of oil can spread out over large areas of the water surface, while spills on land are more likely to be confined locally. It is not known how long the effects of marine oil spills persist. After more than 20 years, wildlife is still being affected adversely by oil from the infamous Exxon Valdez oil spill of 1989 in Prince William Sound, Alaska.[35, 36]

Responsibility is shared by oil for the deaths from outdoor air pollution mentioned above in the discussion of coal. These deaths amount to hundreds of thousands per year, as noted above. Motor vehicle exhaust in large cities creates smog, which has dramatic health effects.[37] Figure 8.2 shows Beijing, China, on days without and with smog in 2005. Modern automobile engines produce far less pollution than engines produced before the 1970s, but the great numbers of gasoline- and diesel-powered vehicles in large cities, especially those in mountainous regions where exhaust fumes tend to collect in valleys, still create urban smog.

---

kilowatt-hours, and the liter is roughly about a quart. A megajoule (MJ) is a million joules. The watt (W) is a unit of power equal to 1 joule per second. A megawatt (MW) is a million watts, and a gigawatt (GW) is a billion (10^9) watts.

Figure 8.2—Beijing without smog and with smog in 2005. The panel without smog shows conditions after rain. This figure was taken from the Wikimedia Commons archive and is reproduced under the terms of the GNU Free Documentation License, Version 1.2.[38] It may be copied freely under the same terms.

Oil sands, tar sands, and oil shale are normally dug up by surface mining operations, including strip mining and open pit mining. The environmental impacts are similar to those in surface mining of coal. A particularly vulnerable ecosystem overlying tar sands is the boreal forest in Alberta, Canada. Reference 39 discusses the ecological significance of this vast forest in detail and both describes and illustrates the devastation imposed by the required surface mining operations.

Another consequence of America's dependency on oil is reduced national security. An unfortunate (for Americans) coincidence of nature and politics is that much of the world's oil reserves are in Middle Eastern countries that are either overtly hostile to the United States, such as Iran, or uneasy partners, such as Saudi Arabia.[40] If we could convert to a petroleum-free transportation infrastructure, as for example with vehicles using hydrogen fuel produced by electrolysis of water, we could be freed of dependency on real and potential enemies.

As demand for oil increases in growing economies like China, all of the impacts on the environment, public health, and political instability are just going to get worse. But projects like the NGNP (cf. Chapter 16) that could lead to our liberation from oil struggle to continue even as paper studies.

8.3 NATURAL GAS

Natural gas[41] is mostly methane (CH_4), but it contains up to 20% of other gaseous hydrocarbons, especially ethane (C_2H_6). The general chemical formula for "saturated" hydrocarbons,[42] or alkanes, is $C_nH_{(2n+2)}$. Figure 8.3 illustrates the structure of a typical alkane, propane. In saturated hydrocarbons, each carbon atom has four single bonds with adjacent atoms, forming a carbon chain surrounded by hydrogen atoms. Butane (C_4H_{10}) is the heaviest alkane that is still gaseous at standard temperature and pressure[43] (20 °C and 1 atmosphere—i.e., 68 °F and 14.696 pounds per square inch in English units). The products of complete combustion of alkanes in oxygen are carbon dioxide and water: $2C_nH_{(2n+2)} + (3n+1)O_2 \rightarrow 2nCO_2 + 2(n+1)H_2O$. Incomplete combustion produces carbon monoxide and particulate matter as well, and combustion in air produces some oxides of nitrogen. All of the gaseous alkanes have higher ratios of hydrogen atoms to carbon atoms than liquid alkanes, so that less carbon dioxide and less particulate pollution are produced per unit of energy released in combustion of natural gas than in combustion of liquid petroleum fuels. Thus, they burn cleaner and produce less air pollution.

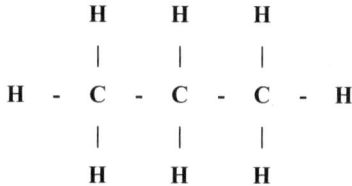

$$
\begin{array}{ccccccc}
\text{H} & & \text{H} & & \text{H} & & \\
| & & | & & | & & \\
\text{H} & - & \text{C} & - & \text{C} & - & \text{C} & - & \text{H} \\
| & & | & & | & & \\
\text{H} & & \text{H} & & \text{H} & &
\end{array}
$$

Figure 8.3—The molecular structure of propane

More complicated hydrocarbons such as alkenes[44] have one or more double bonds between carbon atoms, so that their hydrogen-to-carbon ratio is lower. But these are not significant constituents of natural gas.

Natural gas is usually present in solution in petroleum. As noted above, the petroleum in an underground oil field is under pressure from the overlying rock. When the petroleum is brought to the surface in a drilling operation, the natural gas fizzes out of solution like the carbon dioxide in uncorked champagne. This gas is either burned off, injected back into the oil deposit to repressurize it, or piped to consumers. Natural gas is also found in natural gas fields without petroleum, and in coal beds. Natural gas in all of these forms is called fossil natural gas.

The decay of organic matter such as vegetation, garbage, and manure produces a methane-rich mixture of gases called biogas. These decay processes are caused by methanogenic archaea;[45, 46] archaea are a group of microorganisms formerly considered bacteria, but now classified as a separate "domain" from the bacteria domain and the eukarya domain (which includes us).[47] Biogas as a resource of methane is not heavily exploited yet, but a significant amount of electric power is generated from biogas collected at sewage treatment plants. As of 2017, California has 132 such plants, with a total capacity of almost 1000 MW.[48]

Electric power plants using natural gas employ either gas turbines, in which the gas is burned in the turbine as in a turboprop aircraft engine, or steam turbines, for which the combustion process boils water as in a coal plant or a nuclear plant. Natural gas is also used extensively to heat buildings and provide heat for cooking and other domestic purposes (e.g., gas-powered clothes dryers).

The United States possesses only about 4% of the world's resources of fossil natural gas, yet there is so much natural gas in the ground that this resource might last 100 years at current consumption rates. Most of the world's known fossil natural gas occurs in the Middle East (naturally) and Russia.[49]

Because natural gas is inflammable,[d] its use entails risks of fire and explosion. For example, in September 2010, a gas main in San Bruno, California, ruptured and caused fires and explosions that killed eight people and destroyed 55 homes. Between 2008 and 2017, five natural gas explosions in Pennsylvania killed nine people, injured many more, and damaged or destroyed dozens of homes.[50] In April 2017, a leaking pipeline in Firestone, Colorado destroyed a house and killed two people.[51] In 2018 in northern Massachusetts, a pressure spike in natural gas lines caused leaks that led to dozens of fires and explosion, killing one person and destroying about 80 homes.[52, 53]

Natural gas is fairly easy to store and transport; small tanks for appliances such as outdoor grills are ubiquitous, and many rural people who use natural gas for heating and cooking have large tanks that last for months (I had such a tank when I lived in rural Idaho). In the short term, and perhaps even through the 21st Century, natural gas seems to be the most straightforward alternative to petroleum for automobile and truck fuels. In 2014, there were nearly 20 million natural-gas-powered vehicles operating in the world, mostly in South America and central Asia (Pakistan, India, and Iran).[41] Nevertheless, its combustion does produce carbon dioxide, so if global warming is indeed the looming catastrophe it increasingly promises to be, natural gas is not the ultimate solution.

---

[d] The word "inflammable" means capable of being inflamed. The "in" at the beginning of the word is often misconstrued to mean "not." Therefore, to avoid confusion, the word "flammable" was coined. Here is what language authorities William Strunk, Jr. and E. B. White (in *The Elements of Style*, 2009 edition, Pearson Longman, New York, p. 47) have to say about that: "***Flammable.*** An oddity, chiefly used in saving lives. The common word meaning 'combustible' is *inflammable*. But some people are thrown off by the *in-* and think *inflammable* means 'not combustible.' For this reason, trucks carrying gasoline or explosives are now marked FLAMMABLE. Unless you are operating such a truck and hence are concerned with the safety of children and illiterates, use *inflammable*."

8.4 HYDROPOWER

Hydroelectric power, also called hydroelectricity or hydropower, is electric power obtained from the kinetic and/or potential energy of water.[54] See Chapter 2 for discussions of kinetic and potential energy. Hydropower can be obtained by diverting a river through a channel containing a turbine or by damming a river and passing the high-pressure water at the bottom of the dam through turbines. The former approach, called run-of-the river hydropower, uses the kinetic energy of the flowing river, and the latter approach, called conventional hydropower, uses potential energy. A variation of the conventional approach is the pumped-storage method. In this method, during times of low power demand, some water is pumped to a secondary reservoir at a higher elevation. During times of higher power demand, water from the higher reservoir is released back to the lower reservoir through a turbine. Because it takes energy to pump the water up to the higher reservoir, no additional energy is obtained this way, but more power is available for times of high demand. A schematic illustration of a conventional hydropower system is shown in Figure 8.4.

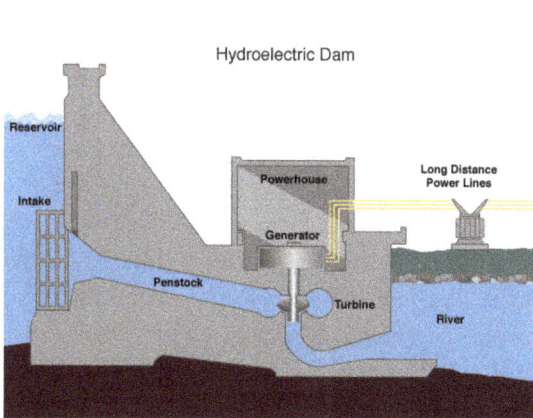

Figure 8.4—A conventional hydropower system. This figure was obtained from the Wikimedia Commons archive and is in the public domain.

As of 2015, hydropower supplies about 20% of the world's electricity. It supplies 100% of Paraguay's electric power, 96% of Norway's, 65% of Venezuela's, 63% of Brazil's, 58% of Canada's, and 47% of Sweden's. It supplies only 20% of China's, 9% of India's, and 6% of that in the United States.[55] The most familiar types of hydroelectric generating facilities are large conventional dams, such as the Hoover Dam near Las Vegas, Nevada.[56] Some of these projects are the largest single sources of electric power in the world. The Three Gorges Dam on the Yangtze River in China, as of 2018 the world's largest power generating station, generates 22.5 GW of power.[57] The Itaipu Dam on the Paraná River on the Brazilian-Paraguayan border generates 14 GW; it supplies 90% of Paraguay's electricity and 20% of Brazil's.[58, 59] The Guri Dam on the Caroni River in Venezuela generates 10 GW.[60] The Hoover Dam generates 2 GW with 17 main turbines. When it first became fully operational in 1939 with only four turbines, it was the largest hydropower plant in the world.[56] These figures are peak generating capacity, available only when the reservoirs are full. In times of drought, the available power is lower. I have driven over Hoover Dam many times in the last three decades, and I have never seen the reservoir (Lake Mead) full.

The potential for hydroelectric development is huge. In many countries, particularly in the developing world, many opportunities for large conventional dams still exist. In many developed countries, however, most sites suitable for large dams have already been used. Even in such countries, however, many opportunities still exist for smaller hydroelectric facilities, including micro hydro and pico hydro plants producing up to 100 kW and up to 5 kW, respectively.[54] Even in the United States, only about 40% of potential hydropower has been exploited.[61]

All this sounds wonderful, but the exploitation of hydroelectric resources comes at a high environmental cost. Whenever a dam is built, the reservoir behind the dam drowns land, often very large areas of it, and the drowned land usually has significant ecological, agricultural, or cultural value. The reservoir also transforms a riverine ecosystem into an unnatural lake and alters the ecological nature of the river below the dam.

Sometimes the impacts of the changes are mixed. Some world-class trout fisheries have been created in rivers below dams. Such river segments, called tailwaters, have uniform cool temperatures year-round, and they are clear because silt in the river upstream of the reservoir falls out in the reservoir. (Eventually, these reservoirs will fill in with silt and cease to be useful.) These tailwater conditions are ideal for the aquatic insects on which trout primarily depend. The White River in Arkansas below Bull Shoals Dam[62] and the Bighorn River[63] in Montana below Yellowtail Dam[64] are two famous examples.

One of the adverse impacts of hydropower comes from the fluctuations in river level caused by releasing large outflows for peak power generation. Such impacts have been particularly severe on the Colorado River below Glen Canyon Dam, especially in the Grand Canyon.[65] Disruptions have occurred in both the aquatic ecosystem in the river and in the riparian ecosystem along the river. On the Bighorn River in Montana, such impacts have been mitigated by the construction of the small Afterbay Dam below Yellowtail Dam; the Afterbay Dam collects water released abruptly for peak power generation and lets it out gradually to keep the flow in the river downstream relatively uniform.[66] Overall, the Bighorn River project is about as environmentally benign as large hydropower projects can get, but it still inundates about 72 miles of river for 250 MW of power.[67]

One of the most ecologically and economically severe impacts of hydropower is its effect on anadromous fish—i.e., fish that are born in inland waters, migrate to oceans to grow up, and return to their natal rivers to spawn. Such fish include most salmons and some trouts. The most dramatic example is the chinook and sockeye salmon of the Snake River system.[68]

Before settlement of the Pacific Northwest by Caucasian settlers in the 1850s, between 8 million and 16 million of these fish ran up the Columbia River each year. These fish moved into every suitable tributary of the Columbia, of which there were hundreds, and fed indigenous Native American tribes and such wildlife species as bears and bald eagles.

The populations of these fish that made the longest journeys were those that spawned in the upper Salmon River system of Idaho. They swam up the Columbia to its confluence with the Snake River, traveled up the Snake to its confluence with the Salmon River, and then moved into the headwaters of the Salmon. The chinook salmon spawned in the upper river and its small tributaries, and the sockeyes spawned in Redfish Lake,[69] a 4.5-mile-long natural lake 6547 feet above sea level[70] and about 900 river miles from the Pacific Ocean. The lake got its name from the red color of the spawning sockeyes, which used to return to the lake in such numbers that they made the water gleam with red.

Many factors have contributed to the decline of these salmon runs. Overfishing had reduced the runs noticeably even by the 1880s, and environmental degradation from mining and logging had also taken a toll. But the runs were still strong until dam construction began on the Columbia and Snake Rivers. Now, four dams on the Columbia and three on the lower Snake bar the salmon's way. Many of the salmon attempting to run up the river are captured and taken to fish hatcheries, but the total number of salmon entering the Columbia was down to 2.5 million by 1993. In 1992, only one sockeye, a male dubbed "Lonesome Larry," made it to Redfish Lake.[71] But Larry did not arrive in vain. He was captured and used in a captive breeding program that has now re-established a small run to Redfish Lake; in 2010, about 1700 were expected to make it there.[72]

The reasons for the impact of the dams on the salmon runs are numerous.[73] The young fish, called smolts, that are ready to go to the sea are programmed naturally to use the higher flows provided by spring snowmelt, or runoff, to carry them quickly. The portion of their trip in the Snake and Columbia Rivers took about five days before the dams were built. Now, they have to swim through 300 miles of slack water to get to the ocean, which takes them up to 30 days.

Instead of being carried by the flowing river, they must use their own energy reserves to swim. The long trip through the reservoirs exposes them to much more predation than they would face in the rivers. Then, many of them go through the dams' turbines and get chopped to death. Overall, the mortality of smolts migrating to the ocean has risen from about 10% before the last two Snake River dams were completed to between 86% and 97%, depending on the conditions in a particular year. A program of barging smolts past the dams has been conducted, but with only limited success. Barged fish do not learn the landmarks on their way downstream that they need to recognize on their way back to find their proper spawning grounds.

The returning fish also face high mortality. To get around the dams, they must ascend fish ladders, which they cannot always find. Also, water at the bottom of the reservoir behind the dam is under high hydrostatic pressure, so that it can dissolve more nitrogen than water at atmospheric pressure. When the water flows through the turbines, it emerges at only a little more than atmospheric pressure, but the nitrogen does not come out of solution immediately: The water becomes supersaturated with nitrogen. The fish breathe the excess nitrogen with their gills, and the nitrogen bubbles out of solution in their bloodstream. This is exactly like what happens when divers ascend from deep water too quickly: The condition is called "the bends." It is fatal to the fish. Some mitigation of this problem has been achieved by placing wings below the spillways to spread the supersaturated water out and dilute the nitrogen, but the fix is not complete.

Salmon advocates have proposed removal of the three dams on the lower Snake River,[74] since the runs were still fairly robust before those dams were completed, but so far the interest groups that benefit from the dams have successfully resisted such proposals.

Another adverse impact of dams is the potential for dam failure. This is not just a hypothetical concern; dams have failed, with consequent loss of life and large-scale property damage. One such failed dam was the Teton Dam on the Teton River in southeastern Idaho.[75] This dam, intended to provide power generation, irrigation water, flood control, and recreation, was built on a site known to have highly fissured and unstable rock. The dam failed on June 5, 1976, shortly after the reservoir was first filled. The complete capacity of the reservoir, about 234,000 acre-feet, went downriver in only 12 hours, with a peak flow rate of over 2,000,000 cubic feet per second (cfs). Typical flows on the river, while varying seasonally, are more like 700 cfs.[76] The dam was completely washed away. The flood killed 11 people (14 according to some sources)[77] and 13,000 cattle, did $194 million dollars in property damage, and wiped two small towns off the map (Wilford and Sugar City). Severe ecological damage was done to the riparian habitat along the river below the dam and to the cutthroat trout fishery in the river itself. The fishery still has not completely recovered, although it is now thriving again.[78] The damage would have been greater had it not happened during the day when people could be more quickly informed of the approaching flood and enabled to evacuate the area in the path of the wave of water. The official investigation into the failure found that geological factors and design decisions both contributed to the collapse.[77]

Another catastrophic failure occurred in France in 1959 with the collapse of the Malpasset Dam.[79] This dam collapsed because a weak clay seam in the rock behind the left abutment of the dam allowed the abutment to shift and the dam to crack right down the middle. A wall of water 40 m high was released, and 421 people lost their lives.

To say the least, hydropower cannot be categorically considered environmentally benign. Sometimes it has significant benefits, but often it is an ecological disaster and a hazard to public safety. Each proposed hydroelectric project should be evaluated on its own merits.

8.5 SOLAR POWER

Sunshine falls on every part of the Earth's surface. It can be used directly for heating, or it can be converted to electricity either by photovoltaic cells or by a thermal cycle. Some see solar energy as the cleanest technology for replacing fossil fuels. Solar energy has been used to power cars, boats, and even light airplanes.[80] However, the ability of sunshine to provide power is limited by the low intensity of the incoming radiation, the efficiency of collecting it, and the cost of the collecting technology. The true cost of the technology should account for the environmental impact of the collection system, including land use, materials extraction, and manufacturing impacts such as waste disposal.

The intensity of the solar irradiation of all wavelengths reaching the outer edge of the Earth's atmosphere when the Earth is at its average distance from the Sun, on a plane perpendicular to the line from the Earth to the Sun, is called the solar constant.[81] Its value varies with solar activity, but the current value is 1.366 kW/m^2. The average value of the intensity of solar radiation of all wavelengths reaching the Earth's surface is about 250 W/m^2. This is the average taken over the whole Earth's surface, including the half of the Earth that is in darkness at any given moment. It accounts for the oblique angle of incidence of most of the incoming radiation and for absorption in the atmosphere, but not for clouds.[82] The actual solar radiation intensity at a particular point on the Earth's surface at a particular time depends on the weather there and on its geographic location. It is greater in the Mojave Desert than it is in Seattle. The total amount of solar energy received in a given time on a given area at a particular location is called the insolation. In Germany, the average annual insolation is about 1000 kWh/m^2,[83] which is equivalent to an average incident solar energy flux of 114 W/m^2.

Figure 8.5 shows the average incident solar energy flux at the outer edge of the Earth's atmosphere and on the Earth's surface. In the United States, the highest value of the flux at the surface, about 300 W/m^2, is in the southwestern desert. For most of the country, the value is closer to 200 W/m^2.

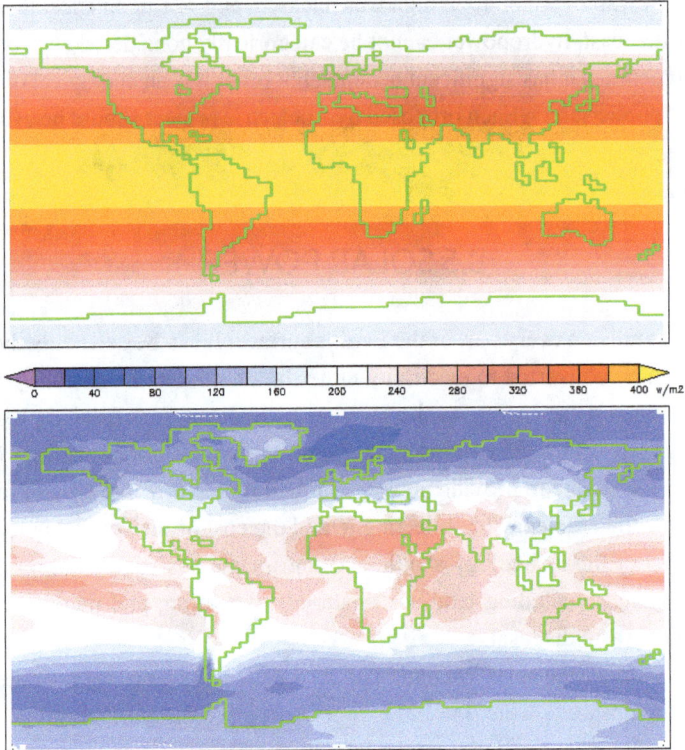

Figure 8.5—Map of solar energy flux at outer edge of Earth's atmosphere and at the Earth's surface. This file was obtained from the Wikimedia Commons and is in the public domain.

Let us do some simple calculations to estimate the land area required to supply U.S. electrical consumption by solar energy. The total U.S. electricity consumption in 2015 was about 3.9 billion MW-hr.[84] That is an average power of about 450,000 MW. Using the national average incident energy flux of 200 W/m^2, we find that the land area required is about 2.23×10^9 m^2, or 860 square miles. That is an area about 30 miles on a side. However, that assumes 100% efficiency at collecting the solar energy. There are several methods for collecting solar energy, none of which is anywhere near 100% efficient. The most familiar are solar photovoltaic (PV) cells, which convert solar energy to electricity directly. Some commercially available cells are 24% efficient,[85] and some experimental cells have reached efficiencies up to 46%.[86] However, the more efficient cells may be too expensive for large-scale application because they use rare and expensive materials. If we assume an efficiency of 20% for large-scale use, we obtain a land requirement of 4300 square miles, which is 66 miles on a side. However, even the most modern systems currently in use have much lower efficiency. A 7.5-acre solar array installed at Denver International Airport in 2008 generates 3 million kW-h/yr,[87] which works out to an average power density of 11.3 W/m^2. This gives an overall effective efficiency of no better than 6%. To supply our 3.9 billion MW-hr per year at that efficiency would require 14,000 square miles, which is 120 miles on a side.

When solar arrays are placed on open land, the habitat under them is essentially destroyed for wildlife or agriculture. Even deserts are habitat for unique and varied plant and animal communities. The environmental impact of wiping out up to 14,000 square miles of wildlife habitat is considerable, and I personally think it would be a tragedy. However, when solar collectors can be put on top of buildings, there is no additional impact on wildlife, and "rooftop solar power" seems to be a very desirable way to generate electricity. Solar heating of home water systems is also an environmentally benign way to use solar energy.[88]

One of the biggest problems with using solar energy is that the sun doesn't shine at night, and some days are so gloomy that not much sunshine reaches the ground even during the day. Therefore, either some means must be provided to store energy at night and for periods of low irradiance, or power from another source must be used to fill in for such times. Now, when solar energy only provides a small fraction of power usage, the latter approach is easy to implement, but if we were to try to supply most or all of our energy from the sun, it would not be possible. Then we would have to rely on batteries, flywheels, or other energy storage technologies, with further decreases in system efficiency and even greater land requirements.

A more efficient way of harnessing solar energy is called concentrated solar power.[89] In this technique, the solar energy falling on a field of mirrors is reflected onto a central collector in which the working fluid for a heat engine is located (see Chapter 7 for a discussion of heat engines). The concentrated sunlight falling onto the central collector can heat the working fluid to very high temperature, permitting much higher efficiency than ordinary steam power plants can attain. Reference 89 shows a graph that gives a peak efficiency of almost 80% for a concentration factor of 5000, a collection temperature of 1500 K (1227 °C), and a heat sink temperature of 300 K (30 °C). But the calculations producing the graph are based on assuming perfect optics and a solar energy flux of 1000 W/m^2, which is not attained anywhere on the Earth's surface (see Figure 8.5). For a solar energy flux of 200 W/m^2, the maximum efficiency becomes only 57%, which would lead to a land area requirement of 1500 square miles (39 miles on a side) for current U.S. needs. Furthermore, the optics will not be perfect, and a rigorous maintenance program will be necessary just to keep the dust off the mirrors. Concentrating solar power is no better at night than solar photovoltaic systems. Finally, any bird flying through the concentrated sunlight near the central collector will be instantly roasted, so this technology adds another environmental concern. This concern is not just speculative. A concentrated solar energy plant has been built in Nevada, and it is indeed roasting birds.[90]

Solar technology is generally perceived as clean, but the manufacture of solar panels produces many kinds of toxic substances that often end up in the environment. These substances include silicon tetrachloride, sulfur hexafluoride, sodium hydroxide, sulfuric acid, hydrochloric acid, nitric acid, hydrofluoric acid, phosphine, arsine, phosphorus oxychloride, phosphorus trichloride, boron bromide, boron trichloride, lead, trichloroethane, ammonia, and isopropyl alcohol, to name just some of them.[91] Most manufacturing processes generate toxic wastes, and the manufacture of electronic components, including solar panels, is no exception. Measures can be taken to capture and often to recycle toxic waste products, but such measures are often not taken, especially in China where many of our electronic products are made. Solar power is not as "green" as you may have thought.

8.6 WIND POWER

Wind power[92] is widely viewed as a panacea for our energy problems. Wind blows everywhere, and in some places it often blows pretty hard. Aside from waste produced in the manufacture of wind systems, wind power produces no emissions. Wind power plants, often called wind farms, are popping up all over the United States, as well as in many other places in the world. Actual electrical energy production has been growing approximately exponentially since about 2000, with a doubling time of about three or four years. In 2015, global electrical energy production by wind worldwide was 833.6 TW-h (a terawatt, or TW, is 10^{12} watts). In 2017, wind generated 4.4%% of the electrical energy used worldwide. However, the wind does not blow all the time anywhere, and the intermittent nature of wind creates difficulties in integrating wind power into the power grid.

Also, wind farms take up a lot of land, and they pose hazards to birds, bats, and aircraft. They create intense low-frequency noise pollution in their vicinity. Some people find them aesthetically appealing, but other people find them aesthetically appalling.

The capacity factor of a wind turbine is defined to be the actual energy generated in a year divided by the energy that would have been generated at the turbine's maximum power output. Because the wind is intermittent, the capacity factor is never anywhere near 100%. Typical capacity factors range from 15% to 50%, with the higher figures attained in particularly windy sites.

The term "penetration" is used by the wind-power industry to mean the fraction of total electrical energy generation that is provided by wind. Thus, in 2017, the global power market penetration by wind was 4.4%. Integrating a power source that fluctuates according to the vagaries of the wind into the power grid is not difficult when only a few percent of the power is supplied by wind, but as penetration increases, integration becomes more difficult. Current studies indicate that penetration of at least 20% can be accommodated into the present power grid. As the wind blows harder and increases the wind system's contribution to the combination of power sources supplying the grid, other sources, such as hydropower and gas turbine plants, can be cut back to meet the demand of the moment.

As penetration increases, it will be more difficult to adjust other power sources to match supply to demand. One approach would be to connect wind farms in distant geographical regions together, so that power from a place where wind is blowing can be sent to places where it is not blowing at any given moment. However, that approach requires long transmission lines, which dissipate electrical energy. Transmission lines that lose a lot of electrical energy are called "lossy."

The wind's intermittency and the need for lossy transmission lines to connect widely separated wind farms mean that for wind to supply a steady, reliable major fraction of our total demand for electricity, much more total rated wind-power capacity must be built than is actually needed. Neglecting transmission losses and taking a middle value of 30% for the capacity factor, we see that the total rated capacity would have to be more than three times the average desired power output. Cristina L. Archer and Mark Z. Jacobson at Stanford University studied the total global potential for wind power from practical sites (those sites where the average wind speed is sufficient), assuming standard wind turbines 80 m in height with blade diameters of 77 m and rated power of 1.5 MW each, at a density of 6 turbines per square kilometer.[93] They found that the world's total wind energy potential is about 72 TW (terawatts, or 10^{12} W). They compared this to the 2001 total global energy consumption and total global electricity consumption rates (about 13 TW and 1.7 TW, respectively). Thus, the total global wind power potential is about five times as great as the total global energy consumption rate and 40 times as great as the total global electricity demand as of 2001. To collect all 72 TW would require wind farms covering about 13% of the Earth's land area that is not covered year-round by snow or ice. To supply 2001 electricity needs would require about 1/40th of this, or 0.325% of the Earth's land area. That is a small fraction, but it is still a lot of land. In the study, the Earth's available land area is taken as 25.4% of the total global area, which is 1.3×10^8 km^2. Thus, 2001 global electricity usage would require 422,500 km^2, or 163,000 square miles, equivalent to a square 400 miles on a side. Current data would be somewhat different, but not qualitatively so.

A key difference between solar power and wind power is that solar collection arrays render the land under them completely useless as ecosystems or farmland, whereas wind farms do not. Only a very small fraction of a wind farm is actually occupied by the bases of the wind turbine towers. The base diameters of typical modern towers are 4.2 m or less.[94] Thus, in a wind farm with 6 turbines per square kilometer, the total area of the turbine bases is less than 83 m^2 per square kilometer, which is only 0.0083%. A larger fraction is taken up by access roads; each wind turbine must be accessible by roads for maintenance. I have traveled quite a bit by personal airplane, and I have often seen wind farms from the air; as one would expect, when the towers are arranged in a grid, the access road system comprises main roads running along grid lines with short spurs leading off the main roads to individual towers. Thus, the total road area is still a small fraction of the total wind farm area, but it is much greater than the area of the tower bases. Where wind farms are located on agricultural land, the impact of the access roads is minor. Where they are located in natural ecosystems, the impact of habitat fragmentation is greater than the impact of land lost to the actual roads. Because of the impact of fragmentation, the U.S. Fish and Wildlife Service has recommended that wind farm development be restricted to land that has already been altered or cultivated and not on intact or healthy natural ecosystems.[95]

Wind turbines kill birds and bats by direct collision. They also kill bats by a process called barotrauma (from the prefix baro-, meaning atmospheric pressure). Bats use echolocation to chase moving objects, and they chase the moving turbine blades. The flow of wind over the airfoil-shaped blades causes the air pressure to drop significantly behind the blades. When bats enter the low-pressure region, the air in their lungs, which is still at the higher pressure of the air where they last inhaled, expands and crushes the capillaries in their lungs. Although bats are feared and persecuted by some people, they are actually enormously beneficial predators of insects, and they are important plant pollinators. A case in point: A colony of 1,500,000 Mexican free-tailed bats that lives under the Congress Avenue Bridge in Austin, Texas, devours up to 30,000 pounds of insects every night.[96] The emergence of the colony from under the bridge is a spectacle that attracts crowds of onlookers;[97] I have had the pleasure of watching it myself. Occasionally bats have contracted rabies and transmitted it to humans, but this has been extremely rare and concerns over it are vastly overblown.[98] At the Judith Gap wind farm in Montana, over a one-year period, wind turbines killed three birds per MW of electricity (406 birds altogether) and 8.9 bats per MW (1206 altogether). This wind farm is located in an area where the environmental impact is relatively low. Other planned wind farms in Montana would be sited in locations of much greater potential impact. Mitigation of the impact of wind farms on birds is a topic of active investigation by government, industry, and conservation groups.[99] The Pennsylvania Biological Survey[95] cites studies that predict an annual mortality by 2020 from wind farms in the mid-Atlantic region of 45,000 birds and 111,000 bats.

Raptors are particularly vulnerable to death by collisions with wind turbines. Most of these birds hunt by looking down for prey on the ground, and they are not accustomed to obstacles 250 feet in the air in open country. They just fly right into the towers or through the disc swept by the rotating blades. A badly sited wind farm at Altamont Pass, California, kills about 4700 birds a year, mainly raptors including golden eagles and burrowing owls, which are both in decline.[100] A study published in 2013 by the U.S. Fish and Wildlife Service revealed that wind farms killed at least 85 bald and golden eagles since 1987, mostly between 2008 and 2012, as wind power was being developed rapidly. Seventy-nine of them were golden eagles, struck by turbine blades. These data do not include the long-standing slaughter at Altamont Pass.[101] The killing of eagles is a violation of federal law, but the Obama administration granted a waiver to the wind industry for the next 30 years, possibly provided that wind farms take measures to reduce the rate of killing.[102] I happen to have a great fondness for birds in general and raptors in particular, especially eagles, and I deplore anything that kills them. Wind farms currently kill about 440,000 birds a year,[102] and it is only going to get worse as the wind industry expands.

Although some people, especially wind developers, may find wind farms visually gratifying, others emphatically do not. In 2009, a proposed offshore wind farm off Cape Cod, called Cape Wind, encountered stiff resistance from local residents who preferred the pristine seascape visible from their locale. Various opinion surveys of residents of Cape Cod and nearby communities showed considerably different results, but a survey in 2009 of Cape Cod Chamber of Commerce members showed 55% against the project. By 2014, opposition had been defeated and the project was proceeding apace. However, after losing some key contracts and suffering licensing and legislative setbacks, the developer cancelled the project in late 2017.[103]

Windmills are expensive to build (they have high capital costs), but cheap to operate (they have no fuel costs). Generally, wind farms are built on borrowed money, and the loans are paid over time from the revenues of plant operations. Comparisons of the costs of various power sources are based on the "levelized cost of energy" (LCOE), which is defined as the average total cost to build and operate a power-generating asset over its lifetime, divided by the total energy generated by that asset over its lifetime. As of 2014, according to a database compiled by the National Renewable Energy Laboratory, the LCOE for land-based wind power was between 40 and 80 USD/MW-h (U.S. dollars per megawatt-hour), while that for solar photovoltaic power was between 60 and 250 USD/MW-h, and that for nuclear power was between 90 and 130 USD/MW-h. The costs of wind and solar power have dropped precipitously in recent years as the relevant technologies have matured; projections by the Energy Information Administration in 2010 for construction to be completed in 2016 were 396.1 USD/MW-h for solar photovoltaic power and 149.3 USD/MW-h for land-based wind power, while its projections in 2018 for construction to be completed in 2022 were 59.1 USD/MW-h for solar photovoltaic power and 48.0 USD/MW-h for land-based wind power.[104] These projections are based on constant dollars so that meaningful comparisons can be made. Based on these projections, wind and solar photovoltaic power are currently competitive with all other power sources. However, there are hidden costs of wind power that are not accounted for in these projections.[105] Wind power is granted huge subsidies in the U.S. by the federal government (most of which go to foreign wind companies); backup power plants, usually coal or natural gas, must be kept on-line to provide power when the wind is calm, imposing operational costs and air pollution; because of the remote siting of many wind farms, wind power suffers more transmission losses than other power sources. Reference 105 gives the best estimate of the true total cost of wind power as 149 USD/MW-h. Thus, wind power still may not really be as economically competitive as it seems.

Despite these economic and environmental issues, wind power is being promoted vigorously. As of 2010, thirty states enacted rules (renewable portfolio standards, or RPSs) requiring that certain fractions of the electricity generated within them must come from renewable sources. In most cases, the RPS is 20% by 2020. A realistic assessment of the challenges posed for the achievement of such standards is given in a report by the Panel on Public Affairs of the American Physical Society.[106, 107] Challenges fall in areas such as wind forecasting, energy storage for times when the wind is not blowing or the sun is not shining, transmission technologies such as superconducting lines, and modernizing the grid to enable long-distance power transmission. Even for a penetration of 20%, the report says that these challenges are greater than many wind and solar power advocates seem to recognize. As of 2017, some states have achieved renewable energy penetration (not counting hydropower) of more than 20% (as high as 46% in Maine), but most states are far from that degree of success. Overall, renewable penetration in the U.S. in 2017 (not counting hydro) was 9.6%.[108]

8.7 POWER FROM BIOMASS

In the context of energy resources, biomass[109] means living or recently living organisms, or materials derived from them, that produce energy. Although fossil fuels originated in plants and animals (e.g., zooplankton), the organisms from which fossil fuels came have been dead for eons, and fossil fuels are not considered biomass. The term biomass is used for renewable energy sources. Biomass includes wood, crop plants such as corn, organic waste material, and even garbage. Gases from decaying organic material in landfills ("landfill gases") can be collected and used for energy production. Biodiesel fuel is obtained from vegetable oils or animal fats, including waste oils.[110] Biodiesel and even gasoline[111] can also be produced by algae.

All the chemical compounds that bear the energy resources in biomass contain carbon, but the carbon in them is obtained from atmospheric carbon dioxide, so burning them simply returns their carbon back to the atmosphere, and thus biomass is not a net source of carbon dioxide in the atmosphere. Energy resources involving such closed carbon dioxide cycles are said to be carbon-neutral. However, when slow-growing plants such as trees are burned, the time required to return the resulting carbon dioxide to new trees is a matter of decades, so the short-term effect of burning them is an increase in atmospheric carbon dioxide.

Biomass can be used to produce electricity, burned directly for heat, or processed into liquids for transportation fuels. As of 2012, biomass accounted for about 1.3% of U.S. electricity production, and 2014 it was projected to approximately double by 2022.[112] The types of biomass used for this electricity included landfill gas, municipal solid waste, agricultural byproducts and waste, sludge waste, wood, wood waste solids and liquids, and other miscellaneous biomass solids, liquids, and gases.[113] The principal wood waste product used for producing energy is "black liquor," a byproduct of the pulp, paper, and paperboard industries. Most of these fuels are essentially free, as they are waste products that need to be gotten rid of. There is some potential for expanding the contribution of biomass to electricity production. However, the largest current use of biomass energy is ethanol for automotive fuels.

Ethanol is simply ethyl alcohol (C_2H_5OH), the alcohol of ethane (C_2H_6). This is the same alcohol found in alcoholic beverages. It has an energy content of 76,000 Btu/gallon,[e] compared to 115,500 Btu/gallon in gasoline.[114] These figures are the so-called lower heating values, which do not account for recovery of the heat of vaporization of the steam component of the exhaust gases. In some applications, such as home heating units, this heat is recoverable, and the total available energy is called the higher heating value, but it is not recoverable in automobile exhaust. The available energy in ethanol is only 76,000/115,500 = 0.658 times, or about two-thirds, as great as that in gasoline; conversely, it takes 115,500/76,000 = 1.52 gallons of ethanol to replace one gallon of gasoline.

Ethanol for fuel[115] is mostly produced by fermentation, just like the ethanol in alcoholic beverages. The process is sugar + yeast → alcohol + carbon dioxide. Note, therefore, that the production of ethanol is a greenhouse gas generator. Ethanol can also be produced by algae directly. About 5% of the world's ethanol production is synthesized from non-renewable resources such as coal or ethylene.[116] Ethylene is usually produced by cracking petroleum,[117] which requires energy, and the resulting alcohol has a lower energy content than the petroleum from which it was made. This route to ethanol is obviously not a net energy producer.

As of 2019, most of the ethanol fuel produced in the United States is produced from corn, but not from the whole plant. Only the kernels are used—the part that is also valuable for food.[118] In Brazil, ethanol fuel is produced from sugarcane. Together, these two nations account for 89% of the world's ethanol fuel production. Brazil has mandated ethanol content of 20% to 25% (depending on the region within the country) for gasoline sold there. Some cities and states in the U.S. have mandated ethanol content of 10%. The U.S. federal government has not mandated the ethanol content of gasoline, but it does mandate production goals for ethanol; for 2011, American motor fuels were required to include 12.6 billion gallons of corn ethanol and 1.35 billion gallons of other biofuels. Nearly 40% of the corn produced in the U.S. is used to make ethanol fuel.[119] As of 2010, about 90 million acres in the U.S. were planted in corn, so the land area devoted to ethanol production was then about 36 million acres.[120] Modern automobiles in the U.S. are capable of running with 10% ethanol (called E10), but there are problems with ethanol-gasoline blends that most consumers are not aware of. Ethanol absorbs water, and the water can cause phase separation of the mixture—i.e., the alcohol-water component separates from the gasoline. This separation damages numerous engine and fuel system components, and can cause

---

[e] The Btu, or British thermal unit, is equal to 1055.05585 J. It is the traditional unit of heat in the British system of units, which is still widely used in the U.S.

engine failure.[121, 122, 123] Many of these problems go away with high ethanol content, such as 85% ethanol fuel (E85), if the proper materials are used for plastic fuel-system components. In October 2010, the U.S. Environmental Protection Agency (EPA), at the request of the corn ethanol industry, issued approval of fuel containing 15% ethanol (called E15) for automobiles manufactured in 2007 or later.[124] This fuel is not approved by automobile and small engine manufacturers, and its use may void engine warranties.[125] A coalition of engine manufacturers filed suit against the EPA for permitting the use of this fuel, which has been shown to cause failures in engines and emissions control equipment.[126] As of 2018, most gasoline pumps in the U.S. contain 10% or less of ethanol, as anyone who fills up the gas tank of a car should know from reading the label on the pump.

The purposes of federal, state, and local mandates for production and use of ethanol are to reduce air pollution and to reduce dependence on foreign oil. Ethanol burns cleaner than gasoline or diesel fuel, producing less carbon dioxide, carbon monoxide, particulates, and other toxic byproducts. These benefits are realized somewhat even by E10 fuel.[127] Assessment of the overall benefits of ethanol must take into account all of the environmental and societal impacts of producing and consuming ethanol fuels. These include the following considerations:

- The land required to grow crops for producing ethanol
- The increased price of food caused by the diversion of food crops to energy production
- The overall carbon balance of ethanol production and use
- The overall energy balance of ethanol production and use
- The impact of fertilizers, herbicides, and pesticides required to produce source crops efficiently
- The increased fuel consumption required to transport larger volumes of fuel to gas stations (since ethanol only contains two-thirds as much energy as gasoline by volume)
- The demands on additional resources such as water.

As noted above, 40% of the U.S. corn crop is devoted to production of ethanol for fuel. This is not a negligible land use. It is a very inefficient use of land, since only the kernels of the corn plant are used for fuel. In the future, however, it is likely that ethanol fuels will be produced from cellulose, which comprises most of the dry weight of plants. There are plants other than corn, such as switchgrass, that are more suitable than corn for the production of cellulosic ethanol, and their use will permit more efficient land use.[128] With corn ethanol, the diversion of so much land to the production of motor fuels has had a large impact on world food prices.[129] A 2008 report by the World Bank[130] claimed that the worldwide price increase was then 75% and had pushed 100 million people under the poverty line. A shift to cellulosic ethanol should substantially abate the impact on food prices of land use for fuel production.

The expanded production of corn for ethanol has led to destruction of land that had been set aside for conservation, including some of the little virgin prairie left in states like Iowa and Nebraska. According to an Associated Press article published in November 2013, five million acres of conservation land—a greater area than that of Yellowstone, Everglades, and Yosemite National Parks combined—have vanished since President Obama's embrace of ethanol, and 1.2 million acres of virgin prairie have been lost just in Nebraska and the Dakotas since 2006. The use of nitrogen fertilizer to boost corn production has led to increased levels of toxic nitrates in Midwestern waterways used for drinking water by millions of people. For example, the Des Moines Water Works, which takes water from the Raccoon and Des Moines Rivers, found excessive nitrate levels in both rivers in 2013: The AP article states, "This year, unfortunately the nitrate levels in both rivers were so high that it created an impossibility for us,' said Bill Stowe, the water service's general manager."[131]

Corn ethanol requires substantial investments of energy. Farming operations consume diesel fuel, and the processes involved in producing ethanol from corn require energy. Roughly two-thirds of the energy required to produce ethanol is consumed in the conversion process from corn to ethanol, and most of the rest is consumed by farming operations. The conversion process consumes both electricity and thermal energy, much of which comes from coal.[132] The ratio of the total energy obtained from burning ethanol to the total energy consumed in producing it is called the ethanol fuel energy balance.[133] The calculation of this balance is complicated, and the resulting figures are controversial. However, claims by some investigators[134, 135] that the energy balance is less than one (i.e., ethanol production consumes more energy than it produces) have recently been discredited. The energy balance for corn ethanol in the U.S. is now believed to be about 1.3; the energy balance for sugarcane ethanol in Brazil is about 8; the potential energy balance for cellulosic ethanol in the U.S. is as high as 36.[133] A 2009 study by the University of Nebraska at Lincoln (UNL) found that the energy balance had recently increased to between 1.5 and 1.8 because of increased efficiency in production processes. Even more important is the ratio of ethanol produced to oil consumed. Since much of the energy used in ethanol production does not come from oil, this ratio is greater than the energy balance ratio. The UNL study found that between 10 and 19 gallons of ethanol are produced for every gallon of petroleum used to produce it.[136] Since ethanol only contains two-thirds the energy of gasoline by volume, that means that the energy contained in the ethanol is between 6.7 and 12.7 times as great as the energy in the petroleum consumed. This gain represents a significant reduction in our need for oil. However, Ted Williams, an environmental muckraker with a regular column in *Audubon* magazine, reports a study by Princeton University's School of Public and International Affairs that claims "producing and burning ethanol creates 93 percent more greenhouse gases than producing and burning fossil fuel."[137]

The phase separation problems noted above for fuels with low ethanol content (e.g., E10 and E15) do not occur with E85 fuel until the water fraction in the fuel is greater than 20% by weight.[138] Ethanol has a higher octane rating than regular gasoline, so engines restricted to running on E85 can have higher compression ratios, which somewhat mitigates the disadvantage in energy content inherent in ethanol. By the (R+M)/2 method of computing octane rating (this is the number posted on the fuel pump at your gas station), E85 has an octane rating between 94 and 96, compared with 87 for regular gasoline and 91 or 93 for premium.[139] The energy content of E85 is 28% less than that in gasoline, but tests on E85 vehicles in Ohio showed only a 25% reduction in fuel economy.[138] Neglecting the amount of petroleum involved in the production process, one can see that about 80% less petroleum would be needed for the U.S. automobile fleet if we shifted to E85 fuel (we would need 1/0.75 = 1.33 times as many gallons, but only 15% of those gallons would consist of gasoline; 1.33x0.15 = 0.2). That shift would greatly reduce our dependence on foreign oil. However, the shift could not be made on corn ethanol. We would need to increase corn production by about 10 times, which would require 400% of the total U.S. farm area currently devoted to corn. (This estimate accounts for both the increased ethanol content of E85 and the lower energy content of ethanol.) Limitations on arable land, as well as the impact on food prices, will ensure that such expanded corn ethanol production will not happen. However, development of cellulosic ethanol may make a transition to an ethanol-based transportation infrastructure possible.

Currently, E85 costs more per unit energy than E10, so the consumer who chooses a "Flex-Fuel Vehicle" (i.e., one that can run on E85 or regular "gasohol") must pay a premium to use it.[140] From the 1980s until 2012, the price of ethanol was artificially reduced by government subsidies,[141] so taxpayers were also paying to push ethanol fuels. The difference in price per gallon between E85 and regular gasoline is tracked on the website http://e85prices.com; as of November 2018, E85 was 20% cheaper by the gallon, but that works out to 8% more per Btu for E85.

In view of the considerations discussed above, one must say that ethanol fuel has great potential, but that it also has great drawbacks that must be overcome before ethanol becomes the primary energy source for transportation, and that as it has been implemented so far, ethanol production has done a great deal of harm.

8.8 GEOTHERMAL POWER

The Earth's interior is very hot. According to geologists' current understanding, the innermost region is the inner core,[142] a sphere about 760 miles in radius, where the temperature is as much as 5700 K (9800 °F). Although it is extremely hot, it is mostly solid because of the high lithostatic pressure at that depth. Next is the outer core,[143] a molten layer of iron and nickel about 1400 miles thick varying in temperature from 4700 K (8000 °F) to 6400 K (11000 °F). Next comes the mantle,[144] about 1800 miles thick. The mantle is mostly solid because of the high lithostatic pressures there, but it is hot enough near its boundary with the core that it would be liquid at atmospheric pressure. The temperature in the mantle varies from about 4300 K (7200 °F) at that boundary to about 1200 K (1650 °F) at its outer limits. Although the high lithostatic pressure makes the mantle solid, the mantle is so hot that it behaves plastically, and because of heating from the core there is convective flow in the mantle on a time scale of millions of years. The outermost layer is the crust,[145] a thin cooler layer of solid rock that floats on the mantle. The oceanic crust, which underlies the oceans, is only 3 to 6 miles thick. The continental crust, which comprises the land masses, is usually between 20 and 50 miles thick. The temperature in the crust varies from the temperature of the interface with the mantle to the surface temperature, which changes with location and season.

When hot rock from the mantle intrudes into the lower-pressure crust, it can liquefy and become magma.[146] Magma is responsible for volcanoes,[147] geysers,[148] and many hot springs.[149] (Other hot springs occur where water emerges onto the surface from a source deep within the crust, where the temperature is relatively high.) Magma usually reaches the Earth's surface or gets close to it at tectonic plate boundaries,[150] but it can also do so at so-called hotspots, where mantle plumes[151, f] melt their way through the crust.

The high temperatures in the Earth's interior are due partly to the original

---

[f] The existence of mantle plumes has recently been called into question. See *Physics Today*, October 2012, pp. 10-12 (letter from Don L. Anderson). Alternative explanations for hotspots are given.

compressive heating associated with the Earth's formation, and partly to decay heat from radioactive elements contained in the Earth's core.[152] Heat flows from the Earth's interior to the surface at a rate of 44.2 TW (terawatts, or 10^{12} watts)[153] and is currently replenished by radioactive decay at a rate of 27 TW (calculated from 860 EJ/yr, where EJ is exajoules, or 10^{18} joules).[154] This radioactive heating is about twice the world's primary energy consumption rate. (Primary energy consumption is the energy consumed from nature before conversion into usable forms.[155]) In the past, the radioactive heating rate was higher according to the exponential time-dependence of radioactive decay (see Chapter 4).

Obviously, there is a lot of heat in the Earth's interior, and it is replenished at a rate that makes its exploitation sustainable, at least at current energy consumption rates. However, it is spread out over a huge volume and flows out over a huge surface area, so most of it is not exploitable. Geothermal energy use relies on locations of volcanism where high temperatures are available close to the Earth's surface or on low-grade heat at shallow depths.[156]

In Chapter 7 it is explained that a heat engine—a device operating in a cycle and converting heat into work—cannot perform this conversion completely, but must reject some of the heat into a heat sink. The efficiency of the conversion depends on the temperatures of the heat source and the heat sink. The hotter the heat source and the colder the heat sink, the higher the efficiency. Although geothermal heat is free once the high-temperature heat source is reached, drilling to reach it is costly, and if the source temperature is too low, the resulting low efficiency makes geothermal electricity generation impractical. Where sufficiently high temperatures are available at depths that can be reached economically, electrical generating plants can be built. In 2015, 24 countries had geothermal electrical generating facilities, with a total rated capacity of 13.3 GW. This capacity may increase to as much as 18.4 GW by 2021 if projects then under consideration are built. In the U.S., there were 64 geothermal electrical power plants in 2014;[157] in 2015, the total rated capacity in the U.S. was 3.7 GW.[158] For comparison, note that a typical large coal or nuclear power plant produces about 1 GW. Because of the cost of drilling to the depths where high temperatures are found in most of the continental crust, geothermal electric projects have limited potential in most countries, although in some relatively small countries with unusual concentrations of suitable sites, geothermal electricity can play an important role. In 2010, geothermal power accounted for about 27% of the electrical energy produced in the Philippines.[156]

Geothermal electricity can be obtained efficiently by taking energy from natural hot water formations such as hot springs at the surface or hot aquifers deeper in the ground, or by injecting water into hot dry rock formations. The latter technique is called hot dry rock geothermal energy or enhanced geothermal energy extraction. To expose sufficient rock area to the injected water, the technique fractures the bedrock hydraulically. This fracturing can create earthquakes; for example, an enhanced geothermal energy project near Basel, Switzerland, triggered 10,000 earthquakes up to 3.4 on the Richter scale within the first six days of water injection and had to be suspended.

Other environmental consequences of geothermal electrical generation include extinction of geysers, emission of toxic gases from deep in the Earth, and transport of toxic chemicals (e.g., mercury and arsenic) to the biosphere. In some locations where geothermal heat sources are near or at the surface, installation of geothermal power plants would inflict unacceptable aesthetic insults (e.g., Yellowstone National Park). On the whole, however, geothermal electric power is relatively benign environmentally where it is practical.

Another use of geothermal energy is direct heating. This may be accomplished by taking heat from hot groundwater, generally where temperatures are too low for generating electricity efficiently; in the geothermal industry, temperatures below 425 K (300 °F) are considered to be low. As of 2007, the total worldwide installed capacity for this kind of heating was about 28 GW.[159]

However, a more widespread type of direct geothermal heating takes heat from the ground at shallow depths by the use of heat pumps.[160] In 2004, there were 15 GW of installed heat pump capacity from 1.3 million geothermal heat pumps worldwide, and the energy production rate from such pumps was growing at 30% per year. A quick look at how these devices work is worthwhile.

The temperature of the Earth's crust varies seasonally down to about 10 m (33 feet), but below that the temperature is steady. Depending on latitude, the ground temperature immediately below the layer of seasonal fluctuation is between 7 °C (45 °F) and 21 °C (75 °F).[161] The temperature increases with depth from that point at a rate of about 25-30 °C (77-86 °F) per kilometer. At shallow depths, the ground temperature in northern or temperate latitudes is lower than the temperature one wants in one's house in winter. However, even at cold temperatures there is a lot of thermal energy in the ground; thermal energy is just the energy of molecular motion, and that doesn't cease until absolute zero (actually, it doesn't quite stop even then, but for practical purposes we can consider it to do so). This thermal energy can be extracted and carried to warmer spaces, such as the inside of your house, by heat pumps.

This transport of heat from a cold place to a warm place is similar to what happens in a refrigerator, which takes heat out of its cold interior and exhausts heat into the warm room. By the second law of thermodynamics (see Chapter 2), it is impossible to build a device that operates in a cycle and produces no other effect than the transfer of heat from a cooler body to a hotter body. The key here is "no other effect." One can transfer heat from a cooler body to a hotter body by a net input of work.

Figure 8.6 shows a heat pump schematically. A working fluid, such as 1,1,1,2-tetrafluoroethane (R-134a),[162] is circulated in a loop. The fluid has the convenient property of changing from liquid to vapor at appropriate temperatures and pressures. In the compressor, saturated vapor (i.e., vapor that is about to condense into liquid) is compressed to high pressure. The compression heats the fluid, so that the fluid remains vapor even though its density increases. This is the stage at which the required work is put into the system. The high-pressure, high-temperature ("superheated"—see Section 17.1) vapor passes through a coil called the condenser in the space to be heated, such as the interior of your house. In the condenser, the fluid pressure is constant, but as heat passes from the fluid to the space to be heated, the fluid temperature falls until the vapor is saturated; as the fluid continues through the coil it continues to transfer heat to the house at constant temperature and pressure and it condenses into saturated liquid (liquid that is just about to boil). From there it goes through an expansion valve, where some of it flashes into vapor. Vaporization requires an input of heat to break molecules loose from the liquid. This heat of vaporization is taken from the thermal energy of the liquid itself, which renders the resulting mixture of vapor and liquid cold. The temperature of this mixture must be less than the temperature of the soil for the heat pump to work. From the expansion valve, the fluid passes into the evaporator coils, which are just a large number of loops

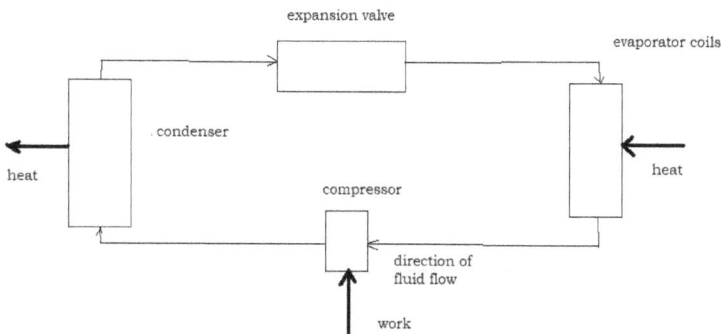

Figure 8.6—Schematic representation of a heat pump.

buried in the ground. Heat passes from the cool or cold soil into the even colder fluid in the coils. Figure 8.7 shows an array of evaporator coils in the ground prior to being covered with soil.[163] Exiting from the evaporator coils, the fluid is once again in the saturated vapor state, ready to enter the compressor.

Figure 8.7—Evaporator coils for a geothermal heat pump. This file is in the Wikimedia Commons and is in the public domain.

The efficiency of a heat pump is called the coefficient of performance, which is the ratio of the heat delivered (i.e., to the interior of the house) to the work required.[164] The coefficient of performance of an ideal heat pump (a Carnot cycle operating in reverse—see Chapter 7) can be shown to be given by the following equation:

$$C_{p,heating} = \frac{T_{hot}}{T_{hot} - T_{cold}},$$

where T_{hot} is the absolute temperature of the high-temperature heat sink (the house interior) and T_{cold} is the absolute temperature of the low-temperature heat source (the ground). This is just the inverse of the Carnot efficiency for an ideal heat engine, as discussed in Chapter 7. Thus, we can see that the coefficient of performance is higher when the temperatures of the heat source and heat sink are close together. That makes sense: It is easier to transfer heat "uphill" when there isn't much of a "hill." When the temperatures are equal, the coefficient of performance is infinite. Practical geothermal heat pumps have coefficients of performance between 4.0 and 5.0. That means that for every unit of electrical energy used in running the compressor, between four and five units of heat energy are pumped from the ground to the house. This is a large gain in energy, and if I were to build a new house for myself, I would consider installing a geothermal heat pump.

A heat pump can be run in reverse for air conditioning. Then the relevant measure of performance is how much heat is taken from the cooler house interior and dumped outside for each unit of work done by the compressor, so the coefficient of performance for cooling is the ratio of these two quantities. For an ideal heat pump in cooling mode, the coefficient of performance is

$$C_{p,cooling} = \frac{T_{cold}}{T_{hot} - T_{cold}} \,.$$

Again, the performance is best when the hot and cold temperatures are close together. A little algebra shows that $C_{p,cooling} = C_{p,heating} - 1$. So for cooling, practical heat pumps have coefficients of performance between 3.0 and 4.0. This is still a substantial energy gain, making geothermal heat pumps a viable choice for home air conditioning.

The potential for this use of geothermal energy is great. Even though ultimately one must consume electricity to use it, the energy gain is large and only a relatively small amount of electricity must be supplied.

8.9 TIDAL POWER

Tidal power[165] is power extracted from ocean tides. In the form of tide mills,[166] tidal power has been used since at least the Middle Ages and possibly since the time of the Roman Empire. Tides can be used to drive turbines for electric power generation, but this application has been limited by high cost and the requirement for sites with high tides or high tidal velocities. However, new technologies promise greater potential for tidal power.

As of 2015, there were at least ten tidal electric power generating stations operating in the world.[167] They are the Annapolis Royal Generating Station on the Bay of Fundy in Canada,[168] the Jiangxhia Tidal Power Station in China,[169] the Kislaya Guba Tidal Power Station in Russia,[170] the Rance Tidal Power Station in France,[171] the Strangford Lough station in Ireland,[172] the Bluemull Sound Tidal Stream Array[173] and the MayGen Tidal Project[174] in Scotland, the Uldolmok[175] and Sihwa Lake[176] Tidal

Power Stations in South Korea, and the Eastern Scheldt Barrier Tidal Power Plant in The Netherlands.[177] Most of these are low-power stations, generating power of the order of 1 MW. However, the MayGen station generates 387 MW, the Sihwa Lake station generates 254 MW, and the Rance station generates 240 MW. Most of these stations are barrage-type facilities, in which a tidal estuary or bay is completely blocked by a dam containing numerous turbines. Strangford Lough, Bluemull Sound, and MayGen are tidal stream generator stations, in which free-standing turbines extract power from the tidal currents flowing through them.

Obviously, a barrage-type facility has the potential to disrupt the movements of aquatic organisms, particularly anadromous fish, as well as the movements of ships and boats. Tidal stream generator stations avoid such impacts, but unless there are many generators their output is limited, and if there are many generators they pose hazards to shipping. A third type of generating station, which may be deployed in the future, is called dynamic tidal power. In this type of station, a barrier as long as 30 miles projects out into the ocean from the coast, and turbines in the barrier extract power as the tidal currents pass through them. Marine organisms and ships can go around the barrier without being fully trapped, although it may not be obvious to the organisms that there is a path around the barrier. Such barriers could generate gigawatts of power.

Several more tidal power stations are currently proposed, most of which would generate power in the GW range. The largest of these is the Penzhinskaya Tidal Power Plant Project in Russia at the northern end of the Sea of Okhotsk,[178] a branch of the Pacific Ocean. This site has the highest tides in the Pacific. One version of the design for the Penzhinskaya plant would generate 87 GW, making it truly huge. However, the proposed plant location is remote, so getting the power to consumers would be costly and entail high losses.

Tidal power thus has considerable potential in some locations, and it is non-polluting, but it does impose environmental problems and some disruption of human activities. As with most other power systems, each proposed project should be evaluated on its own merits and disadvantages.

8.10 PERPETUAL-MOTION MACHINES

I include this section because once in a while someone comes up with a proposal that falls into this category. I will illustrate the category with an example. Early in my career, I enjoyed a stint as an assistant professor at Purdue University, and one year I was asked to be a judge at the local high school science fair. One student, with the approval of his science teacher, presented such a project.

He proposed installing pistons in highway roadbeds that would be pushed down slightly every time a car rolled over them. Pushing the pistons down is a form of work, as we saw in Chapter 2, and work can be converted into other forms of energy. The pistons would be linked hydraulically to a system that would convert the work done on the pistons into electricity. The depression of the pistons would be very small, and the cars would rise out of the depression without significant loss of kinetic energy. Thus, with the accumulated work done by many cars, electricity could be created for free. After passage of the car, the piston would rise by the action of springs to its original level. One-way valving would permit the piston to rise with very little resistance, so that the springs would not have to be very stiff and the work done by the car to compress the springs would be negligible.

My review of the project was not flattering. I was particularly disappointed with the teacher who approved this project, since he should have seen the idea as a perpetual-motion machine. Perpetual-motion machines like this violate the law of conservation of energy, or the second law of thermodynamics, or both. In particular, a machine that would produce more energy than it consumes—essentially, producing net energy out of nothing—violates the law of conservation of energy (also called the first law of thermodynamics), and is called a perpetual motion machine of the first kind. A machine that converts ambient thermal energy (colloquially, the heat in the surrounding environment) completely into work violates the second law of thermodynamics and is called a perpetual motion machine of the second kind.[179] Any scheme that purports to create energy from nothing or to draw net energy from thin air can be dismissed *a priori* as bogus.[g]

The science fair project was a perpetual motion machine of the first kind. The fallacy in the proposal was the claim that, compared to the work done by the car on the pistons, the energy required by the car to climb out of the depression is negligible. In reality, the car gains no kinetic energy (the energy of motion) while depressing the piston, because all of the work done by gravity on the car is absorbed by the piston to generate electricity, but the car loses kinetic energy as it climbs out of the depression, because the piston doesn't rise back up until the car has passed. By the law of conservation of energy (see Chapter 2), the car must lose kinetic energy as it rises out of the depression to regain the potential energy (the energy stored in the Earth's gravitational field at a particular vertical position) it lost when going down into the depression. The

---

[g] You could draw energy from ambient air and convert it to net work (e.g., electricity) if you rejected some waste heat into a colder heat sink such as a lake. However, the efficiency of a device doing this would be so low that the device would be impractical.

driver must add a little gas to regain his original speed. In a perfect world (i.e., one without friction), the energy required by the car to climb out of the depression would be equal to the work extracted from the car to depress the pistons. However, there is friction, and this leads to dissipative losses that cause the car to expend more energy climbing out of the depression than it gave to the pistons. Ultimately, the proposal was a way of stealing gasoline from motorists to produce electricity inefficiently.

One can always find the detailed physical explanation for the fallacy in specific proposed perpetual-motion machines, but sometimes it takes some effort to do so. That is why charlatans can get away with selling such contraptions; you can buy them on the Internet if you are gullible. Don't.

8.11 NUCLEAR ENERGY

The next chapter assesses the issue of anthropogenic climate change, which some people use as an argument in favor of nuclear energy. The subsequent chapters are intended to explain the workings of all aspects of nuclear technology, especially the various forms of nuclear power. We will see that nuclear fuel resources are adequate to last a very long time, especially if spent nuclear fuel is reprocessed to recover unused fuel resources and if nuclear fusion progresses to practical application. We will see that although nuclear power imposes some environmental problems and safety risks, even in the worst nuclear accidents, including the accidents at Fukushima Daiichi, the consequences are surprisingly small compared to the constant toll taken by many other energy technologies.

If the consequences of using any technologies are deemed acceptable, the choices among them will ultimately be economic and practical: Consumers will choose the technology that costs them the least per kilowatt-hour or inconveniences them the least. (Even if electricity is extremely cheap, consumers will not buy electric cars if their range is too short or the recharge time is too long.) In the section on wind power, I briefly noted current comparative costs among some of the available energy technologies. However, the relative costs of energy technologies are subject to considerable change as some resources are depleted and some technologies become more mature. If I made current relative costs a major arguing point, my argument would be obsolete before the book was published. I will simply note from the comparison noted above that nuclear power is not drastically different in cost from other energy technologies. Also, I want to point out that one of the design goals of so-called Generation IV reactors—the next generation of nuclear plants to be deployed—is that they must be economically competitive with other widely available energy technologies.

CHAPTER 9

GLOBAL WARMING

Many of our energy resources are substances such as coal and petroleum that contain carbon. Combustion of any of these fuels produces carbon dioxide. It has been stated often that almost all climatologists believe human-generated carbon dioxide is causing an increase in the Earth's temperature—global warming—that may have serious impacts on human civilization. Proponents of nuclear energy often cite that consensus to support nuclear energy over other energy sources.

However, there is a strong voice in the popular media, supported by a very few scientists, that doubts that opinion. I am not a climatologist, and this is not a book on climatology. Nevertheless, I owe it to my own scientific integrity and to the trust my readers place in me to examine the claims of climatologists objectively, so that I can decide whether nuclear energy deserves support as a mitigator of global warming.

Those claims may be distilled into the following points:

- Carbon dioxide acts as a greenhouse gas
- The concentration of carbon dioxide in the atmosphere has been increasing since the beginning of the Industrial Revolution
- Human activity is responsible for much, probably most, of that increase in carbon dioxide
- The Earth's temperature has been increasing, particularly in the latter part of the 20th Century and the 21st Century
- The increase in atmospheric carbon dioxide and the increase in the Earth's temperature can be shown by computational models to be a cause-and-effect relationship via the greenhouse effect
- The same computational models predict continued increases in the Earth's temperature if human-generated carbon dioxide continues to accumulate in the atmosphere

- The predicted future increases in the Earth's temperature have a high potential to disrupt global ecology and human civilization.

I will briefly summarize my assessment of each of those points. Some of the discussion applies concepts from the science of thermodynamics and heat transfer; these concepts are discussed in some detail in Chapters 2 and 7.

9.1 THE GREENHOUSE EFFECT

The greenhouse effect is well understood; it is the reason greenhouses work, and it has been experienced by everyone who has been in a closed automobile. A brief explanation of it in the latter context will be instructive.

Imagine that your car is parked in the sun with the windows rolled down. Suppose that a nice breeze is flowing through the car. Sunlight falling on the windshield glass is partially reflected from it and partially transmitted through it. The transmitted sunlight falls onto the interior surfaces. The interior heats up because the interior surfaces absorb the incident sunlight and the heated surfaces transfer heat into the air in the interior. However, the air flowing through the car carries away heat from the interior (this kind of heat transfer is called convection), so that the interior does not heat up too much. The temperature of the air flowing through the car increases slightly as the air picks up heat from the interior, but not much, because the brisk wind passes through the car quickly. The interior radiates heat back outwards, but at a relatively low rate because it is not very hot.

Now suppose that you return to your car, get in, and sit in it to wait for a friend. Your car is not air-conditioned, so you leave the motor off. There is a lot of annoying traffic noise, so you roll the windows up.

The loss of convective cooling allows the interior to heat up much more than it did with the windows down. Eventually, a new equilibrium temperature will be reached, in which the amount of heat entering the car as sunlight is rejected by thermal radiation, or heat radiation (radiative heat transfer takes place across space by electromagnetic waves, as discussed in Chapters 2 and 4), which is in the infrared portion of the electromagnetic spectrum. (For simplicity, we are neglecting the heat conducted through the roof of the car and transferred to the air by convection.) Radiative heat transfer is governed by the Stefan-Boltzmann Law (see Chapter 2): The emitted heat flux varies as the fourth power of the absolute temperature. The car's interior will heat up until it becomes so hot that the portion of the infrared radiation escaping out through the window glass is equal to the total radiation absorbed by the interior. Also, the nature of the silicate window glass causes a problem. The transparency of the window glass is highly dependent on the wavelength of the radiation impinging on it. Silicate glass is very transparent to visible radiation, but much less transparent to infrared radiation.[1] The portion of the incident sunlight in the visible part of the spectrum passes through the glass into the car almost without attenuation. However, the interior re-radiates the absorbed energy in the infrared regime, to which the glass is much less transparent. The glass absorbs a certain fraction of the energy radiated from the seats and re-radiates it again in all directions, also in the infrared regime. Some of this energy re-radiated from the glass goes back into the car interior and is reabsorbed. Because more of the energy entering as sunlight remains in the car than would remain if the glass were transparent to infrared radiation, the car heats up even more than it would in that case. On a hot day, if you do not give up and roll the windows down, the heat in the car will kill you. This effect—the greenhouse effect—is inevitable. And it is not a matter of conjecture or scientific dispute. What is often not noted in the popular press is that the loss of convective cooling has a greater effect on heating up the car than the translucency of the glass to infrared radiation.[2]

Convection carries heat from place to place on the Earth, but all heat transfer to and from the Earth as a whole occurs solely by radiation. Greenhouse gases in the atmosphere act like the window glass. They are fairly transparent to incident solar radiation in the visible and ultraviolet regimes. They are much less transparent to infrared radiation.[3] Sunlight absorbed by the Earth's surface is re-radiated as infrared radiation. The greenhouse gases absorb some of it and re-radiate some of that back to the Earth's surface. The Earth's surface is warmer because of greenhouse gases than it would be without them. This effect is also not a matter of conjecture or scientific disagreement.

The contribution of a particular gas to the greenhouse effect depends on its potency and its concentration. The most important greenhouse gas in the atmosphere is water vapor. Carbon dioxide is second, followed by methane and ozone. A National Aeronautics and Space Administration (NASA) study finds that water vapor and clouds contribute 75% of the total greenhouse effect, carbon dioxide contributes 19%, and all other constituents contribute 7%, on average (the total doesn't add up to 100% because of rounding). On a clear day, these numbers change to 67%, 24%, and 9%, respectively.[4]

The physics of infrared transmission and absorption in the atmosphere is discussed in detail in Reference 5.

9.2 THE HISTORY OF CARBON DIOXIDE
IN THE EARTH'S ATMOSPHERE

Critics of the global warming theory rightly claim that the concentration of carbon dioxide in the atmosphere has varied greatly during the Earth's history. Carbon dioxide concentrations 500 million years ago are believed to have been 20 times more than today's concentration, and in the Jurassic Period they are believed to have been 4 to 5 times greater.[6] The critics argue that in view of such large fluctuations in concentration, during which life on Earth got along just fine, we should not be concerned about changes on the order of 100%, whether they are human-caused or not.

That is a specious argument. Conditions millions of years ago, or hundreds of millions of years ago, were much different, and nobody in his right mind would want to see the Earth return to such conditions. Furthermore, the sun was several percent dimmer 500 million years ago.[7] Plants didn't even appear on land until the Silurian Period, which began about 444 million years ago. In the Jurassic Period, 200 million to 145 million years ago, there were no ice caps and much of what is now Europe was under water.[8] The last time the atmospheric carbon dioxide concentration was as high as it is today was about 2.5-3 million years ago. At that time, the temperature in the Arctic was between 15 and 20 °C greater than it is today, and the sea level was 20 meters higher than today.[9] All this is interesting, but it is not relevant to current concerns about climate change. The relevant question is not whether carbon dioxide concentrations were different in the distant past, but whether changes from the recent concentration might have caused, and might continue to cause, climatic changes that could disrupt human civilization.

The concentration of carbon dioxide in the atmosphere between the present time and 800,000 years ago has been measured by analyzing the composition of bubbles of air trapped in polar ice. A new method based on the ratio of boron to calcium in the shells of ancient algae allows the concentration to be measured as far back as 20 million years.[10] The findings of the new method agree with the ice core measurements back to 800,000 years ago. When two independent methods agree, the validity of both methods is strongly supported (this principle is formally named "consilience," or "convergence of evidence"[11]).

Based on these measurements, we know that the concentration of carbon dioxide has varied cyclically during the last half-million years in concert with the advance and recession of glaciers in the Ice Age. During the last 800 years, the concentration of carbon dioxide only varied about 2.5% until the beginning of the Industrial Revolution about 200 years ago; during this period the concentration was about 280 parts per million (ppm). Since then the concentration has increased to 413 ppm as of April 2019, an increase of about 48%.[12] Figure 9.1 shows these variations up to the year 2000.

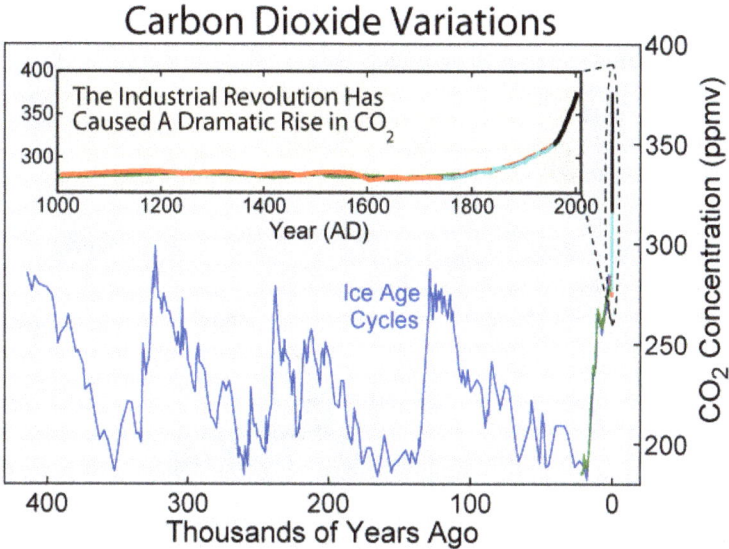

Figure 9.1—Variation in atmospheric carbon dioxide during the past 400,000 years. The label *ppmv* means parts per million by volume. This figure was created by Robert A. Rohde from publicly available data and is reproduced under the terms of the GNU Free Documentation License, Version 1.2.[13] It may be copied freely under the same terms, or downloaded from http://en.wikipedia.org/wiki/File:Carbon_Dioxide_400kyr.png.

The new method finds that the current atmospheric concentration of carbon dioxide is greater than it has been at any other time in the past 15 million years.[10]

Natural sources of atmospheric carbon dioxide include metabolic processes of plants and animals, fires, and geologic processes such as volcanoes. Volcanoes produce less than 1% of the amount produced by human activity, whereas all natural processes combined produce about 10 times as much as human activity.[6]

9.3 ANTHROPOGENIC CARBON DIOXIDE

As noted above, the carbon dioxide content of the atmosphere has increased by 48% in about the last 200 years. Is it possible that human activity has caused this? To answer this question, we must compare the amount of carbon dioxide generated by human activity to the amount by which carbon dioxide in the atmosphere has increased.

This is a question of great interest, so other people have already looked into it. The CO_2 concentration varies seasonally and from year to year; Reference 12 gives the average concentration in 2010 as 410 ppm by volume (0.041%), or 623 parts per million by mass (0.0623%). The total mass of the Earth's atmosphere is 5.15×10^{18} kg,[14] so the mass of carbon dioxide in the atmosphere is 3.21×10^{15} kg, or 7.06×10^{15} pounds. These figures were 45% greater in 2010 than the corresponding values 200 years ago; thus, the earlier figures are 283 ppm by volume, 430 ppm by mass, and 4.87×10^{15} pounds. The atmospheric carbon dioxide content has increased by 2.19×10^{15} pounds during this time.

According to *Science Daily*, as of 2007, power generation emits nearly 10 billion tons of carbon dioxide per year worldwide, and this is about one-fourth of total human-caused carbon dioxide emissions.[15] Thus, total annual carbon dioxide emissions from human activities are 40 billion tons, or 8×10^{13} pounds. This is 1.1% of the total mass of carbon dioxide in the atmosphere in 2010, and 1.64% of the mass of carbon dioxide that was present 200 years ago, and it is 3.7% of the net increase. If all the carbon dioxide released into the atmosphere stayed there, human carbon dioxide generation at the present rate could supply the entire observed increase in only 27 years. It doesn't all stay there—about a third of it is promptly dissolved in the oceans[6]—but it is obvious that human activity could easily be responsible for the entire increase.

9.4 EARTH'S TEMPERATURE HISTORY

Reconstruction of Earth's temperature history is a challenging subfield of climatology, but many kinds of evidence have been applied to obtain estimates going back hundreds of millions of years. A fascinating review is given by John Baez.[16] Most records show that the Earth's temperature in the last half of the Twentieth Century has been warmer than most periods in the last 2000 years, but that there was another comparable warming, the Medieval Warm Period. Figure 9.2 shows several reconstructions for this time period.[17] The colored lines are smoothed by averaging over the ten years around the data time, and the black line for recent years shows one-year averages of actual measurements. Thus, the black line is not directly comparable with the colored lines. One can see that the Medieval Warm Period was as nearly warm as our current times.

Compared to the Twentieth-Century average, global temperature has risen by about 0.84 °C as of 2017.[18]

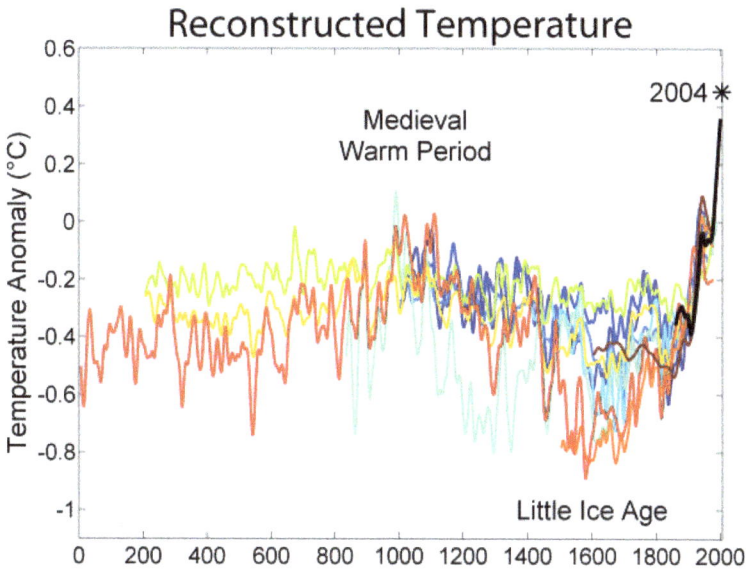

Figure 9.2—Ten-year temperature averages for the last 2000 years obtained by different methods, plus actual one-year average measurements in recent history. This figure was obtained from the Wikimedia Commons archive and is in the public domain.

9.5 CORRELATION BETWEEN ATMOSPHERIC CARBON DIOXIDE AND EARTH'S TEMPERATURE

J. R. Petit, et al.,[19] studied ice cores from the Russian scientific station in Antarctica. They measured the concentration of carbon dioxide in the Earth's atmosphere over the past 420,000 years, along with records of dust, sodium (from marine aerosols), oxygen-18, methane, glacial ice volume, and the ratio of deuterium to ordinary hydrogen in ice. The latter two records are indicators of temperature in the region where the ice formed. They found that the temperature and the carbon dioxide content are closely correlated. This correlation does not show whether changes in atmospheric carbon dioxide caused the temperature changes or vice versa, but it strongly indicates that there is a cause-and-effect relationship one way or the other.

There have been numerous claims in recent years that there has been no net global warming since about 1998, e.g., Reference 20. Rebuttals to this claim have also been made, e.g., References 21 and 22. The National Oceanic and Atmospheric Administration (NOAA) reported in 2010 that the first decade of the 21st Century was the hottest on record.[23] Reference 18 provides clear evidence that the Earth's temperature his been steadily increasing since 1910, except for a few cold years between the mid-1940s and the mid-1970s. I do not have the data on which these claims are based, but my own purely personal experience suggests that summers during this period have been hotter than they used to be, and winters have been milder. The rather brutal winter of 2013-14 seems to be an exception. But while North America shivered and suffered storm after storm, Australia was in the throes of another record-setting heat wave.[24]

In fact, the Ice Age fluctuations in Earth's temperature were probably caused by variations in the intensity of the solar radiation reaching the Earth, resulting from a wobble in the Earth's rotation on its axis, a related precession of the equinoxes, and changes in the eccentricity of the Earth's orbit.[25] Cooler temperatures would permit more carbon dioxide to be dissolved in the oceans, so that the carbon dioxide concentration in the air would follow the ocean temperature, which would follow the air temperature. The correlation between carbon dioxide concentration and air temperature is complex.

More recent records of Earth's temperature may also show some correlation with solar activity. Figure 9.3 shows the Earth's temperature history since 1950 along with two "phenomenological solar signatures" constructed from different measurements of total solar irradiance.[26] In the figure caption, the initialism TSI means total solar irradiance and the initialism PSS means phenomenological solar signature. Earth's temperature fluctuated in phase with the solar cycles, but it increased overall more than either solar signature curve would suggest. The authors of Reference 26 believe that increased solar irradiance could account for up to 69% of the observed temperature increase, which leaves at least 31% that could be explained by the greenhouse effect.

However, the Intergovernmental Panel on Climate Change (IPCC)[27] considers the effect of solar variability to be negligible.[26] (The IPCC is an international body assembled by the United Nations Environment Programme and the World Meteorological Association. Its mission is to review and assess worldwide scientific, technical, and socio-economic information relevant to climate.)

In the absence of feedback processes—i.e., responses to the initial warming that tend to produce a cooling reaction (negative feedback) or further heating (positive feedback)—it would be easy to calculate the temperature increase from a given increase in atmospheric carbon dioxide.

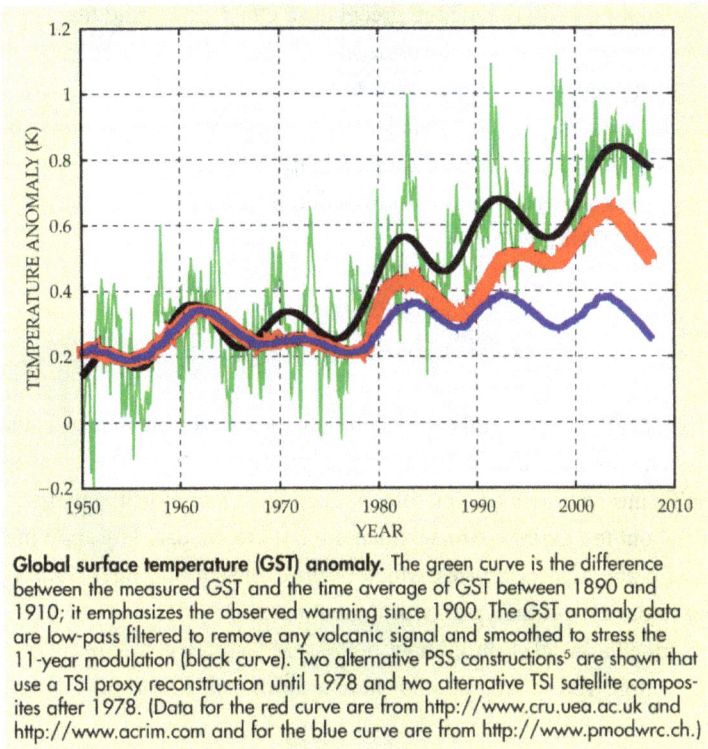

Global surface temperature (GST) anomaly. The green curve is the difference between the measured GST and the time average of GST between 1890 and 1910; it emphasizes the observed warming since 1900. The GST anomaly data are low-pass filtered to remove any volcanic signal and smoothed to stress the 11-year modulation (black curve). Two alternative PSS constructions[5] are shown that use a TSI proxy reconstruction until 1978 and two alternative TSI satellite composites after 1978. (Data for the red curve are from http://www.cru.uea.ac.uk and http://www.acrim.com and for the blue curve are from http://www.pmodwrc.ch.)

Figure 9.3—Global surface temperature for 1950-2008, along with curves representing the effect of solar variability. Reprinted with permission from Nicola Scafetta and Bruce J. West, "Is climate sensitive to solar variability?", *Physics Today*, Vol. 61, No. 3, p. 51, March 2008, Copyright 2008, American Institute of Physics.

Accurate results are obtained for the present climate because we know the present values of the Earth's albedo (which measures the amount of incoming solar radiation that is reflected back into space) and emissivity (which measures how much incoming radiation is absorbed and how much thermal radiation is emitted). We can't be so confident about the values those parameters will take in an environment with more carbon dioxide in the air. In reality, there are strong feedback processes that follow an initial increase in the Earth's temperature.

As an example, one important feedback effect is increased evaporation from the oceans, causing increases in the water content of the atmosphere. Water vapor is a greenhouse gas, so increased water vapor in the atmosphere would tend to exacerbate the greenhouse effect. However, increased water vapor would also lead to more condensation in the form of clouds. Clouds reflect incoming sunlight, so they would reduce solar heating and mitigate the greenhouse effect. But—and this is a very important point—unless there is an increase in temperature in the first place, there will be no increased evaporation. How much mitigation increased cloud cover will provide cannot be predicted without detailed quantitative analysis—it might be quite a bit. Such quantitative analyses are discussed in the next section.

Other feedback mechanisms are discussed in Section 9.8 below on future consequences of global warming.

9.6 CLIMATE MODELS

In the physical sciences, you can't claim to understand a phenomenon unless you can construct mathematical theories that make quantitative predictions that agree well with available measurements. Climatologists construct theoretical models for making predictions about the Earth's climate. Such models are validated by applying them to what is known about past climates. Much of the criticism of global-warming theory by skeptics is directed at the validity of these models.

Two Wikipedia articles on climate models are sufficiently informative to bear reading.[28, 29] I urge you to read both of these articles. But I want to summarize them very briefly.

There is a hierarchy of models, starting with a simple "zero-dimensional" model and increasing in complexity. The zero-dimensional model does not account for variation of properties in any spatial direction or in time, but simply summarizes the radiative energy balance of the Earth in radiative equilibrium, where the incoming radiative heat flow is equal to the outgoing radiative heat flow. This model uses single numbers for the Earth's albedo and the Earth's emissivity. The key to getting good results from this model is to evaluate the albedo and emissivity accurately. Estimating these parameters properly gives an average Earth temperature of 15 °C, or 59 °F, which is generally taken as the standard value. The emissivity accounts for greenhouse gases.

Zero-dimensional models won't account for feedback mechanisms, but they should give fairly accurate answers unless some time-dependent phenomenon such as a feedback loop dominates the physics. The zero-dimensional climate model shows the consequences of the greenhouse effect clearly in the absence of such phenomena.

Increasingly complex models include increasingly greater detail in spatial variation and time dependence. The most complex models are atmosphere-ocean global climate models (or atmosphere-ocean general circulation models) (AOGCMs), which account for water vapor in the atmosphere, transfer of water between the atmosphere and the oceans, polar ice, and the carbon cycle, which is the biogeochemical cycle by which carbon is circulated through the Earth's various systems.[30] These models are highly computation-intensive, and they can only be run on the most powerful available computers.

AOGCMs agree reasonably well with climate records over the past hundred years, with some differences in details. Figure 9.4 shows annual and five-year averages for global mean surface temperatures since 1880.[17] Reference 29 shows a comparison of observed temperatures with the predictions of two such models, and it also discusses uncertainties in the models. This comparison is reproduced below in Figure 9.5. Room for improvement exists, but compilation of the results of a wide spectrum of AOGCMs is regarded by the authors of Reference 29 as a good tool for prediction of future trends.

Note that the observations show that the Earth's temperature increased about 0.75 °C during the Twentieth Century, with the greatest increase occurring in the final twenty years.

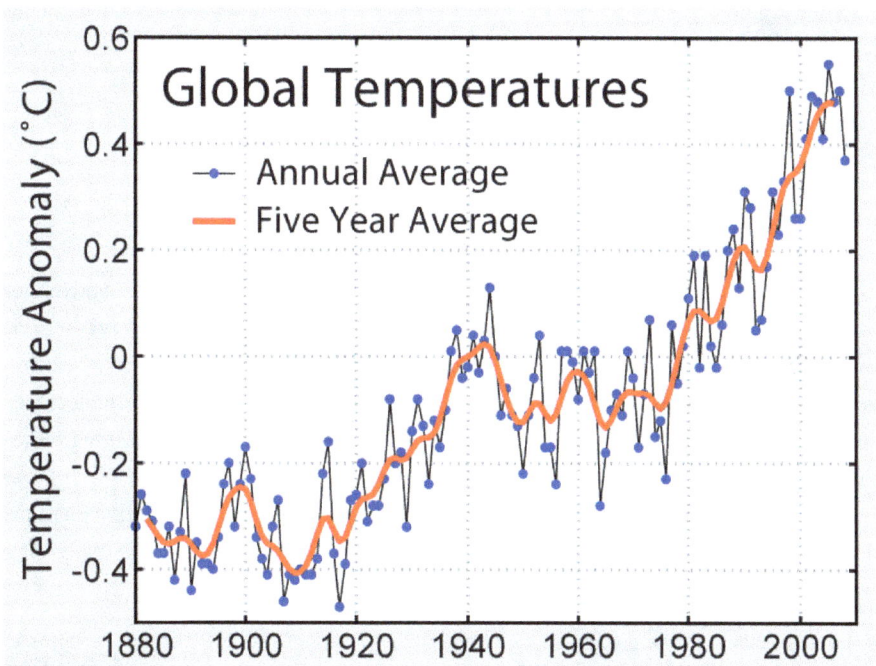

Figure 9.4—Annual and five-year average global mean temperatures since 1880. This figure was obtained from the Wikimedia Commons archive and is in the public domain.

Mean Temperature Anomalies (Global)

— Observations — Canadian Model — Hadley Model

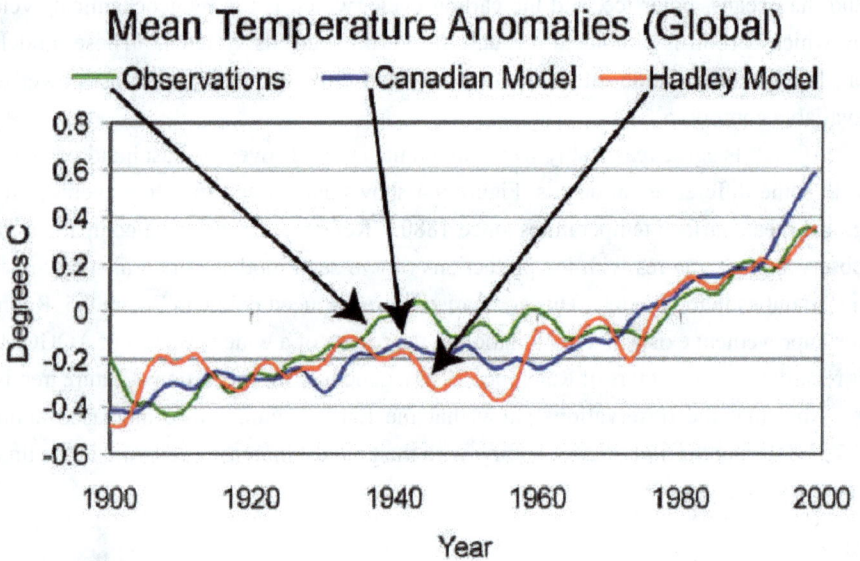

Figure 9.5—Observed Earth temperature and predictions by two AOCCMs. This figure was obtained from the Wikimedia Commons archive and is in the public domain.

9.7 PREDICTIONS OF FUTURE CLIMATE CHANGE

One great uncertainty in the AOGCMs is their treatment of clouds and upper tropospheric[a] humidity. This uncertainty leads to significantly different predictions of future temperature increase by different models for the same scenarios of carbon dioxide emissions. However, all the models predict significant temperature increases (from 1.3 to 4.5 °C, or 2 to 7.2 °F) by the year 2100. A graph of some of these predictions is shown in Figure 9.6, taken from Reference 29.

As explained in Chapter 1, before scientific work gets published in scientific journals, it is subjected to a rigorous process of peer review. Peer-reviewed research expressing

---

[a] The troposphere is the portion of the atmosphere where most weather occurs; it varies in thickness, but it extends from the surface to as much as 60,000 feet. The word comes from the Greek *tropos*, denoting turbulence, as the troposphere is turbulent while the layers above it have smoother airflow except when penetrated by thunderstorms. The layer above the troposphere is the stratosphere, which extends up to about 160,000 feet (30 miles). Above that are several very rarefied layers, which peter out at about 400 miles (http://en.wikipedia.org/wiki/Troposphere, http://en.wikipedia.org/wiki/Stratosphere).

concern about anthropogenic global warming began to appear in the 1970s. IPCC was formed in 1988 and has issued summaries of the state of climate science every five or six years.[31] By 2007, the consensus among climate scientists was that the predictions of global temperature increase by climate models are at least qualitatively correct. Reference 32 summarizes this consensus. The article cites the 2007 report of the IPCC:

> In February 2007, the IPCC released a summary of the forthcoming Fourth Assessment Report. According to this summary, the Fourth Assessment Report finds that human actions are 'very likely' the cause of global warming, meaning a 90% or greater probability. Global warming in this case is indicated by an increase of 0.75 degrees in average global temperatures over the last 100 years.

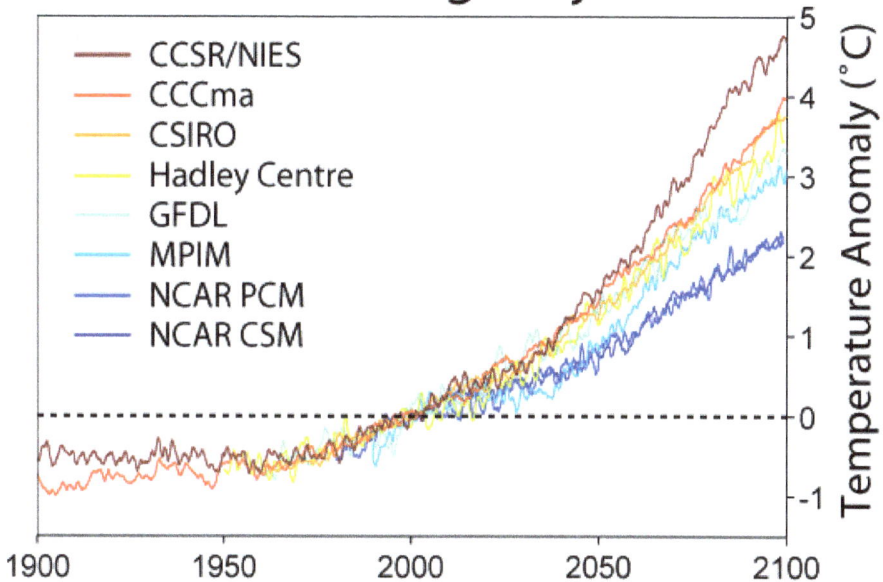

Figure 9.6—Predictions of future Earth temperature changes by various AOGCMs. This figure was taken from the Wikimedia Commons archive and is reproduced under the terms of the GNU Free Documentation License, Version 1.2.[13] It may be copied freely under the same terms, or downloaded from http://en.wikipedia.org/wiki/File:Global_Warming_Predictions.png

The article summarizes the declarations of dozens of organizations of scientists in fields related to climatology, along with statements of less related organizations, supporting the IPCC position. The article states, "Since 2007, no scientific body of national or international standing has maintained a dissenting opinion. A few organisations hold non-committal positions."

Since 2007, the consensus among climatologists has only become stronger. Numerous reports have concluded with almost complete certainty that human activity is responsible for measured global temperature increases. In 2018, the second National Climate Assessment, a report by 13 agencies in the administration of anti-climate-science President Donald Trump, predicted costs to the U.S. economy of $141 billion from heat-related deaths, $118 billion from sea-level rise, and $32 billion from infrastructure damage, along with other costs, by the end of the 21st Century. This report also ascribes increasingly severe weather events and wildfires that have already been happening to anthropogenic climate change.[33] The 2018 IPCC report also predicts increasingly dire consequences from anthropogenic climate change.[34]

9.8 FUTURE CONSEQUENCES

If the global temperature increases even a few degrees Fahrenheit, much if not all of the Arctic sea ice will melt. Extensive melting of this ice is already being observed, and it is reported frequently in the popular news media.[e.g., 35] However, contrary to popular misconception, the added volume of liquid water will not raise the sea level. Sea ice floats, and like any floating object it displaces its weight of water already. The water now in sea ice will increase in density on melting to occupy exactly the volume that it now occupies below the waterline, and no more. If the Greenland ice cap melts, that will raise the sea level. If the ocean becomes warmer, it will expand, and that will also raise the sea level. Projected temperature increases in the 21st Century will not be enough to melt the Antarctic ice cap, except for sea ice shelves. These shelves are cantilevered over the ocean and not supported by buoyancy; therefore, when they melt, the sea level will rise.

In addition, climate on land will change in various ways, making some areas dryer and other areas wetter, causing disruptions in agriculture. Ecosystems will be affected worldwide, with some species expanding in their range and others contracting in their range or even disappearing.

Proposals for coping with predicted climate changes range from drastic reductions in carbon dioxide emissions immediately, through measured responses designed to obtain the most benefit per dollar invested, on to complete denial that there is even a problem. In the following, I first discuss these approaches, attempting to give a fair hearing to the claims of climate science deniers. Then I give my own view of the future.

The first approach is exemplified by the Kyoto Protocol.[36] This is an international treaty in which 184 countries (as of October 2009), notably excluding the United States, agreed to reduce their carbon dioxide emissions by 5.2% from 1990 levels by 2012. Most nations have failed to meet their commitments; instead, so far most have increased their emissions.

Bjorn Lomborg, a Danish political scientist, argues that efforts made now to reduce carbon dioxide below present levels will be extremely expensive and not very effective in lowering global temperatures.[37] He thinks that we should use non-carbon-dioxide-emitting technologies to supply increased demand in the future, and that we should focus now on research and development to make such technologies more affordable in the future. A major assumption in his thesis is that it will be much easier to pay for non-emitting energy sources in the future when average per capita incomes are much greater than they are now. He asserts that the worldwide average per capita income will be $100,000 (in current U.S. dollars) in the year 2100, whereas it is about $5000 now. I find that assertion very dubious.

The most extreme opposing view that has any scientific basis is provided by Ian Plimer.[38] He compares climate history to records of human prosperity and finds that, contrary to the predictions by mainstream climate prognosticators that global warming will bring famine and death to huge numbers of people, warmer periods have brought prosperity and colder periods are the times of suffering. He points particularly to the Medieval Warming that occurred from 900-1300 AD, during which worldwide temperatures were 1-3 °C warmer than they were in 2009, the date of publication of his book, and civilizations flourished almost everywhere. (However, one place where the Medieval Warming Period did not cause increased prosperity was in the American Southwest, where prolonged drought caused the abandonment of Puebloan communities that had been occupied for more than a thousand years; see Reference 39.) Plimer disputes assertions that the Earth's temperature has increased as much as climatologists claim, citing the urban heat island effect as a distorting influence on temperature records: It is known that, because of the heat they generate, cities are warmer than the surrounding countryside, and urban growth has changed the environments of many traditional temperature-recording sites from countryside to urban environments. He correlates climate history with solar activity and concludes that carbon dioxide has little effect as a driving force on climate. Plimer criticizes climate models for not accounting for solar variability. However, the best way to assess the effects of carbon dioxide concentration is to examine these effects in isolation. If models predict that temperature will increase when carbon dioxide increases, and if the effects of solar variability are greater, then the effects of carbon dioxide will be superimposed on the greater effects of solar variability. Furthermore, some of Plimer's scientific arguments are simply wrong. For example, he claims that the atmosphere already contains so much carbon dioxide that the greenhouse effect is "saturated," and adding more carbon dioxide won't cause the Earth's temperature to rise any more. However, this claim is not consistent with the predictions of sophisticated radiative transport models—models that are extremely well validated.[40]

Patrick Moore[41] has asserted that the carbon dioxide content of the atmosphere has been decreasing steadily over the last 540 million years as marine invertebrates have withdrawn carbon from their environment to build shells of calcium carbonate. He says that at the beginning of this period there was 17 times as much carbon dioxide in the atmosphere than there is now, but that the carbon dioxide in the atmosphere has been drawn down to its present level by such organisms, along with the trees that became coal beds and the organisms that became petroleum. For the last 150 million years, he says, biological processes have removed 37,000 tons of carbon from the air every year. He claims that the lowest level of carbon dioxide in the air at which plants can survive is 150 ppm, and that at the present rate of carbon dioxide withdrawal by organisms, without anthropogenic addition of carbon dioxide, the atmospheric carbon dioxide content would fall below the level needed to sustain plant life within 2 million years.[42] In his view, restoring carbon dioxide to the atmosphere by industrial processes will save the world in the long run! However, human civilization has adapted to the mostly stable climate the Earth has enjoyed since the end of the last Ice Age. If climate change makes marginally inhabitable areas of the Earth uninhabitable, people who live there will have to move, and the people in the areas they move to will not welcome them. The time frame of our concern is not 2 million years, but 20-100 years.

The most egregious opposing view is the claim, repeated loudly and often by some contrarians, that the whole global warming scenario is a hoax. That is a bogus claim for two reasons.

First, scientists compete for both recognition and funding. If you prove everyone else wrong, you gain fame and glory, and you are also more likely to be considered favorably for funding. (I would personally love to prove all the global warming predictions wrong. I would become very famous and possibly even rich; also, if the predictions are correct, some things that I love dearly will be lost, so I fervently hope they are wrong.) Furthermore, the whole peer review process is based on skepticism: The author of a proposal or a research paper has to make a convincing case for his research program or his findings. It is extremely far-fetched to think that essentially the entire community of climatologists and Earth scientists would collude to create the impression of a crisis in the public mind when there is none in reality.

But more serious is the allegation that the entire community of climatologists and Earth scientists is engaged in a massive deliberate fraud. This is a slander on the integrity of the entire scientific enterprise, and by extension the integrity of all scientists. It is true that once in a while individual scientists commit scientific fraud. However, the process of checks, reviews, and repetition of measurements ensures that scientific fraud is eventually revealed, sometimes very quickly. When that happens, the career of the perpetrator is ruined forever. It is really stupid to commit scientific fraud! I do not believe that there is any such collusion among climatologists.

Of course, one can be honestly wrong, and there is no shame in that. It is possible that the models used by climatologists are missing some important details (mostly feedback effects) that render their predictions erroneous. But that is extremely unlikely. The more frequent extreme weather events—hurricanes, droughts and resulting wildfires, and storms—that we have been seeing in recent years are exactly what predictions from climate models would lead one to expect. Furthermore, the feedback effects omitted from AOGCMs are almost all positive—i.e., they will make things worse. A comprehensive treatment of the state of the art in climate science for laypersons, including detailed descriptions of the likely feedback effects, is given in Reference 43. The most important feedback effects are summarized next.

One of the most disruptive consequences of global warming will be sea level rise, which will occur because of thermal expansion of ocean water and because of melting of glaciers, the Greenland ice cap, and Antarctic ice shelves. Melting of ice is a positive feedback mechanism. Ice is very reflective, but the darker ground exposed by the vanished ice absorbs much of the sunlight that falls on it. This absorption causes warming, which causes more ice to melt, and so on and on. This increased heating in the Arctic is called Arctic amplification; temperature increase in the Arctic is expected to be three to four times as great as the average temperature increase in the Northern Hemisphere. Remember the observation cited above from Reference 9 that the last time carbon dioxide concentrations were as high as they are now, the sea level was 20 meters—over 65 feet—higher than it is today. You should take no comfort in noting that it is not that high now, despite the temperature increase we have already seen. Consider what happens when you take an ice cube out of the freezer and lay it on the kitchen table. Of course, it will melt, but not instantly. It takes time. Even if no more carbon dioxide were added to the atmosphere after today, the ice would continue to melt. We could very well end up with the amount of ice the Earth had 2.5 million years ago, with sea levels 20 meters higher. But we continue to add carbon dioxide to the atmosphere. Things are likely to be much worse than that eventually.

The Arctic permafrost[44]—permanently frozen ground beneath the layer of dirt that melts and thaws annually—is currently beginning to melt. Permafrost contains a great deal of methane, an even more potent greenhouse gas than carbon dioxide. As methane is released into the atmosphere by melting permafrost, it will cause increased warming, which will cause more melting of the permafrost, etc.

Hotter climate promotes more wildfires, as we have seen recently in California and elsewhere. Plants, by photosynthesis, remove carbon dioxide from the atmosphere and release oxygen, using the carbon in the extracted carbon dioxide to build plant tissue. When plants burn, they release the stored carbon into the air as carbon dioxide, increasing the greenhouse effect, which will cause increased warming, which will cause more fires, etc. The boreal forests of Siberia and northern Canada often rest on peatland; peat fires are difficult to stop and release much more carbon dioxide than the forests above them. The feedback loop goes on and on.

Global warming will cause some areas to experience multidecadal "megadroughts." Die-off of plant cover causes greater absorption of sunlight in the soil, causing more heating, thence more drying out, etc. Large areas of arid but vegetated land will become desert as carbon dioxide emissions continue to heat the Earth up.

There are many other adverse effects to be expected from global warming, as explained at length in Reference 43.

My own view is that mankind is facing changes that could make parts of the Earth uninhabitable, force unwelcome migrations, and cause widespread agricultural failure and collapse of marine resources. I think it is still possible to prevent the worst impacts of global warming, but we have to do some very unlikely things. We need to stop emitting carbon dioxide, which will mean completely abandoning fossil fuels. Not only that, but we must also find ways to remove carbon dioxide from the atmosphere. Instead of depleting the world's forests, we need to restore them, although that in itself will not be enough. "Carbon sequestration" is a subject of considerable discussion but no practical solutions yet.[45] In Chapter 8, the environmental liabilities of other non-emitting power sources are discussed; I think nuclear power should be the primary means of replacing fossil fuels, directly generating electricity and indirectly producing hydrogen for internal combustion engines, as discussed in Chapter 16 below. However, the necessary rapid shift away from fossil fuels would cause a great deal of economic hardship for a great many people, so it's not likely to happen. Most of all, people need to be persuaded that human population growth is the root cause of almost all environmental problems and that human population must be stabilized, which is not likely to happen soon enough. The potential consequences of continuing human population growth, even without climate change, have been known for many years, but there has been a great deal of cultural and religious resistance to the message. Dire predictions made long ago by people such as Paul Ehrlich[46, 47] were not realized because increases in productivity outpaced population growth for many years. But Ehrlich et al. were not wrong, their timetable was just too pessimistic. Currently, however, agricultural productivity has begun to decline for many reasons, including global warming.[48] This topic needs a whole book for a comprehensive treatment, so I'll just state that I'm convinced of the "population connection"[49] and leave it at that.

The last part of this book deals with individual nuclear systems.

PART III
NUCLEAR SYSTEMS

CHAPTER 10
NUCLEAR MEDICINE

One of the benefits of nuclear technology that most people don't think much about until they need it is nuclear medicine.[1] Nuclear medicine can be classified into two broad categories: diagnostics and treatment. Most of these procedures apply one or more of the nuclear reactions introduced in Chapter 5.

10.1 DIAGNOSTIC PROCEDURES

Over 13 million Americans undergo some sort of diagnostic or therapeutic procedure each year that involves nuclear technology.[2] This section describes diagnostic procedures. Some of these involve the injection, ingestion, or inhalation of radioactive materials, and others impose radiation on the body from an outside source.

Radionuclides administered internally are usually attached to "tracer" compounds that tend to collect in specific body parts of interest. Positron emission tomography[3] (PET) and single-photon emission computed tomography[4] (SPECT) scans are examples of such techniques. "Tomography" refers to techniques in which images of "slices" (Greek *tomos*), or planar sections through regions of interest, are constructed. Radionuclides suitable for diagnostic procedures must have short half-lives, so that they will not continue to irradiate the patient and those around him and create cancer risks by irradiation for protracted periods of time.

Most diagnostic procedures that impose external radiation use X-rays. Besides conventional X-rays, computed tomography[5] (CT, formerly CAT, for computed axial tomography) falls into this category. However, magnetic resonance imaging[6] (MRI) exploits an interaction between atomic nuclei and externally imposed magnetic fields.

We now look at some of these techniques in greater detail.

10.1.1 Positron emission tomography

PET scans use a positron emitter, such as fluorine-18 (F-18). Positrons have a range of only a few millimeters in tissue; after slowing down to low energy by electrostatic deflections, a positron combines with an atomic electron in an annihilation reaction that produces two gamma rays of 0.51 MeV each, which take off in nearly opposite directions. These gamma rays generally pass out of the body with only minor energy loss, and they are recorded on a scanning device that encircles the region of interest in the patient. Only coincident pairs of gamma rays (i.e., pairs that hit the scanning device within a few nanoseconds of each other) are recorded. The time difference between the impacts of the two gamma rays in the scanning device gives the location of the annihilation event along the chord between the two impact points.

The resolution of a PET scan is limited by the range of the positrons—i.e., a few millimeters—since the positron could have been emitted in any direction from the emission site, and by the deviation in the relative angle between the two gamma rays from exactly 180°, which may result either from scattering of the gamma rays or from the incoming momentum of the positron. A typical PET scan uses F-18 in the compound fluorodeoxyglucose (FDG), which acts like glucose metabolically. FDG concentrates in areas of higher metabolic activity, so an increase in gamma ray hits from a particular area of the body might indicate some abnormality, such as a tumor. The half-life of F-18 is 110 minutes, which is short enough that the F-18 concentration in a patient is negligible in a day or so, but long enough that F-18 can be produced in centralized locations that can serve several hospitals. Fluorine-18 is produced in cyclotrons by bombarding oxygen-18 with protons (using the $^{18}O\ (p,n)\ ^{18}F$ reaction); O-18 is naturally occurring, with an abundance of 0.2% in natural oxygen.

From the records of tens of thousands of coincidence events, complex computer codes construct images of the positron-electron annihilation sites, in which areas of concentration show up as hot spots, or, in some cases in which the tracer is excluded from diseased tissue, as cold spots.

Figure 10.1 shows an image of a human brain produced by a PET scan.[7] Yellow and red areas are regions of greater radionuclide accumulation.

Typical radiation doses to patients in PET scan procedures are about 700 mrem, which is several times the normal annual background radiation dose in the U.S. Clearly, the decision to administer radioactive tracers is a matter of benefit from the procedure versus the small possible risk of adverse effects (mainly delayed cancers) from the increase of radiation. Also, the people around the patient will receive an increased radiation dose from the gamma rays that pass out of the patient and into those other people.

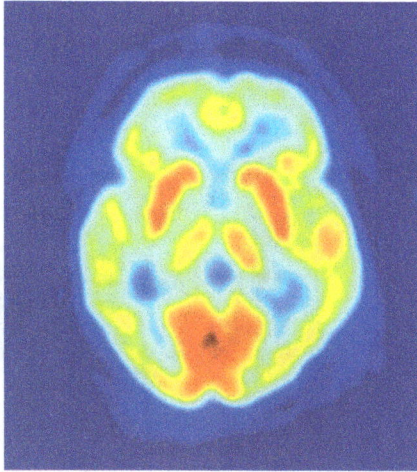

Figure 10.1—PET scan image of a human brain. This image, created by Jens Langner, has been released into the public domain and put in the Wikimedia Commons.

Medical personnel may limit their time near the patient and/or wear protective clothing (lead aprons, etc.), and the patient may be isolated for several hours while the radionuclide decays away.

10.1.2 Single-photon emission computed tomography

SPECT scans create tomographic images by using a gamma camera[8] at multiple angles to detect the locations of gamma-ray emitters in the body. The gamma camera is rotated around the body part of interest and images are taken every few degrees of rotation. The term "single-photon emission" means that a radioactive decay event produces a photon directly, rather than a pair of photons as in PET scans. The gamma-ray emitters are attached to pharmaceuticals that concentrate in particular organs or types of tissue. For example, technetium-99m can be attached to drugs that concentrate in bone, the heart, the brain, or other organs.

Figure 10.2 shows lung SPECT scan images.[9]

Depending on the radionuclide used and the amount of it needed for adequate imaging of particular organs, the radiation dose to the patient in SPECT scans can vary widely, from as little as 0.6 mrem to 3.7 rem.

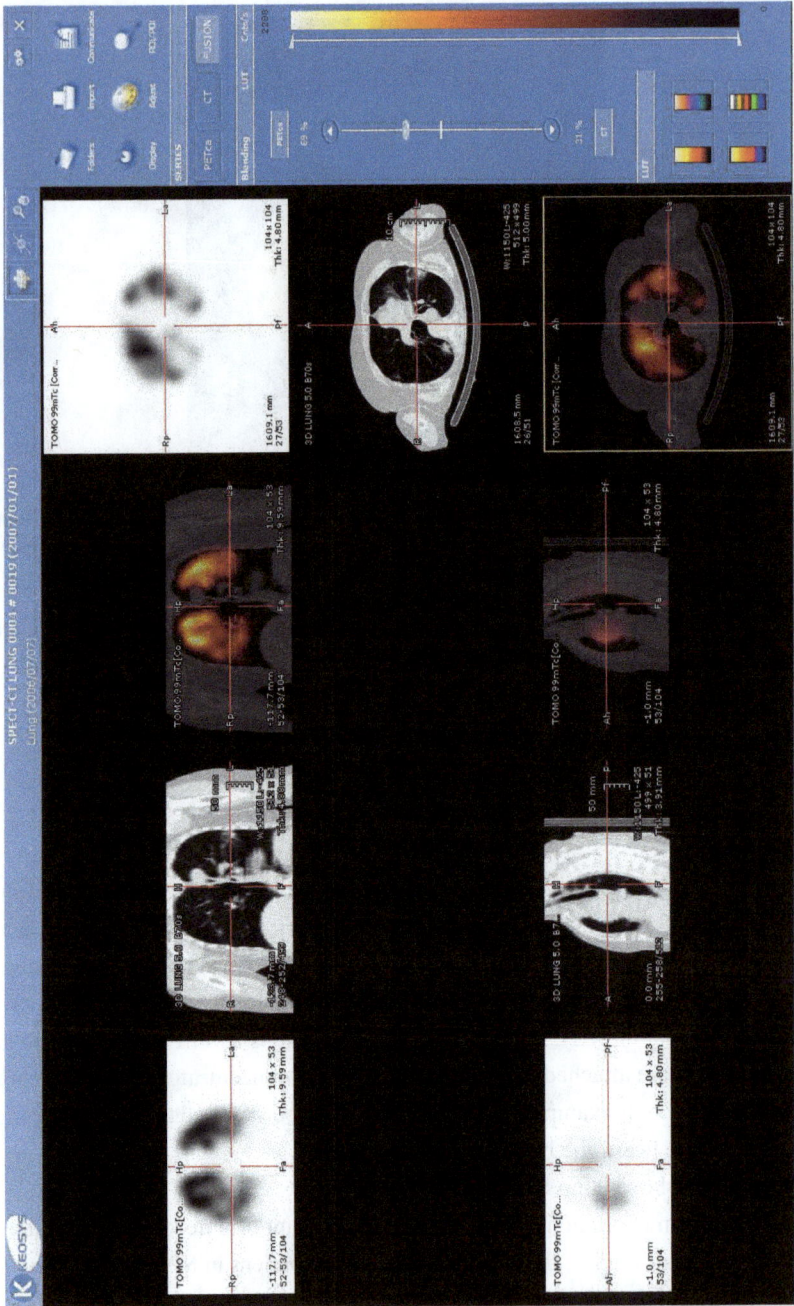

Figure 10.2—Lung images incorporating SPECT technology. This file was taken from the Wikimedia Commons and is in the public domain.

Technetium-99m (^{99m}Tc or Tc-99m) is almost ideal for diagnostic imaging, because it has a short half-life (6 hours) and it emits a gamma ray of 141 keV, which is strong enough for good imaging but soft enough to cause limited tissue damage. Accordingly, it is used for up to 90% of all radioisotopic diagnostic imaging procedures performed in the U.S., which amount to over 10 million Tc-99m scans per year.

Technetium does not occur in nature. Its name is derived from its status as a product of technology, having first been discovered, by Perrier and Segré in 1937,[10] in molybdenum (Mo) that had been irradiated in a cyclotron. For nuclear medicine, it is produced indirectly from nuclear reactors. One of the fission products of uranium fission is Mo-99, which is radioactive and decays with a half-life of 66 hours into Tc-99 and Tc-99m. In the production process, Mo-99 is extracted from uranium targets irradiated in the reactor, and it is shipped to hospitals in "Moly cows" that are "milked" of the Tc-99 and Tc-99m as they are formed from the decay of the Mo-99. The half-life of Tc-99 is 213,000 years, so Tc-99 is radiologically unimportant, but it reduces the amount of the tracer pharmaceutical that carries the useful Tc-99m. Therefore, the ratio of Tc-99 to Tc-99m needs to be limited (to values below about 20). In the decay of Mo-99, only 11% of the technetium is Tc-99, but Tc-99m decays into Tc-99, so the ratio grows continuously.

In North America, until recently only one nuclear reactor was used to produce Mo-99, the National Research Universal (NRU) reactor in Chalk River, Ontario, Canada which began operations in 1957. The proposed replacement reactor, MAPLE-1, was built, but a design flaw discovered during startup testing prevented it from receiving an operating license.[11] The NRU reactor ceased production in October 2016.[12] As of December, 2016, the Nuclear Regulatory Commission has recommended approval of a system whereby university research reactors in several locations would irradiate low-enriched uranium targets, which would be shipped to a central facility at the University of Missouri's Discover Ridge Research Park at Columbia for extraction of Mo-99.[13]

Technetium-99m can also be produced by electron accelerators.[14] A beam of 40 MeV electrons strikes a tungsten converter plate, in which the slowing down of the electrons produces Bremsstrahlung gamma rays. These gamma rays enter a molybdenum target 96% enriched in Mo-100, an isotope that comprises 9.63% of natural molybdenum. By the photoneutron reaction ^{100}Mo$(\gamma, n)^{99}$Mo, the molybdenum-99 parent of technetium-99m is produced in the target. After irradiation, the targets are placed into "goats" for milking of the technetium-99m product at the production facility. (They are called goats because goats are also milked but they are distinguished from cows. The research team was amused by this Jeff Foxworthy redneck joke: You might be a redneck if you know how to milk a goat.) This process has been demonstrated on the laboratory scale, but it has not yet been commercialized. It would have the advantage that technetium-99m would be produced in many facilities distributed near population centers and shipped to nearby hospitals, so that the entire North American supply would not be dependent on a single facility. Other accelerator-based production schemes are possible, as also discussed in Ref. 14. Some of these are being pursued as replacements for reactor-based production. This is a rapidly changing field of research and development; up-to-date accounts can be found on the pertinent DOE website.[15]

10.1.3 Magnetic resonance imaging

Also called nuclear magnetic resonance imaging (NMRI), magnetic resonance imaging[16] (MRI) is a non-invasive technique that provides extremely detailed, high-resolution images of structures in the body. Some nuclei, such as the hydrogen nuclei (which are just protons) in water molecules, have magnetic dipole moments. That is, they act like little bar magnets, so that they can be aligned with magnetic fields. The patient is placed inside a large, powerful magnet, and the protons align with its strong magnetic field. A secondary magnetic field, which oscillates at radiofrequencies in a plane perpendicular to the main magnetic field, is superimposed on the main magnetic field. The protons in different types of tissue respond differently to the oscillating field, returning modified radiofrequency signals that can be detected and distinguished. Figure 10.3 shows an MRI scan of a human knee.[17]

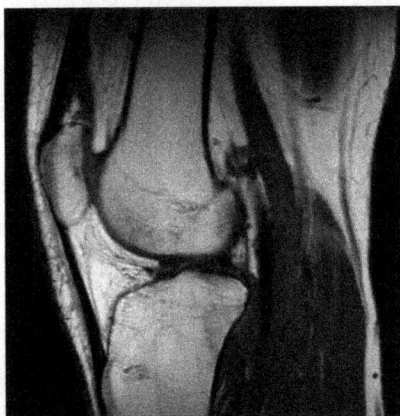

Figure 10.3—MRI image of a human knee. This file from the Wikimedia Commons has been released by its author under the GNU Free Documentation License, version 1.2,[18] as updated by the Creative Commons Attribution-Share Alike 3.0 unported license.[19] See the indicated references for conditions.

Because MRI does not apply ionizing radiation, the patient suffers no radiation damage from an MRI scan. However, the powerful magnetic fields can interfere with cardiac pacemakers, and they can drag metal implants or artifacts around in the body, causing mechanical damage. Such artifacts might include tiny metal fragments in the eyes, which are often carried by welders. Therefore, if you have ever done any welding, you might have your eyes X-rayed before receiving MRI. For such reasons, some patients are ineligible for MRI.

For most people, however, MRI is an extremely safe and effective technique, and accordingly it is one of the most common diagnostic imaging procedures for conditions in which X-rays are inadequate. It has not completely replaced X-rays because it is time-consuming and expensive.

10.2 THERAPEUTIC PROCEDURES

Therapeutic procedures in nuclear medicine[20] primarily employ the introduction of radioisotopes into the body. For some cancers, gamma rays from an external source are directed at the tumors. A promising experimental technique, neutron capture therapy, irradiates disease sites with a beam of neutrons.

10.2.1 Radioisotopic therapy

Radioisotopes are used both to treat cancer and to provide pain relief. Cancer cells are more sensitive to radiation than normal cells, so radiation received locally in tumors preferentially kills cancer cells. Most healthy cells damaged by radiation repair themselves, but cancer cells damaged by radiation do not. Radioisotopes are delivered to particular types of tissue, where alpha or beta radiation from the radioisotopes is deposited locally. Radioisotopes can be delivered to a specific type of tissue by attachment to a pharmaceutical that is injected into the blood stream or ingested, and then taken up by that type of tissue. For example, iodine is taken up by the thyroid, so the radioisotope I-131 concentrates there and treats thyroid cancer very successfully. Strontium concentrates in bone, so Sr-89 is used to relieve pain from bone cancer. Radioisotopes can also be placed mechanically. For example, "seeds" containing palladium-103 are implanted directly into the prostate gland to treat early-stage prostate cancer. Implantation of radiation sources is called brachytherapy.

10.2.2 Teletherapy

Teletherapy is the application of gamma rays from an external source to a tumor. Cobalt-60 is normally used for this purpose. However, linear accelerators producing high-energy X-rays are more versatile and therefore preferred for such treatment.

10.2.3 Neutron capture therapy

Neutron capture therapy (NCT), or boron neutron capture therapy (BNCT),[21, 22] is an experimental technique in which epithermal neutrons (neutrons with slightly higher energies than thermal energy) are aimed at cancerous tissue to which boron-10 has been delivered by a borated pharmaceutical formulated to deposit preferentially in cancer cells. Boron-10 has a very high absorption cross section for thermal neutrons, in the $^{10}B(n,\alpha)^7Li$ reaction. Both the alpha particle and the lithium-7 nucleus have very short ranges, so they deposit their energy in very small volumes, on the scale of a single cell. Thus, the technique has promise for treating cancers that infiltrate normal tissue. Such cancers include glioblastoma multiforme, an inoperable brain cancer, and metastatic melanoma. The neutrons are slowed from epithermal energy to thermal energy as they pass through tissue between the skin and the tumor. For melanomas on the skin, the neutron beam is composed of thermal neutrons, since penetration, with its concomitant internal neutron thermalization, is not necessary.

The neutron source for BNCT is usually a collimated beam (i.e., a beam in which the particles travel without spreading out) from an appropriate nuclear reactor, such as the research reactor at the Massachusetts Institute of Technology. Particle accelerators can also produce neutron beams from high-energy proton beams (e.g., beams with proton energy of about 50 MeV) directed into lithium or beryllium targets, for example by the $^7Li(p,n)^7Be$ reaction.[23] Neutron beams produced this way tend to contain a lot of fast neutrons, even after passing through moderating materials.[24]

Clinical trials at the Brookhaven National Laboratory defined tolerance limits for exposure of healthy tissue and showed limited therapeutic response. A key to achieving success with BNCT is to find pharmaceuticals to which boron can be attached that become highly concentrated in the target cancer cells. The ratio of pharmaceutical concentration in tumor tissue to that in surrounding healthy tissue needs to be at least 10 for therapeutic response without excessive damage to healthy tissue. Better pharmaceuticals are needed to provide sufficient concentration of boron in cancer cells.

However, funding for development of effective pharmaceuticals and pursuit of further clinical trials has not been available. Recent research has been limited to studies in vitro and with small animals such as rats.[24]

CHAPTER 11

LIGHT-WATER REACTORS

Nuclear reactors are categorized in several different ways. One categorization depends on the neutron energy spectrum. If most of its free neutrons are thermal neutrons, a reactor is called a thermal reactor; if most of its free neutrons are fast neutrons (i.e., if they have energies well above thermal energy), a reactor is called a fast reactor. Another categorization depends on the moderator. If its moderator is light (ordinary) water, a reactor is called a light-water reactor (LWR). If its moderator is heavy water, a reactor is called a heavy-water reactor (HWR). (See below for an explanation of light and heavy water.) If its moderator is graphite, it is called a graphite reactor. Yet another categorization depends on the coolant. Thus, there are liquid-metal reactors, molten-salt reactors, and gas reactors. Reactors designed to produce more fuel than they consume are called breeder reactors. (Such reactors do not violate any physical laws; they consume a fissile fuel, such as Pu-239, and produce more of it from a so-called "fertile" material such as U-238. Breeder reactors are discussed in Chapter 17.) Basically, reactors are categorized according to their distinctive features. Thus, categories can be combined in reactor names: For example, the liquid-metal fast breeder reactor, or LMFBR, is a liquid-metal-cooled breeder reactor with a fast neutron spectrum.

Most operating power reactors in the world are LWRs. Light water is ordinary water. Almost all water molecules comprise one atom of oxygen and two atoms of ordinary hydrogen, whose nucleus consists only of one proton. Light water is abundant and cheap, and it has excellent heat-transfer and moderating properties. It is an excellent moderator because the mass of each of its ordinary hydrogen nuclei is almost exactly equal to the mass of a neutron. Recall from Chapter 2 that when a moving object collides elastically with a stationary object of equal mass, the incoming object can transfer all of its kinetic energy at once to the stationary object. This only happens when the collision is head-on, but even glancing collisions between objects of equal mass are more efficient at energy transfer than collisions in which the mass of the incoming particle is much different from that of the stationary particle.

The drawback to using light water as a moderator is that ordinary hydrogen has a significant thermal neutron absorption cross section (the reaction is $^1H(n,\gamma)D$, where D is deuterium, described below), so it removes some of the thermal neutrons from the free-neutron population and makes it difficult to sustain the nuclear chain reaction. That is, light water is slightly poisonous to thermal neutrons. For this reason, it is impractical to fuel LWRs with natural uranium, in which only 0.72% is the fissile isotope U-235. Therefore, the fuel in LWRs is enriched in U-235 to a concentration in the range of 2-5%. The enrichment process is discussed in Chapter 21.

There are three naturally occurring isotopes of oxygen, but 99.762% of natural oxygen is oxygen-16. There are two naturally occurring isotopes of hydrogen. Different isotopes of all the other elements are denoted only by stating their atomic numbers (e.g., oxygen-16, oxygen-17, and oxygen-18 or ^{16}O, ^{17}O, and ^{18}O for these stable isotopes), but for hydrogen the symbols D for deuterium and T for tritium are also used for the isotopes that have one and two neutrons, respectively, in addition to the proton. Tritium, or ^3H, is radioactive with a half-life of 12.33 years, so it is not naturally present in significant quantities on Earth (a little is produced in the upper atmosphere by cosmic rays).[1] Deuterium, or ^2H, comprises 0.015% of natural hydrogen atoms.[2] If water is enriched in deuterium until it is essentially D_2O, it can serve as an effective moderator in an HWR. HWRs are discussed in Chapter 15.

Reactors are also classified according to the maturity of their technology.[3] Early experimental and proof-of-principle reactors are now regarded as Generation I reactors. Most currently operating power reactors are refinements of Generation I concepts, and they are classified as Generation II reactors. Generation III reactors are further refinements of Generation II reactors, with many of the improvements pertaining to reactor safety. Generation II reactors use engineered safety systems, such as emergency core cooling systems, that require some sort of actuation in order to operate in accident scenarios such as loss-of-coolant accidents. Generation III reactors apply the principle of passive safety, in which no system actuation or other operator action is required to prevent core damage—or at least not immediately. Some Generation III designs would require operator intervention after three days.[4] However, this time interval is long enough to allow careful analysis of the accident and well considered actions in response. Such deliberation is valuable in order to prevent ill-considered response actions that could make the situation worse. Some Generation III reactors have been deployed already. Generation IV reactors are even more advanced designs that are currently undergoing research and development. Evolutionary advances in Generation III designs are sometimes called Generation III+ designs. Because of their advantages in safety and efficiency, and because their designs have been standardized for easier passage through the licensing process, all future reactors built anywhere will be Generation III or III+ reactors until Generation IV designs become sufficiently mature to be commercialized.

This chapter discusses LWR designs, both in the usual Generation II configurations and in the new Generation III and III+ configurations.

All LWRs use light water as the moderator. But water is also an excellent heat transfer medium because of its combination of heat capacity, heat of vaporization (i.e., it absorbs a lot of energy in passing from liquid to vapor), and convenient boiling temperatures at practical pressures. Therefore, LWRs use the same stream of water as both the moderator and the coolant. Most nuclear power reactors are basically sources of heat for boiling water, and for that reason they are sometimes denoted as nuclear steam supply systems (NSSSs).

11.1 BOILING-WATER REACTORS

LWRs are further divided into the categories of boiling-water reactors (BWRs) and pressurized-water reactors (PWRs). The difference is that in BWRs the water is boiled in the reactor core and the same stream of steam that emerges from the core goes into the turbine to generate electricity. In PWRs, the water emerges from the core as a high-pressure liquid and passes through heat exchangers called steam generators where it transfers its heat to a secondary coolant loop at lower pressure. The water in the secondary loop boils in the steam generators and flows into the turbine.

BWRs are slightly simpler than PWRs because they do not use steam generators and all the associated piping in the secondary loop. However, sending the steam directly from the reactor into the turbine introduces radioactive contamination in the form of activation products into the turbine, which makes turbine maintenance more difficult. Also, because boiling causes dissolved solids to precipitate out, soluble poisons are not used in BWRs for reactivity control: Caked boron compounds in the coolant channels and the upper plenum would quickly make the reactor inoperable.

Thus, PWRs and BWRs each have advantages. The choice between them is not clear-cut, and different utilities have chosen each type. In the United States, all Generation II BWRs have been built by the General Electric Company (GE), which is now GE Hitachi. First, we look at the typical GE Generation II BWR.

11.1.1 Generation II BWRs

The general operating scheme of a Generation II BWR is described in Section 11.1.1.1. Reactor safety systems are discussed in Section 11.1.1.2. General Electric built six evolutionary models of Generation II BWRs, designated as BWR/1 through BWR/6. Details of the reactor and safety system designs differ among these models. There were also three different containment structure designs, denoted as Mark I, Mark II, and Mark III. Some differences among these models are described below, but this discussion does not attempt to identify all the differences.

11.1.1.1 Reactor and power plant layout

Figure 11.1 shows the steam cycle schematic diagram for a General Electric Generation II BWR, and Figure 11.2 shows the layout of the reactor vessel interior.

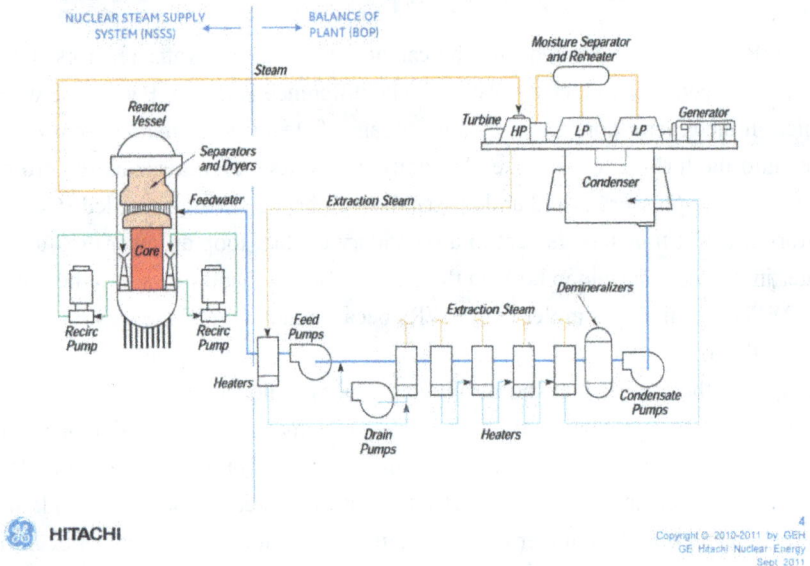

Figure 11.1—Typical GE BWR layout. Reproduced by courtesy of GE Hitachi.

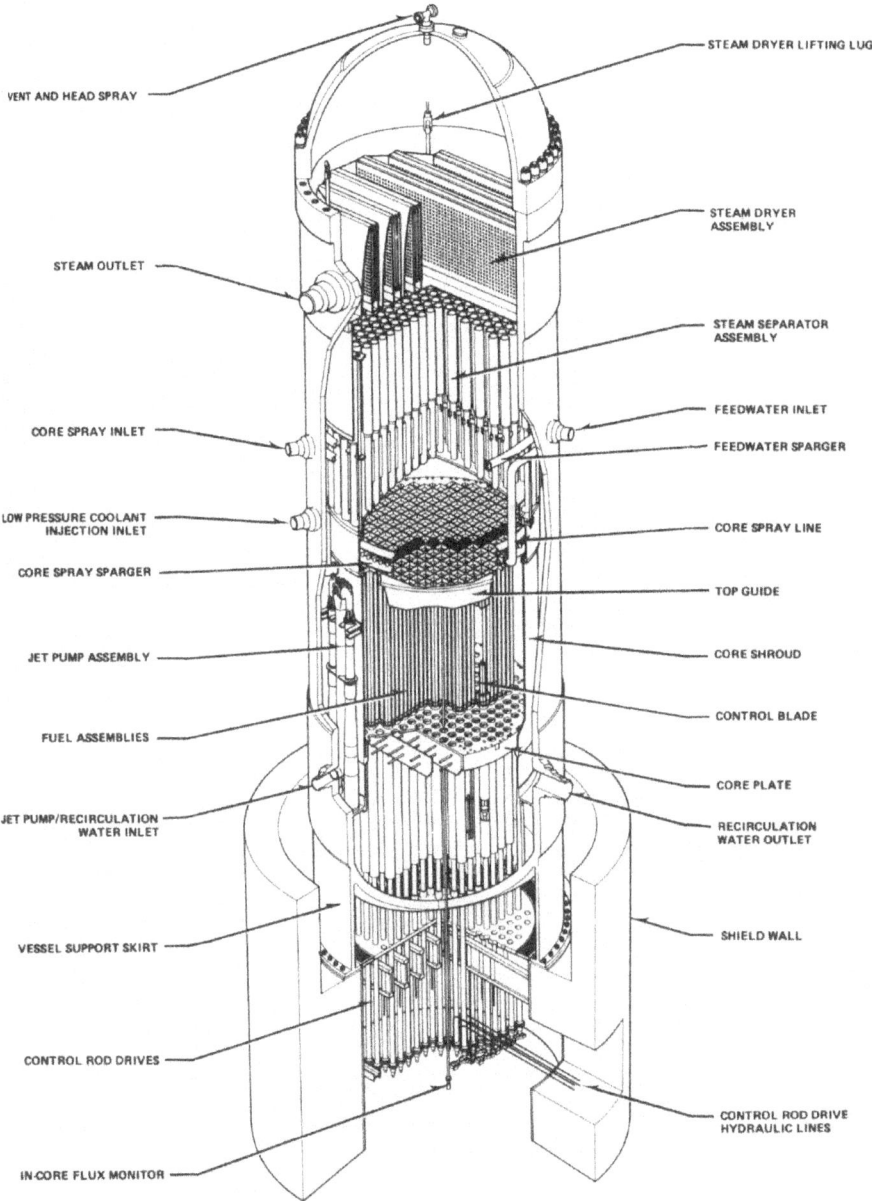

VENT AND HEAD SPRAY

STEAM OUTLET

CORE SPRAY INLET

LOW PRESSURE COOLANT
INJECTION INLET

CORE SPRAY SPARGER

JET PUMP ASSEMBLY

FUEL ASSEMBLIES

JET PUMP/RECIRCULATION
WATER INLET

VESSEL SUPPORT SKIRT

CONTROL ROD DRIVES

IN-CORE FLUX MONITOR

STEAM DRYER LIFTING LUG

STEAM DRYER
ASSEMBLY

STEAM SEPARATOR
ASSEMBLY

FEEDWATER INLET

FEEDWATER SPARGER

CORE SPRAY LINE

TOP GUIDE

CORE SHROUD

CONTROL BLADE

CORE PLATE

RECIRCULATION
WATER OUTLET

SHIELD WALL

CONTROL ROD DRIVE
HYDRAULIC LINES

**Figure 11.2—Typical GE Generation II BWR reactor vessel
and contents. Reproduced by courtesy of GE Hitachi.**

Recall the discussion of Newton's law of cooling in Chapter 2. The rate of heat transfer between a solid surface and a fluid depends on the heat transfer coefficient characterizing the surface and the fluid and on the temperature difference between them. Steam is a much less effective heat transfer medium than liquid water (i.e., it has a much lower heat transfer coefficient), so the coolant cannot be allowed to boil completely to vapor in the core. If it did, in order for all the heat generated in the all-vapor regions to be transferred into the steam, the fuel temperature would have to become too high in those regions for the materials to withstand. This condition is called "dryout." But the coolant entering the turbine must all be steam. Therefore, the upper plenum in a BWR contains steam separator and steam dryer components above the core to send the steam to the turbine and redirect the remaining liquid water to another pass through the core. This redirected, or "recirculated," water passes downward in an annular channel outside of the core to the lower plenum. It would be too complicated, and it would interfere with the function of these components in the upper plenum, to provide penetration channels for control rods from the top, so the control rods are inserted from the bottom.

The circulation of liquid water from the steam separator and steam dryer components back to the lower plenum, along with the feedwater flow from the primary coolant pumps, is driven by jet pumps, as shown in Figure 11.3. Figure 11.3 (a) shows

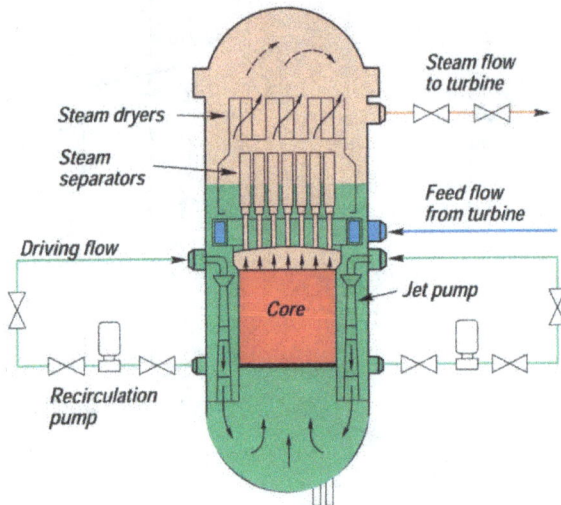

(a) **Jet pump recirculation plumbing and pressure vessel flow
pattern. Reproduced by courtesy of GE Hitachi.**

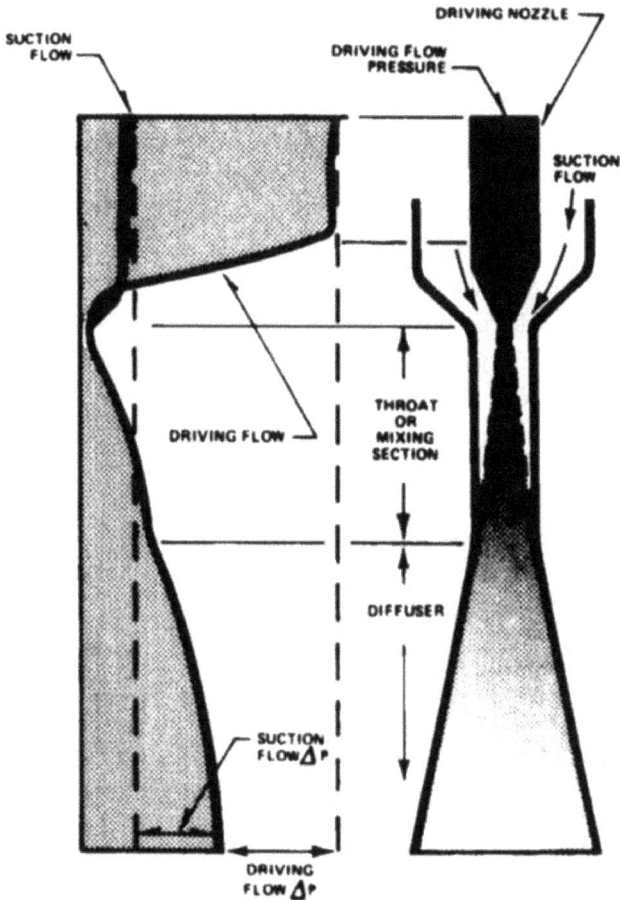

(b) **Operation of jet pump. Reproduced by courtesy of GE Hitachi.**

Figure 11.3—Jet pumps in a BWR.

how a small part of the recirculating water passes through external loops from which it is pumped at high pressure into the jet pumps. Figure 11.3 (b) shows how this small high-pressure flow creates a suction in the jet pump that draws a much larger volume of water with it to create the overall recirculating flow. The right-hand side of Figure 11.3 (b) shows the hardware and flowing fluid, and the left-hand side shows a graph of driving pressure and suction pressure versus position. This suction is an example of the Venturi effect discussed in Chapter 2. The feedwater to replace the steam sent to the turbine enters the pressure vessel above the jet pumps, as shown in Figure 11.3 (a).

Since the void fraction, or the fraction of the cooling water that has been boiled into steam, must be carefully controlled throughout the core of a BWR in order to prevent dryout, the coolant cannot be allowed to wander transversely from its principally upward flow direction through the core. To prevent such "crossflow," fuel assemblies, or groups of fuel rods, are arrayed within long open-ended boxes called fuel channels. The array of fuel rods within a modern BWR fuel channel is typically 10x10, although in earlier fuel assembly designs it was 8x8; the arrangement is preserved by grid plates with holes for coolant flow and fuel rod penetration. Each set of four fuel assemblies is associated with a cruciform control element in a fuel module. Figure 11.4 shows the layout of fuel modules in a typical BWR core; the crosses represent the cruciform control rods. Figure 11.5 shows a fuel module with 8x8 fuel assemblies. Some of the fuel modules have in-core flux monitors to keep track of the power distribution in the core.

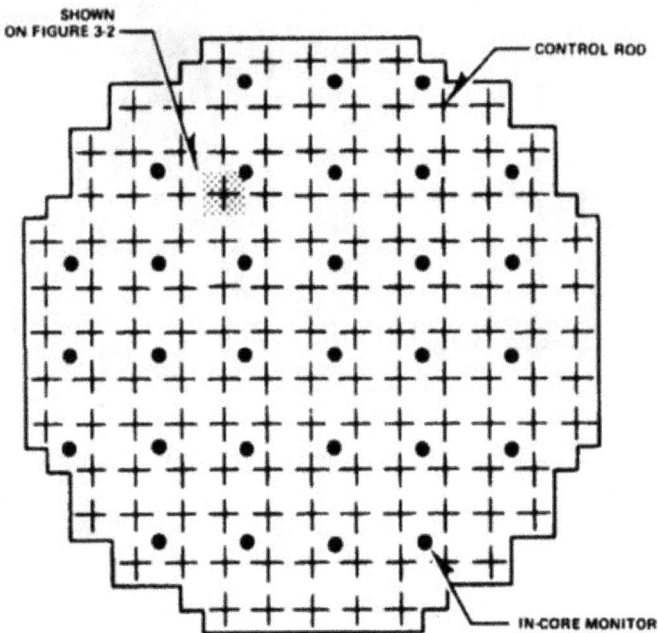

Figure 11.4—Arrangement of fuel modules in a BWR. The crosses represent the cruciform control rods. The reference to Figure 3-2 applies to the original GE document from which this figure was taken. Here, the shaded area is enlarged in Figure 11.5. Figure reproduced courtesy of GE Hitachi.

**BWR/6 FUEL
ASSEMBLIES
& CONTROL
ROD MODULE**

1. TOP FUEL GUIDE
2. CHANNEL FASTENER
3. UPPER TIE PLATE
4. EXPANSION SPRING
5. LOCKING TAB
6. CHANNEL
7. CONTROL ROD
8. FUEL ROD
9. SPACER
10. CORE PLATE ASSEMBLY
11. LOWER TIE PLATE
12. FUEL SUPPORT PIECE
13. FUEL PELLETS
14. END PLUG
15. CHANNEL SPACER
16. PLENUM SPRING

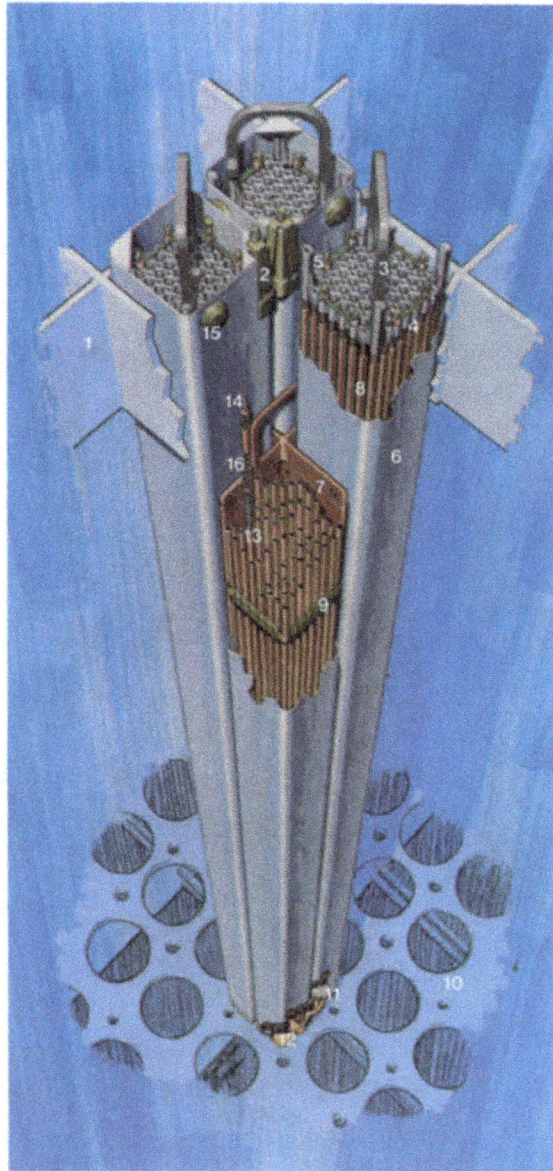

GENERAL ⊛ ELECTRIC

Figure 11.5—Fuel module in a BWR. Reproduced by courtesy of GE Hitachi.

The fuel rods themselves are sealed cylinders of a zirconium alloy called Zircaloy, chosen for its low neutron absorption cross section. Zircaloy has the drawback that at high temperatures it reacts exothermically with water and produces potentially explosive hydrogen ($Zr + 2H_2O \rightarrow ZrO_2 + 2H_2$), but its use allows reactors to operate with uranium of about 1% lower enrichment than stainless steel fuel rods.[5] In order to operate safely, reactors with Zircaloy fuel rods must be protected from reaching temperatures in accidents at which the Zircaloy-water reaction could take place. Both the Three Mile Island accident and the reactor damage incurred in the Fukushima Daiichi reactors in Japan show that this required protection cannot always be achieved. However, Generation III LWRs may be able to guarantee that the cladding is never exposed to excessive temperatures. Also, the development of accident-resistant fuel is an area of active research now.[6]

Within each fuel rod is a stack of ceramic fuel pellets, usually uranium dioxide (UO_2) enriched to 2-5% U-235. During reactor operation, the fuel pellets release fission-product gases and swell, so room must be provided within the rods for these gases and the enlarged fuel pellets. This room is provided by the upper plenum. A schematic diagram of fuel rods is shown in Figure 11.6.

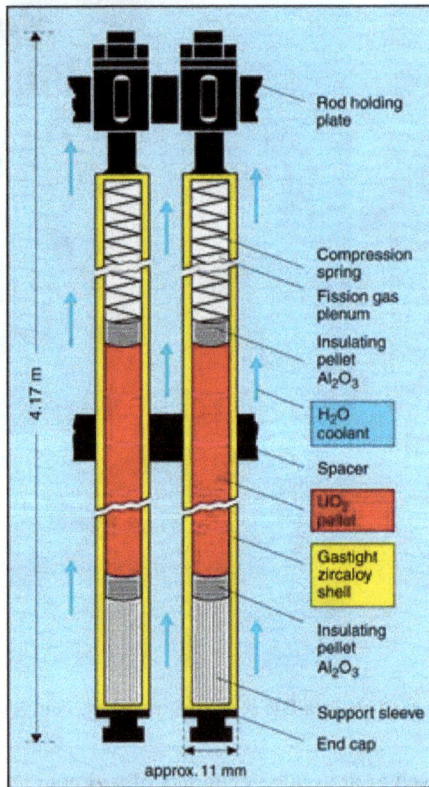

Figure 11.6—Typical LWR fuel rod and pellets.[7] Figure reproduced by courtesy of Informationskreis KernEnergie, Berlin (www.kernenergie.de)

Each fuel rod extends from the bottom to the top of the core. The active length (i.e., the length of the portion with fuel pellets) of a fuel rod in a typical BWR core is 3.66 m. The external diameter of the fuel rod is typically 1.24 cm, the thickness of the tube material is about 0.86 mm, and there is initially a gap of about 0.23 mm between the inner surface of the tube and the pellet. So the pellet diameter is about 1 cm.[8]

11.1.1.2 Safety systems

Safety of nuclear reactors in normal operations and accident scenarios is vitally important to the public and to the owners of the reactors. Even an accident that causes no physical harm to the public can be financially devastating to the reactor's owner, as the Three Mile Island and Fukushima Daiichi accidents have been, and the potential exists for accidents that take human life, as the Chernobyl accident did. These accidents each have chapters of their own: Chapter 12 for Three Mile Island, Chapter 13 for Chernobyl, and Chapter 14 for Fukushima Daiichi.

Safety in normal operation is provided by such natural and automatic processes as stabilizing reactivity feedback phenomena, as discussed in Section 6.3.1 and Appendix I. However, abnormal events such as earthquakes can exceed the ability of normal processes to accommodate them. Therefore, additional safety systems are provided to deal with abnormal occurrences.

There is an entire subfield of nuclear engineering devoted to imagining every plausible occurrence that can happen to nuclear reactors. This subfield is called probabilistic risk assessment (PRA). PRA engineers quantify risk as the product of the probability of an occurrence and the consequences: Risk (consequences per unit time) = frequency of events (events per unit time) x consequences per event.

Reactors must be able to sustain without core damage such normal occurrences as disturbances to the electric power distribution grid caused by severe weather, but also events of low probability, such as earthquakes that rupture major coolant pipes, that would have the potential to release significant amounts of radioactive materials. Nuclear regulatory agencies such as the U.S. Nuclear Regulatory Commission (USNRC) define design-basis accidents as postulated accidents that a nuclear facility must be designed and built to withstand without loss to the systems, structures, and components necessary to ensure public health and safety.[9]

The U.S. Department of Energy classifies expected event frequencies at a reactor site as anticipated ($>10^{-2}$/yr, or one event every hundred years or less), unlikely (10^{-2}-10^{-4}/yr, or one event in a time interval between a hundred and ten thousand years), extremely unlikely (10^{-4}-10^{-6}/yr, or one event in a time interval between ten thousand and a million years), credible (>10^{-6}/yr, or more than one event every million years), and incredible (<10^{-6}/yr, or fewer than one event in a million years).[10] Reactors must be able to withstand all credible events, so design-basis accidents are those with expected occurrence frequencies of 10^{-6} per year or greater. However, Generation III, III+, and IV reactors are designed to withstand even less probable events. In Canada, design-basis accidents are defined as those with expected occurrence frequencies between 10^{-5} and 10^{-2} per year.[11] For perspective, note that your individual chance of being hit by lightning is a little less than 10^{-6}/yr.

Obviously, the validity of a PRA analysis depends on how accurately the analysts identify the probabilities of the various occurrences that might threaten the integrity of a nuclear system. In the recent case of the Fukushima Daiichi reactors destroyed by an earthquake and the ensuing tsunami, a beyond-design-basis accident happened after only about 40 years of service. Thus, one should suspect that the actual occurrence frequency of such events may be considerably greater than the 10^{-6} per year previously assumed for them.[a] As happened after the Three Mile Island accident, considerable changes in reactor design and operations have been implemented as the course of the Japanese accidents was analyzed and understood. The Fukushima Daiichi accidents are discussed in detail in Chapter 14.

The worst design-basis event for most reactors is a large-break loss-of-coolant accident (LOCA).

In this accident, a large coolant pipe entering the pressure vessel suffers a complete ("guillotine") break in two locations (a "double-ended break"), so that a section of the pipe falls completely away and all inflow from that pipe is lost. Such a break might happen in a

---

[a] One must be careful here. If the probability of a magnitude-9 earthquake at a particular site is one in a million per year, that earthquake could still happen at any time in that million years, including right away, or not at all. Only in a sufficiently large number of million-year intervals will the average rate be one every million years. Consider the analogy of rolling dice. The probability of rolling, say, a five is one in six on each throw. But you can throw the die six or more times without rolling a five even once. The probability of rolling two fives in a row is one 36. But once the first five has been rolled, the probability of getting another five on the next throw is one in six. Such probability rules can be applied strictly when the probabilities of events are known. However, in estimating the probability of an earthquake close enough to a reactor site to cause a design-basis accident, one must search the geological record. Especially for undersea earthquakes, such records may be hard to search. Worldwide, earthquakes causing tsunamis are fairly common. There have been 23 notable tsunamis since 1992 (https://en.wikipedia.org/wiki/List_of_historical_tsunamis). The Sumatran tsunami of 2004, triggered by a magnitude-9 earthquake, killed 150,000 people (http://news.nationalgeographic.com/news/2004/12/1227_041226_tsunami.html). I think it is very likely that the probability of large undersea earthquakes with devastating tsunamis has been substantially underestimated.

severe earthquake. The coolant in the pressure vessel is blown out and the coolant depressurizes. This "blowdown" is not instantaneous, but it is finished in less than a minute. The mass of coolant in the core is greatly reduced, and the void fraction is greatly increased.

Because the reactor is designed with a negative void coefficient of reactivity (see Section 6.3.1), the nuclear chain reaction stops immediately. This is an inherent physical response that does not require any action by reactor operators. But the core continues to generate heat at a reduced rate because the fission products are radioactive and the radioactive decay events release heat. Therefore, some means must be provided to remove this heat in order for the core temperature to remain low enough that core damage will not occur. The zirconium-water reaction must be prevented, and the core components must not be allowed to melt.

The reactor is equipped with shutdown control rods, which are automatically inserted immediately when signals from various sensors indicate that operating limitations are exceeded. Such automatic shutdown is called a reactor "SCRAM" (see Section 6.3.4). These control rods are not necessary for reactor shutdown in a LOCA, but they help prevent recriticality when subsequent actions reflood the reactor core and restore moderating water.

A potential occurrence unique to BWRs is the "pressure transient," which can happen if the pressure in the core increases. Then some of the steam in the core will condense into liquid, and neutron moderation will increase. (Since PWRs have no steam in their primary cooling system, they are not subject to this kind of event.) This will lead to an increase in power. To prevent pressure transients from occurring, the reactor is provided with safety relief valves that open when the pressure exceeds a preset limit and thus relieve the pressure in the core. The valves open channels that lead to a "wetwell," which is a reservoir of liquid water that condenses escaping steam from the core into low-pressure liquid. This reservoir is called the pressure suppression pool. In earlier BWR designs, the wetwell is in a toroidal (doughnut-shaped) tank beneath the reactor pressure vessel. Figure 11.7 shows a cutaway view of Mark I (the earliest) BWR containment structures, including the toroidal wetwell. The region around or above the wetwell is called the drywell. In the cutaway portion of the torus in Figure 11.7, one can see downward-directed pipes leading out of a toroidal pipe within the larger torus. These downward-directed pipes end in spargers, or spray heads, which vent the steam from the safety relief valves into the pressure suppression pool below the water line. In later BWR designs, the wetwell is an annular region between a steel containment building and a weir wall, as shown in Figure 11.8 for the Mark III containment design.

The principal cooling of the post-LOCA core is provided by the emergency core cooling systems (ECCSs). There are actually several different ECCSs in a BWR.[12] Figure 11.9 illustrates the ECCSs in a BWR schematically.[13] One of the ECCSs (the high-pressure core spray system) injects cooling water at high pressure and at a flow rate up to 5000 gal/min into the pressure vessel above the core. This system can begin working even before the blowdown is completed, or in accidents that do not depressurize the core. The second system is the automatic depressurization system; this system keeps the reactor pressure within prescribed limits or, if the high-pressure injection system cannot deliver enough water to cool the core, it reduces pressure enough for the third system to operate. This third system comprises the low-pressure systems, including the low-pressure core spray system, which can deliver up to 12,500 gal/min, and the low-pressure coolant injection system, which can deliver up to 40,000 gal/min of water to maintain an adequate water level and cooling in the core. There is also a containment spray system that sprays water into the containment building atmosphere to condense steam and onto the outer surface of the pressure vessel to keep it cool. The pumps in all of these systems are powered by redundant on-site diesel generators that turn on automatically if off-site power is lost in an accident.

DRYWELL TORUS

GENERAL ⬡ ELECTRIC

Figure 11.7 Cutaway view of Mark I BWR containment structures, showing toroidal pressure suppression pool (wetwell) located within the drywell. Reproduced courtesy of GE Hitachi.

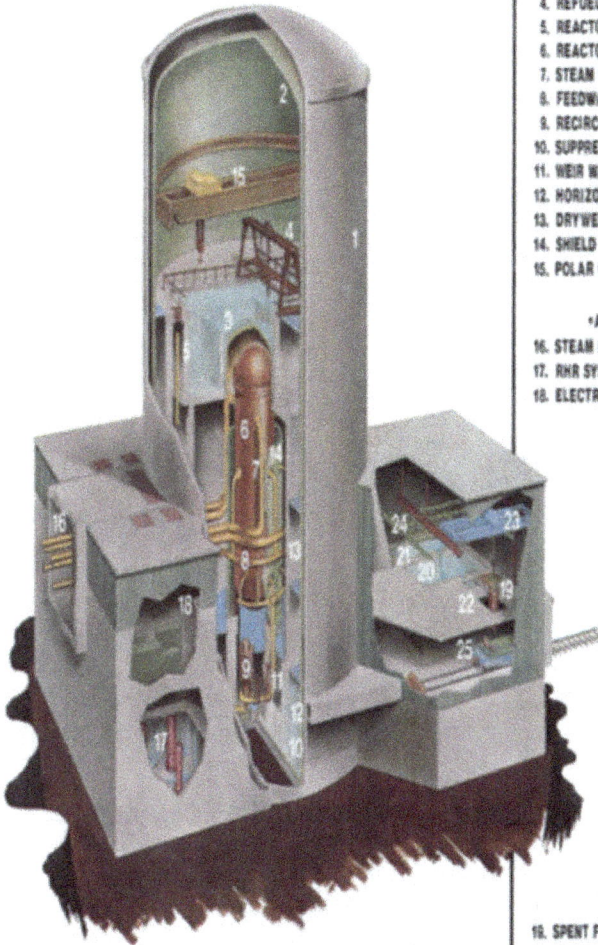

MARK III CONTAINMENT

•REACTOR BUILDING•
1. SHIELD BUILDING
2. FREESTANDING STEEL CONTAINMENT
3. UPPER POOL
4. REFUELING PLATFORM
5. REACTOR WATER CLEANUP
6. REACTOR VESSEL
7. STEAM LINE
8. FEEDWATER LINE
9. RECIRCULATION LOOP
10. SUPPRESSION POOL
11. WEIR WALL
12. HORIZONTAL VENT
13. DRYWELL
14. SHIELD WALL
15. POLAR CRANE

•AUXILIARY BUILDING•
16. STEAM LINE TUNNEL
17. RHR SYSTEM
18. ELECTRICAL EQUIPMENT ROOM

•FUEL BUILDING•
19. SPENT FUEL SHIPPING CASK
20. FUEL STORAGE POOL
21. FUEL TRANSFER POOL
22. CASK LOADING POOL
23. CASK HANDLING CRANE
24. FUEL TRANSFER BRIDGE
25. FUEL CASK SKID ON RAILROAD CAR

GENERAL ✷ ELECTRIC

Figure 11.8 Cutaway view of Mark III BWR containment structure, showing wetwell annulus. Reproduced by courtesy of GE Hitachi.

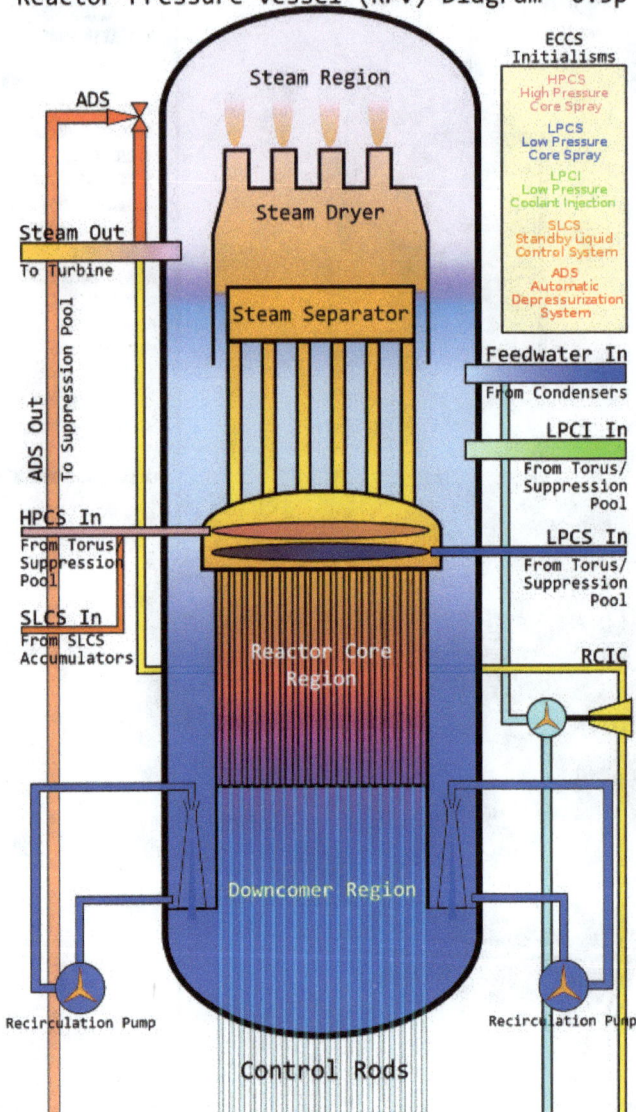

Figure 11.9—Emergency core cooling systems in a BWR. This figure was created by David C. Synnott and placed in the Wikimedia Commons; it is reproduced here under license CC-BY-3.0 (http://creativecommons. org/licenses/by/3.0); any further reproduction of this illustration must attribute the creator and comply with the same license.

Besides removing steam from the containment atmosphere, the containment spray system also contains additives such as sodium hydroxide or sodium thiosulfate, which can combine with some fission product gases in the containment atmosphere and put them into a liquid solution. There are additional measures for the removal of radionuclides from the containment atmosphere, such as HEPA filters, charcoal filters, and the use of paints containing substances that combine with halide fission products.

Steam vented into the space between the pressure vessel and the containment building wall condenses and pours into the pressure-suppression pool through the spargers. The wetwell also collects water from the ECCSs for reuse. There is also a residual heat removal heat exchanger, not shown in the figures, that ultimately dumps the radioactive decay heat from the core into the environment.

The steam that enters the interior of the containment building from the pipe break will be radioactive, and it is ultimately the containment building itself that prevents the release of radioactive substances into the environment. In early Generation II BWRs, the containment building is a concrete structure over 1 m thick. In later models, there is also a steel containment shell inside the concrete building. These containment structures are designed not only to withstand large pressure pulses from the LOCA itself, but also to endure extreme external forces, such as those from tornadoes, impacts from large aircraft, and other unlikely or extremely unlikely events.[14]

Figure 11.9 also shows a channel labeled RCIC, which is not identified in the legend. This is the reactor core isolation cooling system.[15] It is not actually one of the emergency core cooling systems, but it is able to replace water boiled off by residual decay heat or lost through small leaks. It is driven by high-pressure steam from the pressure vessel, and needs no electric power other than battery power to operate the valves in its piping. The valves can also be operated manually. It only works, however, if the pressure in the primary cooling system remains high. Not all Generation II BWRs have reactor core isolation cooling systems; the BWR/3 design uses isolation condensers, which are discussed in Section 11.1.2 below.

The foregoing discussion draws heavily on the previously referenced textbook by Knief and on the text by Glasstone and Sesonske,[16] as well as on Reference 12.

The principal weakness in Generation II BWRs is that emergency core cooling depends on the mechanical actuation of systems. For example, the diesel generators that power the ECCSs must turn on. To counter this weakness, the principle of "defense in depth" is applied. There are three ECCS systems instead of one, and there are backup diesel generators. Nevertheless, it would be better if no actuation were needed for engineered safety systems. Thus, the idea of "passive safety" was developed and applied in Generation III concepts.

11.1.2 Generation III BWRs

In the past, nuclear reactors were licensed individually, so that each license application was treated as an entirely new entity. Changes in regulatory policy have allowed standardized designs to be licensed as a category, so that only site-specific differences among plants need to be addressed in individual reactor license applications. GE developed the Advanced Boiling Water Reactor (ABWR) in the 1990s and received initial U.S. Nuclear Regulatory Commission (NRC) approval for the design in 1997.[17] (Whenever a reactor vendor makes design changes or updates, the NRC reviews them, so final NRC approval is not given until the final reactor design is established.) As of 2005, four ABWRs in Japan were completed and operating; in 2018, another ABWR in Japan was scheduled for completion and startup in 2021. ABWRs have been licensed in several other countries, including the United States. Construction was begun on a fifth ABWR in Japan and on a two-reactor project in Taiwan, but those projects were suspended. For the licensed project in the U.S., at Bay City, Texas, as of 2018 construction has not been scheduled.[18]

The biggest improvement in the ABWR over the Generation II BWR is the replacement of the external recirculation system and the jet pumps by internal recirculation pumps. The removal of the external piping eliminates the possibility of breaks in those pipes, which were up to 24 inches in diameter. The external piping remaining in the ABWR is 2 inches in diameter or less. The greatly reduced potential coolant loss rate reduces the demand on the ECCSs. The ECCS reservoirs can supply emergency core cooling water for up to three days before they need to be replenished. The ECCSs are actuated automatically without operator intervention, but they do require actuation, so that the ABWR does not possess true passive safety. However, the calculated maximum core damage frequency—i.e., the expected frequency of design-basis accidents—is between 5 and 50 times less than in Generation II BWRs.[19] That is, the events the reactor is designed to withstand are more severe events regarded as 5 to 50 times less probable than the events Generation II reactors are designed to withstand. Figure 11.10 shows the layout of the GE Hitachi ABWR, and Figure 11.11 shows the interior of the pressure vessel.

Reference 20 presents a fairly comprehensive overview of the ABWR.

The ESBWR, for Economic Simplified Boiling Water Reactor, is a Generation III+ reactor, with truly passive safety.[21, 22] The reactor uses no recirculation pumps at all, but depends on natural convection, an inherent physical process, to drive the coolant flow. In an accident, decay heat would be removed from the shut-down core by heat exchangers called isolation condensers, which are heat exchangers located in open pools of water above the containment. Evaporation of the clean water in the open pool takes the heat away into the atmosphere. (Some older BWRs, notably the BWR/3 model used in Fukushima Daiichi Unit 1, also have isolation condensers.[23]) Makeup coolant would be provided from pools in the gravity-driven cooling system, and natural circulation would continue to drive the coolant through the core. Not only would the maximum core damage frequency be ten times less than even in the ABWR,[19] but the cost of construction is expected to be less than the cost of other LWRs. A very detailed discussion of the ESBWR is presented in Reference 24.

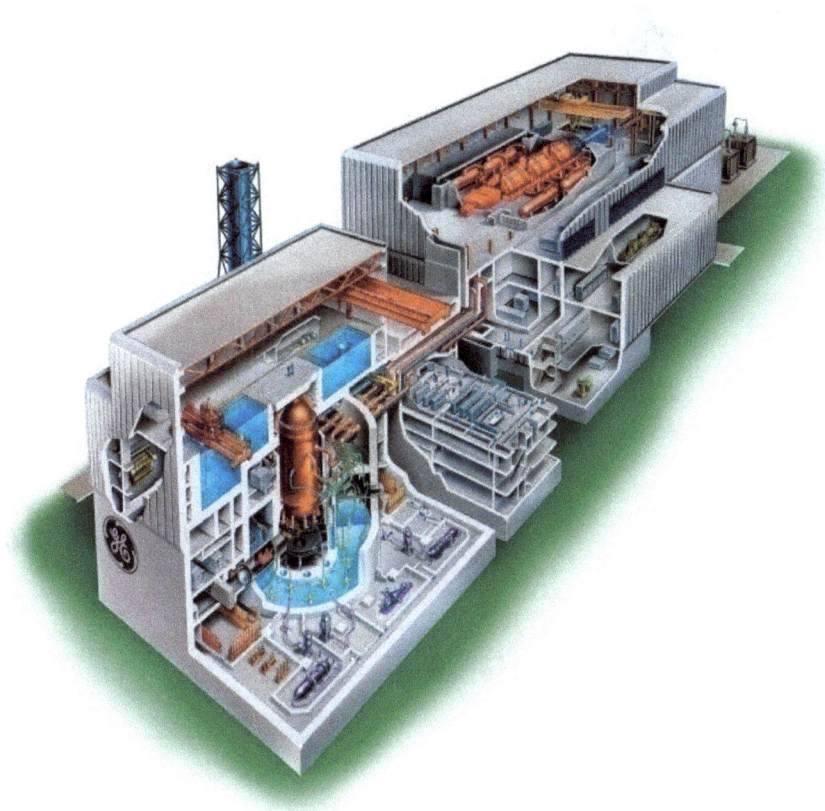

Figure 11.10—Layout of the GE Hitachi ABWR power plant.[25] Reproduced courtesy of GE Hitachi.

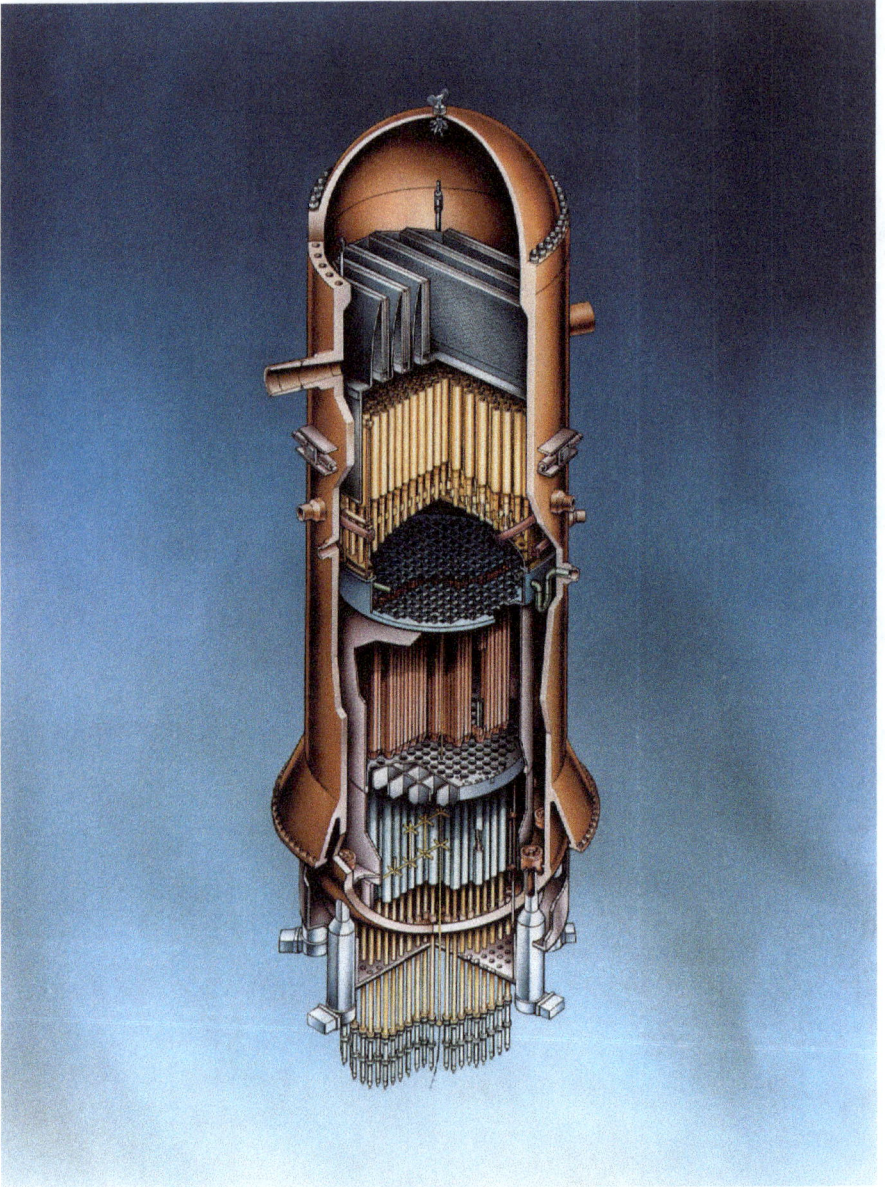

**Figure 11.11—Cutaway view of ABWR pressure vessel.
Reproduced by courtesy of GE Hitachi.**

Table 11.1 presents some parameters of interest for several different BWRs.

Table 11.1—Design parameters for BWRs

BWR Version	Late Gen II BWR[26]	ABWR[17, 20]	ESBWR[21, 22, 24]
Primary loop coolant pressure (MPa/psi)	7.17/1040	7.17/1040	7.17/1040
Core inlet temperature (°C)	278		
Core outlet temperature (°C)	288	287	
Net electric power (MW)	1178	1350-1460	1575-1600
Efficiency (%)	32.9	34	35
Maximum core damage frequency (/yr)	1×10^{-6}	1.6×10^{-7}	3×10^{-8}

11.2 PRESSURIZED-WATER REACTORS

Most PWRs in the U.S. up to now have been produced by the Westinghouse Electric Company, but some have been produced by Babcock and Wilcox and Combustion Engineering. Other countries have their own nuclear industries that produce PWRs, such as Mitsubishi in Japan and Areva (which was formerly Framatome) in France. Westinghouse, Areva, and Mitsubishi have also developed Generation III or III+ designs.

11.2.1 Generation II PWRs

Most operating power reactors are Generation II PWRs.[27] Not only are they widely used by utilities, but they are also used in almost all nuclear-powered ships. The general operating scheme of a Generation II PWR is described in Section 11.2.1.1. Reactor safety systems are discussed in Section 11.2.1.2.

11.2.1.1 Reactor and power plant layout

Figure 11.11 shows the major components of a typical PWR power plant schematically. Figure 11.12 shows the pressure vessel and its interior.

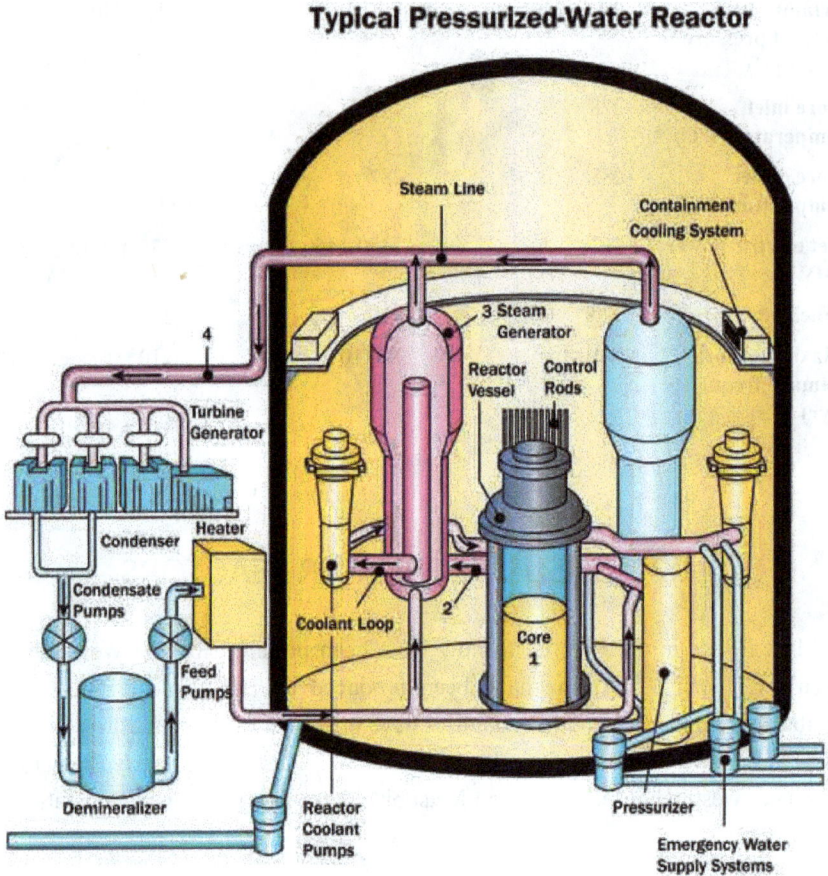

Figure 11.11—Typical PWR power plant components. This figure was taken from Reference 28, a work of the Nuclear Regulatory Commission. As a work of the United States Government it is in the public domain under the terms of Title 17, Chapter 1, Section 105 of the U.S. Code.

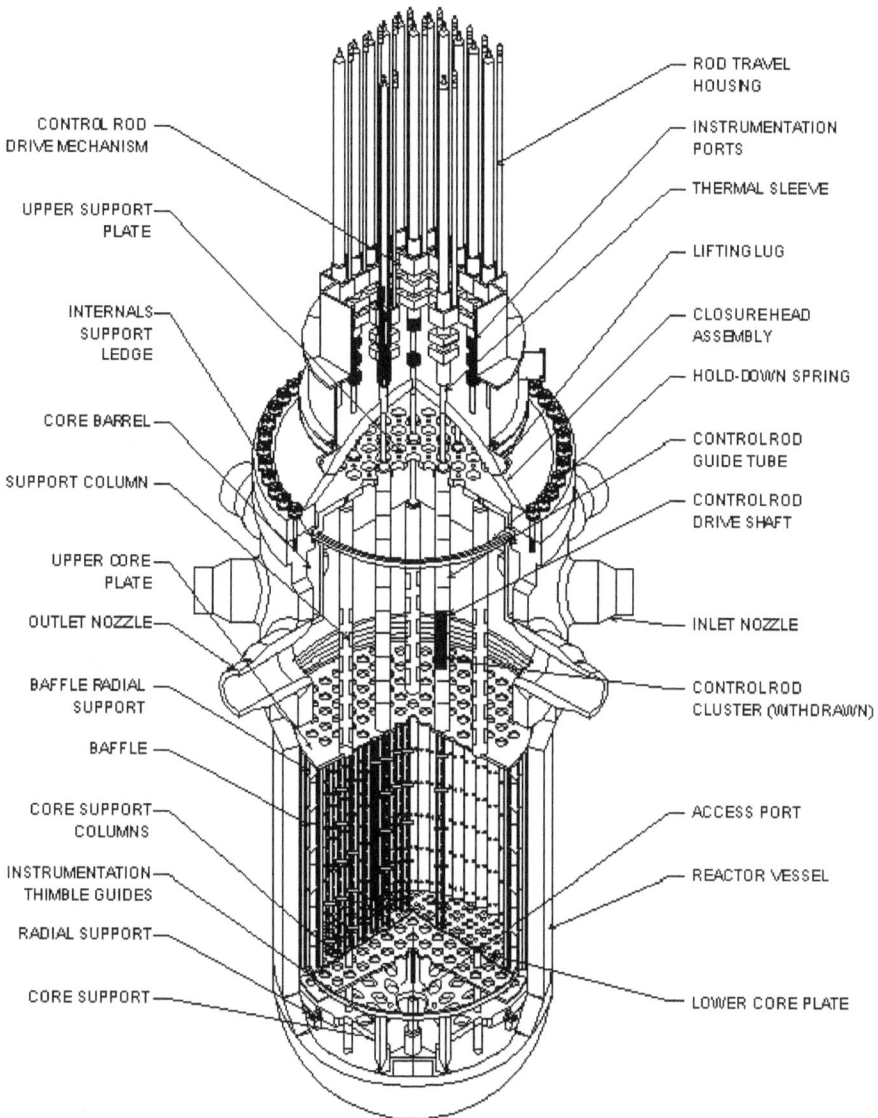

CONTROL ROD
DRIVE MECHANISM

UPPER SUPPORT
PLATE

INTERNALS
SUPPORT
LEDGE

CORE BARREL

SUPPORT COLUMN

UPPER CORE
PLATE

OUTLET NOZZLE

BAFFLE RADIAL
SUPPORT

BAFFLE

CORE SUPPORT
COLUMNS

INSTRUMENTATION
THIMBLE GUIDES

RADIAL SUPPORT

CORE SUPPORT

ROD TRAVEL
HOUSING

INSTRUMENTATION
PORTS

THERMAL SLEEVE

LIFTING LUG

CLOSURE HEAD
ASSEMBLY

HOLD-DOWN SPRING

CONTROL ROD
GUIDE TUBE

CONTROL ROD
DRIVE SHAFT

INLET NOZZLE

CONTROL ROD
CLUSTER (WITHDRAWN)

ACCESS PORT

REACTOR VESSEL

LOWER CORE PLATE

Figure 11.12—Interior of pressure vessel of a Westinghouse PWR.[29, 30] **This file was obtained from the Wikimedia Commons and is in the public domain.**

As noted above, the PWR contains steam generators where heat from the primary coolant loop is transferred to a secondary coolant loop that actually supplies steam to the turbine. The primary loop contains a component called a pressurizer that controls the pressure in the loop. The pressurizer is filled partly with liquid water and partly with steam. A spray nozzle in the pressurizer injects liquid water to condense some steam and reduce the pressure when the pressure in the primary loop (and thus in the pressurizer) increases. An electrical heater generates more steam in the pressurizer, and thus increases the pressure in the pressurizer, when the pressure in the primary loop (and thus in the pressurizer) decreases. These two systems keep the pressure in the primary loop at the desired value.

Because there are no steam separator and steam dryer components in the pressure vessel of a PWR, the control rods in a PWR can be inserted from the top, so that the shutdown rods can fall into the core by gravity in a SCRAM.

As shown in Figures 11.11 and 11.12, the core inlet nozzles are located above the core, and the coolant flows downward outside of the core barrel before reversing direction in the lower plenum and flowing upwards through the core. Placing the core inlets above the core prevents all the coolant from flowing out of the core in case an inlet pipe is broken in an earthquake or other severe accident. The outlet nozzles are also located above the core.

The fuel rods in a PWR are similar to those in a BWR, as illustrated in Figure 11.6. However, because the pressure in a PWR does not permit the coolant to boil, crossflow is not the problem in a PWR that it is in a BWR. Therefore, the fuel assemblies are not enclosed in a boxlike fuel channel as in a BWR, but are held instead in an open lattice by grid plates. The lattice is usually either 16x16 or 17x17, with some of the lattice positions occupied by control rod guide tubes and instrumentation. Lattices may also be 14x14, 15x15, or 18x18. The control rods are operated in clusters as shown in Figure 11.13.[31] The central control rod drive mechanism shown at the top of the figure ends in an array of fingerlike projections, each of which holds a control rod at its top. When the central mechanism is raised or lowered, all of the control rods attached to it are raised or lowered with it. Most of the grid plate locations shown in the figure are empty for purposes of illustration, but in reality they are all occupied by fuel rods, as suggested at the bottom grid plate, except in locations occupied by control rod guide tubes or instrumentation.

The control rods contain a neutron poison, which may be B_4C or an alloy of silver, indium, and cadmium. Most of the control rods contain the poison material along the entire active length of the core, but in some of them the poison material is contained in only the lower end of the rod. These rods may be inserted partially in order to absorb neutrons in specific axial locations. This technique allows the axial neutron flux profile to be shaped more uniformly, so that the fuel consumption can also be more even axially.

As noted in Reference 27, a PWR does not contain enough excess reactivity to permit prompt supercriticality. However, it must contain a sufficient positive reactivity margin to permit desired increases in power. As the fuel is depleted, this positive reactivity margin is maintained by the simultaneous depletion of soluble boron (in the form of boric acid dissolved in the coolant) and fixed "burnable poisons" contained in rods occupying selected positions in the fuel lattice.

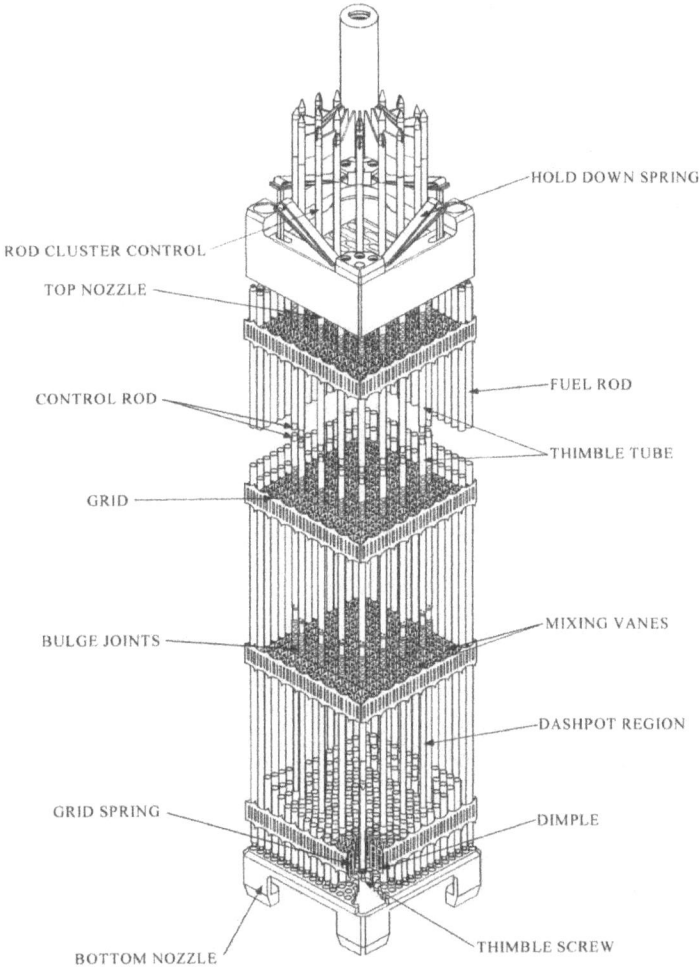

Figure 11.13—PWR fuel assembly. This figure was taken from a work of the U.S. Nuclear Regulatory Commission.[31] As a work of the United States Government it is in the public domain under the terms of Title 17, Chapter 1, Section 105 of the U.S. Code.

11.2.1.2 Safety systems

The same general principles of safety analysis and reactor protection that were discussed for BWRs in Section 11.1.1.2 also apply to PWRs. The differences in safety systems are due to the differences in the design details. The ECCSs in a PWR are shown schematically in Figure 11.14.

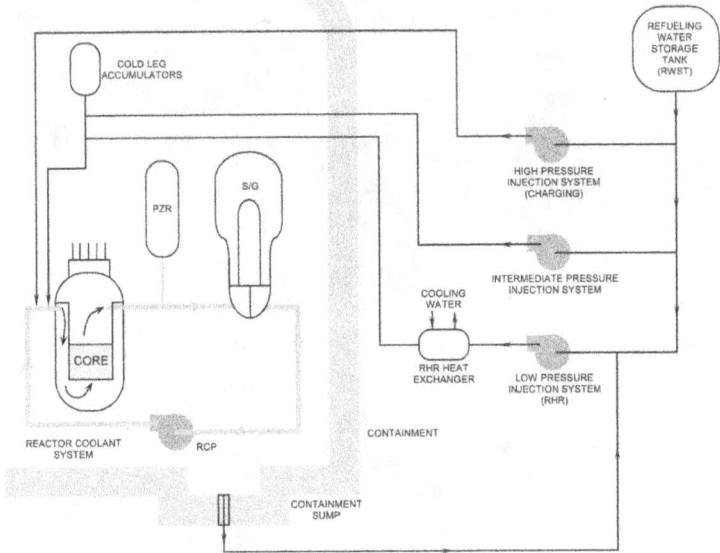

Figure 11.14—Emergency core cooling systems in a PWR. This figure was taken from a work of the U.S. Nuclear Regulatory Commission.[31] As a work of the United States Government it is in the public domain under the terms of Title 17, Chapter 1, Section 105 of the U.S. Code.

The instrumentation tubes located throughout the reactor monitor numerous parameters that characterize the reactor's performance, such as temperature, pressure, and neutron flux. All of these parameters have normal operating ranges, and when any parameter falls outside its normal range, the reactor is scrammed. The full-length control rods in the reactor are suspended above the core by electromagnets, and when the SCRAM signal interrupts the electric current in the magnets, the control rods fall into place by gravity.

As with the BWR, the most severe credible accident is a double-ended large-break LOCA in a coolant inlet pipe. A blowdown phase ensues, in which the reactor is depressurized as coolant blows out of the break into the containment building. Loss of moderation from the coolant shuts down the neutron chain reaction, but the control rods are inserted anyway to prevent recriticality as the ECCS floods the core again.

Three independent ECCSs, which can be powered by several alternative sources including backup diesel generators, supply borated cooling water to carry away the decay heat released by the core after shutdown. The high-pressure injection system and the accumulator system maintain coolant flow during the blowdown phase. The high-pressure injection system pumps water from the refueling water storage tank into the core. The accumulators are pressurized tanks from which water flows into the core when the pressure signals from the core cause accumulator isolation valves to open. After the blowdown phase is complete, the low-pressure injection system pumps water from the refueling water storage tank first. Once that tank is emptied, the low-pressure injection system takes water from the floor of the containment building, where it has collected from the blowdown and from a containment spray system, which sprays cooling water into the air within the containment building to condense steam from the air. The low-pressure injection system is part of the residual heat removal system, which takes the decay heat from the core and exhausts it to the environment.

The containment spray system and other measures similar to those in BWRs remove radioactive fission products from the containment atmosphere. The containment buildings for PWRs and BWRs are similar, with steel inner vessels and concrete buildings designed to withstand high internal pressures and violent external impacts.

As with BWRs, the safety systems in Generation II PWRs are mechanically actuated, with pumps and valves to drive coolant flows and open flow channels. These systems are highly reliable and redundant, applying the principle of defense in depth. However, systems that do not rely on mechanical actuation are inherently more foolproof. Thus, passively safe designs have been developed in Generation III and III+ PWRs.

11.2.2 Generation III PWRs

The United States, France, and Japan have developed Generation III and III+ PWR designs. In the U.S., Westinghouse has designed the AP600 and AP1000 reactors.[32, 33, 34] In France, Areva has introduced the EPR.[35] The acronym originally stood for European Pressurized Reactor; to appeal to a global customer base, Areva changed the meaning to Evolutionary Power Reactor. Now they just call it EPR. In Japan, Mitsubishi has developed the Advanced Pressurized Water Reactor (APWR) and its U.S. version, the US-APWR.[36] More recently, Areva and Mitsubishi have collaborated on a Generation III+ reactor design, the ATMEA1.[37] The acronym is based on a combination of the Areva and Mitsubishi company names. South Korea has developed the APR-1400 advanced PWR.[38]

All of these competing Generation III and III+ PWRs apply the principles of passive safety, redundancy, and defense in depth to achieve similar safety performance. These principles are described below for the Generation III+ AP1000 reactor.

The AP1000 is an advanced, higher-power development from the AP600 design, which was never built. The AP1000 received license approval from the U.S. Nuclear Regulatory Commission (NRC), but some design improvements introduced since licensing required review from the NRC before the license could be extended to the improved design. By December 2010, the NRC had reviewed the improved design and issued a Final Safety Evaluation Report (FSER). Before the NRC can issue a "rule" certifying the design and granting a license for an improved design, the FSER must by law be presented to the public for review and comments.[39] As of December 2011, the public review and comment phase had been completed and the NRC had certified the revised design. Final NRC approval for construction and operation, which depended on site-specific safety issues and environmental impacts, was given for four AP1000 reactors in the U.S., two each at the Vogtle Electric Generating Plant in Georgia and the Virgil C. Summer Nuclear Generating Station in South Carolina.[32] However, in March 2017, Westinghouse Electric Company filed for bankruptcy as a result of large cost overruns at all four plants, together with slowing growth in demand for electricity, falling prices for natural gas, and maturing solar and wind power technologies.[40] These factors led to the cancellation of the Virgil C. Summer plants after partial construction, but the Vogtle plants are still under construction as of February 2019. In China, four AP1000 reactors have been constructed, and two of them are in operation as of January 2019.[32]

The AP1000 is much simpler in design than Generation II PWRs, containing 50% fewer safety-related valves, 80% less safety-related piping, 85% less control cable, and 35% fewer pumps. The building volume requiring resistance to seismic events is 45% smaller. Its Passive Core Cooling System (PCCS), functioning by gravity, stored compressed gases, and natural circulation, replaces the ECCS of earlier PWRs. This system is shown in action in Reference 41. The animation in Reference 41 shows only one PCCS, but there are actually two independent PCCS systems, one for each of the two independent primary coolant loops. In a LOCA, core cooling is immediately provided by high-pressure water injection from the pressurizer, the core makeup tanks (CMTs), and the accumulators. As the CMT water level drops, the primary cooling system depressurizes in three stages. High-pressure steam from the core passes backwards through the pressurizer and into a large pool called the in-containment refueling water storage tank (IRWST), where it is condensed into liquid. Water from the containment building sump and the IRWST is pulled into the core by natural convection, cooling the core without the use of pumps. The IRWST is the thermal reservoir for residual decay heat removal. The Passive Containment Cooling Water Tank is located above the containment building, providing gravity-driven cooling of the containment building atmosphere and the pressure vessel. The passive safety systems of the AP1000 permit the reactor to respond to a LOCA without operator intervention for 72 hours.[42] The components of the AP1000 safety systems mentioned above are shown in Figure 11.15.[43]

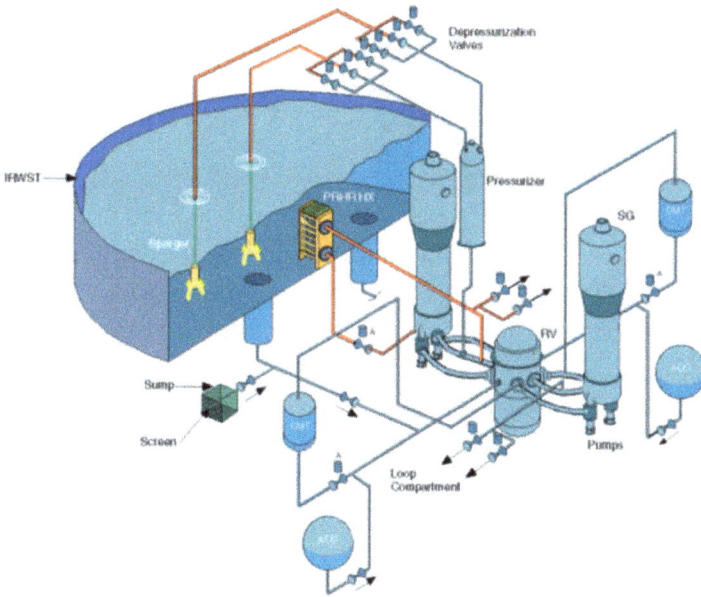

Figure 11.15—Safety systems in the AP1000. From Reference 43, reproduced courtesy of T. L. Schultz, Westinghouse Electric Company

EPRs are under construction in Finland, China, and France. APWRs and US-APWRs are planned for Japan and the U.S., respectively. South Korea has four APR-1400s under construction and two more planned.[44]

Table 11.2 gives important parameters for Generation II and Generation III and III+ PWRs.

Table 11.2—Design parameters for PWRs

PWR Version	Gen II PWR	AP1000[45]	EPR[35, 46]	US-APWR[47, 48]
Primary loop coolant pressure (MPa/psi)	15.5/2235	15.5/2235	15.8/2279	15.5/2235
Core inlet temperature (°C)	292	281	295.9	288
Core outlet temperature (°C)	325	321	327.2	309.3*
Net electric power (MW)	1150	1000	1650	1700
Efficiency (%)	33.7	29.4	36.7	38
Maximum core damage frequency (/yr)	10^{-6}	2.41×10^{-7}	6.1×10^{-7}	
Fuel enrichment (%)	>2.6		5	

*Core average temperature

The next three chapters are devoted to analyses of the three large nuclear power reactor accidents that have occurred so far: Three Mile Island, Chernobyl, and Fukushima Daiichi. While it was not technically an LWR, because it was graphite-moderated, the Chernobyl reactor was cooled by light water, so it can logically be grouped with the Three Mile Island and Fukushima Daiichi reactors for its cooling medium as well as by its accident history.

CHAPTER 12
THE THREE MILE ISLAND ACCIDENT

On 28 March 1979, a partial core meltdown occurred in Unit 2 of the Three Mile Island (TMI) Nuclear Station near Middletown, Pennsylvania, about 10 miles southeast of the capital city of Harrisburg. Ironically, this accident occurred while the movie *The China Syndrome* was playing; this movie dramatizes a hypothetical scenario in which a malfunctioning pump leads to the threat of the core melting its way through the containment building and on down into the ground, metaphorically all the way to China. The Three Mile Island accident was a pivotal event for the future of nuclear power in the United States. It caused deployment of nuclear power to come to an abrupt halt: Orders for power plants were cancelled, completed plants were never operated, and some existing plants were shut down.[1,2] No further orders for new nuclear plants were placed until very recently, although now some Generation III and III+ designs have been licensed by the NRC, some construction and operating license applications have been submitted for Generation III or III+ plants, and construction was begun for four AP1000 reactors,[3] although two of them have been cancelled, as noted in Chapter 11. The entire design and operating philosophy for nuclear reactors was reviewed and modified, with much greater emphasis on safety. Existing plants were modified, especially in their control systems and displays, and much more rigorous training programs were instituted for operators. The concept of passive safety was made a primary goal for future power plant designs. Because of the extent of the damage, the size of the reactor, and the potential for greater consequences, this accident is generally regarded as the most serious accident that has ever occurred in a U. S. nuclear facility.[a]

---

[a] No fatalities resulted from the Three Mile Island accident, whereas three people were killed by a steam explosion and core meltdown in the SL-1 reactor, a small experimental reactor at the

A fairly technical summary of the accident is provided in the nuclear engineering textbook by Ronald Allen Knief, who worked as manager of training activities at the TMI facilities after the accident.[4] Most of the following discussion is taken from his summary.

The TMI accident was a loss-of-coolant accident (LOCA), but since no pipe break was involved, it was not immediately recognized as such. A gauge that displayed information crucial to understanding the event was inconspicuously located and not checked when it should have been, so that the nature of the event was not understood in time and the actions of the operators were inappropriate and led to the melting of part of the core.

Figure 12.1 shows the schematic layout of the TMI-2 reactor. For simplicity, the figure only shows one primary coolant loop, but the reactor actually had two independent primary loops. The pressurizer was shared by the two loops.

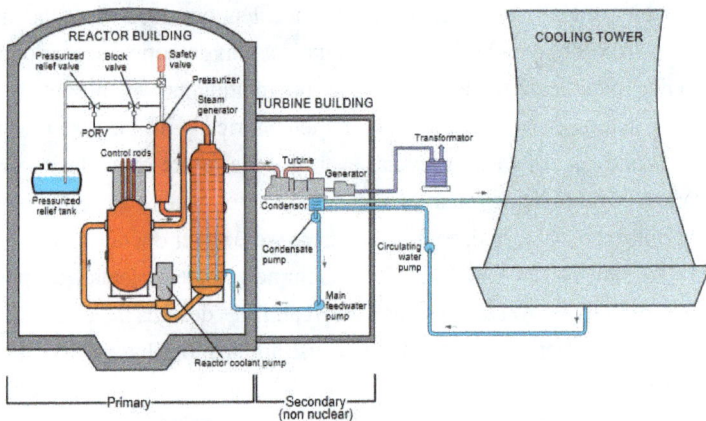

Figure 12.1—Simplified schematic diagram of the TMI-2 nuclear power plant. This figure was taken from the Wikimedia Commons.[5] It was originally created by the U. S. Nuclear Regulatory Commission; as a work of the U.S. federal government, the figure is in the public domain under the terms of Title 17, Chapter 1, Section 105 of the U.S. Code.

National Reactor Testing Station (as the present Idaho National Laboratory was then known) in Idaho in 1961. (See https://en.wikipedia.org/wiki/SL-1.) The SL-1 reactor only produced 3 MWt (megawatts of thermal power), so even though the damage was as complete as it could have been, it was contained within the reactor building. Still, one can argue that any accident with a loss of life is more serious than an accident that only damages property.

In the upper left region of the reactor containment building, the figure shows a valve called the power-operated relief valve (PORV), which Knief calls the "electromatic relief valve." The line in which this valve is located empties into the pressurized relief tank, or drain tank, shown in the left of the containment building. The purpose of this valve is to prevent excessive pressure buildup in the primary coolant system by any possible cause of increased pressure. This valve can be opened and closed manually, but it opens or closes automatically in response to signals from a pressure monitor. The relief tank contains a relief valve and a rupture disk to prevent excessive pressure build-up in the tank. In response to a pressure increase caused by actions described below, the electromatic relief valve opened automatically, but when the pressure subsided, the valve stuck open. Loss of coolant through this valve was the direct cause of the LOCA.

There were several abnormal conditions affecting the reactor before the accident occurred. First, the primary cooling system was losing coolant through one or more of the valves associated with the pressurizer. Besides the electromatic relief valve itself, there were a block valve and a safety valve in the pressurizer piping, as shown in the figure. The coolant loss rate exceeded regulatory limits, but the operators believed at the time that it did not. The lost coolant entered the drain tank, which was therefore partly filled when the accident began.

Second, two valves in the emergency feedwater lines were closed, although they should have been open. The emergency feedwater lines (not shown in the figure) supply coolant in the secondary coolant loop to the steam generators. When the main feedwater pumps "trip"—i.e., shut down manually or on a signal from plant protective monitoring sensors—the turbine also trips, and the emergency feedwater pumps start automatically to continue providing a means of transferring heat from the primary coolant loop to the secondary coolant loop. The closed valves in the emergency feedwater lines were opened manually 8 minutes into the accident. The effect of these closed valves on the severity of the accident is debated.

Third, a blockage existed in a transfer line in the demineralizer system in the turbine building, also not shown in the figure. The demineralizer system maintains the purity of the coolant in the secondary loop; it works by passing the coolant through ion-exchange resins that absorb impurities. Operators had been attempting for about 11 hours to transfer spent resins from the demineralizers to a resin regeneration tank, but the blockage in the line thwarted their efforts. Attempts to clear the blocked line caused the condensate pump in the secondary loop to trip. This trip was the initial event in the sequence of events that led to melting in the core.

The chronology of these events is presented in Table 12.1. The subsequent discussion explains the events.

Table 12.1—Chronology of events in the TMI-2 reactor accident

Approx. time	Event
0	Condensate pump trip
1 s	Main feedwater pumps and turbine trip
3 s	Electromatic relief valve opens
8 s	Reactor trips
13 s	Pressure drops below set point for electromatic relief valve, but valve sticks open
14 s	Emergency feedwater pumps reach normal discharge pressure
38 s	Emergency feedwater directed to steam generators, but flow is blocked by closed valves
2 min	High-pressure injection ECCS starts automatically; also, drain tank relief valve lifts
3-5 min	Operators throttle high-pressure injection ECCS
7 min	Coolant transfer from containment to auxiliary building begins
8 min	Closed valves in emergency feedwater lines are opened
15 min	Drain tank rupture disk lifts
73 min	Main coolant pumps in Loop A are tripped
100 min	Main coolant pumps in Loop B are tripped
1.5-3.5 h	Water level in core falls; core exposed and partially melts; zirconium-water reaction produces hydrogen
142 min	Block valve on pressurizer line is closed manually
150 min	In-core thermocouple readings go off scale
3 h	Site emergency is declared
3.5 h	General emergency is declared
9.5 h	Hydrogen detonation in containment building produces pressure spike
13.5 h	First main coolant pump in Loop A restarts to restore forced-convection cooling
16 h	Second main coolant pump in Loop A restarts

As noted above, the actions of the maintenance crew attempting to clear the blocked resin transfer line led to a trip of the condensate pump, which in turn immediately caused trips in the turbine and main feedwater pumps. The momentary loss of heat removal by the secondary coolant loops from the primary coolant loops in the steam generators caused an increase in the primary coolant temperature and pressure, which caused the electromatic relief valve to open. This pressure increase also triggered a reactor SCRAM—the control rods were automatically dropped into the core to shut down the nuclear chain reaction.

A few seconds after the reactor SCRAM, the primary coolant pressure fell enough for the electromatic relief valve to close automatically. However, although the voltage that holds the valve open was cut off, the valve stuck in the open position and steam flowed out through the top of the pressurizer into the drain tank. This flow gradually reduced the amount of coolant in the primary cooling system and eventually uncovered the core.

The emergency feedwater pumps came on automatically to restore cooling flow in the secondary loops through the steam generators, but the closed valves in the emergency feedwater lines prevented that flow. As noted above, the contribution to the core damage by the lack of secondary flow is debated, but the loss of secondary flow certainly didn't help matters. The closed condition of the valves was noticed 8 minutes into the accident, and the valves were opened manually.

Two minutes into the accident, the two high-pressure-injection ECCSs turned on automatically in response to low primary coolant pressure. This injection would have made up for the loss through the stuck electromatic relief valve if it had been allowed to continue.

At this point, a phenomenon occurred that led to misunderstanding of the situation by the operators and induced them to make disastrous errors.[6] As the steam flowed out of the pressurizer, it drew liquid into the pressurizer and raised the liquid level too high. If it contains insufficient vapor volume, the pressurizer cannot perform its function of pressure regulation, so an excessive liquid level is to be avoided and in fact violates operating regulations. Seeing indications of an excessive liquid level in the pressurizer, operators inferred incorrectly that the primary cooling system was full, so they turned off one of the high-pressure ECCS injection pumps and throttled back the other. The rate of high-pressure injection was no longer enough to make up for the loss through the stuck electromatic relief valve.

A clue to the true nature of the problem could have been provided by a meter showing the pressure in the drain tank. At about the same time that the high-pressure-injection ECCSs came on, the drain tank relief valve opened, and the meter would have shown that the drain tank was full. However, this meter was located on a panel behind the reactor console, out of view of the operators. It was not checked during the frantic efforts of the operators to respond to the confusing indications from the main instrument clusters. After 15 minutes, the drain tank rupture disk ruptured, and thereafter the meter indicated atmospheric pressure in the drain tank, which is the normal reading for an unfilled tank. The poorly chosen location for the drain tank pressure meter caused the operators to miss information that could have saved the reactor.

As explained in Section 11.1.1.2, continued cooling must be provided even after the nuclear chain reaction has been shut down, because fission products continue to generate heat by radioactive decay. In the TMI-2 accident, the amount of coolant available in the primary cooling system to remove this heat was continuously decreasing, so the coolant temperature increased until the coolant began to boil. The steam voids in the coolant caused vibrations in the primary coolant pumps. The operators shut down the pumps in Loop A 73 minutes into the accident, and they shut down the pumps in Loop B at 100 minutes. If the primary cooling system had been full of liquid water, as the operators believed it was, natural convection would have been sufficient to remove the decay heat from the core. If the pumps had been allowed to continue running while vibrating violently, the pump seals could have failed and coolant could have flowed out of the valves onto the containment building floor.

When coolant flow in the primary system ceased, the steam-water mixture separated, and the upper portion of the core was exposed. The temperature of the exposed region increased, first enough for the zirconium-water reaction to take place (1200 °C), and then enough for the core to melt (2850 °C).[7] About 50% of the Zircaloy cladding reacted with steam and disintegrated, and about 45% of the fuel melted.[6] This was a very bad thing indeed, but it should be noted that the melted core remained within the reactor pressure vessel and never made it to China.

At 142 minutes into the accident, the loss of coolant from the pressurizer was noticed, and the block valve above the pressurizer was closed manually. However, the zirconium-water reaction had produced hydrogen gas that was blocking the flow of water, and all efforts to restore flow in the primary cooling system failed until 13.5 hours into the accident. At that time, coolant flow began to be restored in primary system Loop A, and the situation was brought under control.

Hydrogen is produced normally in reactor operation by radiolysis of water: The intense radiation in the reactor core separates water into hydrogen and oxygen. The appearance of excess hydrogen facilitates the recombination of hydrogen and oxygen, so that the amount of free oxygen remains low and the hydrogen poses no threat of explosion. However, some of the hydrogen escaped into the containment building atmosphere through the open electromatic relief valve. When the core became uncovered and the zirconium-water reaction ensued, much larger amounts of hydrogen were released, and some of it formed a bubble at the top of the reactor vessel, above the primary coolant inlet and outlet nozzles, while some of it was released into the containment atmosphere. Nine and a half hours into the accident, the concentration of hydrogen in the containment atmosphere became high enough for ignition to occur, and the hydrogen exploded. However, the peak pressure in the containment building was only 0.19 MPa (28 psi), which was well within the design limits of the building. The hydrogen bubble in the reactor vessel did not explode, because the concentration of oxygen in the bubble was too low to permit combustion.

Overflow through the drain tank relief valve and rupture disk spilled onto the containment building floor and ran down into the containment building sump, which is a sort of collection well below floor level. This coolant was highly contaminated with radionuclides, and some radioactivity made its way from there to the outside environment through the ventilation system. Also, water collecting in the containment building sump was automatically pumped to storage tanks in the auxiliary building. A blown rupture disk on one of these tanks allowed radioactive water to spill onto the floor on the auxiliary building, and some of it leaked from there into the Susquehanna River. The amount of radioactivity released was about 10 MCi, mostly in the form of Xe-133. This radionuclide is chemically inert and passes through the human body without being retained. A substantial portion of the local population evacuated the area voluntarily (about 80,000 of the 200,000 living within 21 miles of the plant), and all pregnant women and preschool children living within five miles of the plant were ordered to evacuate.[8] It has been estimated that 2 million people were exposed to an average of 1.5 mrem, and the maximum possible off-site dose was 83 mrem. The highest estimated dose associated with a particular individual was 37 mrem. A dose of 100 mrem increases one's chance of acquiring cancer by one in 50,000, compared to the one in seven chance from all causes. Therefore, the public health consequences of the accident were negligible; the most serious effect was mental stress on the people living near the reactor.[9]

Following the accident, President Jimmy Carter appointed a commission, led by John Kemeny, to investigate the causes of the accident and to recommend measures to prevent future accidents. The Kemeny Commission found numerous deficiencies in training, oversight, attitudes, and design (especially in the design of human-machine interfaces such as control room layout), emergency planning, and public access to information.[9] Measures to address these deficiencies have been implemented in all operating U. S. reactors. However, the requirement in Generation II PWRs and BWRs for safety systems to be actuated by valves, which have the potential to stick in the wrong position, spurred attempts to design passively safe reactors. Great simplification and enhanced reliability have been introduced in Generation III and III+ designs, as described in Chapter 11, but true passive safety, in which no action is needed at all to prevent core melt, is only offered by Generation III+ reactors and by Generation IV reactors such as the high-temperature gas-cooled reactors discussed in Chapter 16.

CHAPTER 13
THE CHERNOBYL ACCIDENT

Chernobyl Reactor Unit 4 was one of four reactors operating at the Chernobyl Nuclear Power Plant in Ukraine, near the city of Pripyat, on 26 April 1986. At that time, Ukraine was part of the Soviet Union. These reactors were of a type called RBMK, for *reaktor bolshoy moshchnosti kanalniy* (in the Russian Cyrillic alphabet, Реактор Большой Мощности Канальный), meaning "high-power channel-type reactor."[1] The channels are long (7 m) vertically oriented pressure tubes in which the cooling water is allowed to boil. The core of the Chernobyl reactor was 7 m (23 feet) high and 14 m (46 feet) in diameter, and the reactor produced 3000 MW of thermal power (MWt) and 1000 MW of electric power (MWe).[2] At the time of the accident, the core contained 1659 fuel assembles with a total of 200 metric tons of uranium dioxide fuel.

RBMK reactors are graphite-moderated, but water-cooled. Graphite has a very low neutron absorption cross section. The reason for using this combination is that it enables natural (unenriched) uranium to be used as fuel. When the temperature of the graphite moderator increases, the thermal expansion of the graphite reduces the moderation rate per unit volume, so the thermal neutron population decreases and thus so does the fission rate. This effect is stabilizing. However, when the void fraction in the water coolant increases (as by an increase in the boiling rate in a pressure tube), the usual competition between reduced moderation and reduced absorption takes place in the water. (Recall the discussion in Section 6.3.1 on reactivity feedback.) The large amount of graphite creates an overmoderated condition, so increased coolant boiling gives rise to a positive feedback loop, and the reactor is unstable.

The inherent instability of the RBMK design is manageable under most circumstances, but it is an accident waiting to happen. On that awful April day in 1986, it happened.

Before following the sequence of events in the accident itself, we examine the culture in the bureaucracy that administered the Soviet nuclear program. This bureaucracy, along with a very detailed account of the accident, is described from an insider's viewpoint by Grigori Medvedev in *The Truth About Chernobyl*.[3] Most of the account in this chapter is taken from Medvedev's book; the rest comes from References 1 and 2. At the time of the accident, Medvedev was deputy director of the department in the Soviet Ministry of Energy that dealt with the construction of nuclear power plants. Fifteen years earlier, in 1971, after having worked for many years as a shift foreman at a nuclear power plant, he was assigned to the Chernobyl site as a deputy chief engineer for operations at Chernobyl Unit 1. At that time, the first Chernobyl plant was under construction. So Medvedev knew the plant intimately, from the ground up, and he also knew most of the people personally who had positions of responsibility at the Chernobyl plant as well as in the Ministry of Energy. He was also a technically astute nuclear reactor physicist, unlike most of the cast of characters involved in the accident. Medvedev was sent to the accident site on 8 May, twelve days after the accident, to attempt to figure out what had gone wrong and what to do about it.

13.1 SECRECY AND INCOMPETENCE IN THE SOVIET BUREAUCRACY

Appearances are important to nearly everyone, but in the Soviet nuclear bureaucracy they assumed such importance that accidents and incidents were concealed not only from the general public and higher-ups in the bureaucracy but also from colleagues in the nuclear industry. This was actually official policy; for example, on 19 May 1985, Anatoly Ivanovich Mayorets, the national Minister of Energy and Electrification, issued an order stipulating that "Information about the unfavorable ecological impact of energy-related facilities...shall not be reported openly in the press or broadcast on radio or television."[a] In this atmosphere, complacency was natural, and in fact nuclear reactors became regarded as foolproof.

Quite to the contrary, the Soviet nuclear power program had a long series of serious reactor accidents, many of which involved core damage and release of radionuclides into the environment. These accidents continued unabated right up to the time of the Chernobyl accident; the last previous accident listed by Medvedev took place in June 1985 and killed fourteen people. (Medvedev also lists accidents and incidents in the U.S. nuclear program, but in comparison to the Soviet experience most of these

---

[a] After the accident, on 18 July 1986, Mayorets issued an order forbidding his subordinates to tell the truth about Chernobyl in the press or on radio and television. Medvedev reasonably infers that this order was intended to protect Mayorets from losing his job.

incidents were minor, and in any case they were not covered up.) Medvedev himself was the victim of one of those accidents, suffering acute radiation sickness from which he had not fully recovered when he went to Chernobyl in 1971.

The policy of secrecy prevented any focused efforts to address the design and operational flaws that led to the accidents, and the attitude that reactors were foolproof encouraged the assignment of fools to the administration and staffing of the reactors, as well as to the bureaucracy that administered the Soviet nuclear program on a national level. Assignment to nuclear facilities was regarded as prestigious and was sought by bureaucrats seeking advancement in the bureaucratic hierarchy. Without understanding the requirements of nuclear systems, administrators set unrealistic deadlines for construction and operational objectives. Few reactor operating or administrative personnel had any significant background in nuclear technology, and when things started to go wrong at Chernobyl the operators and administrators were oblivious to what was happening and had no idea what to do about it.

The chain of events that culminated in the Chernobyl accident began with the design of a rather routine test by the Chernobyl chief engineer, Nikolai Maksimovich Fomin, an electrical engineer whose background had been in coal-fired power plants.

13.2 THE FLAWED EXPERIMENT

Electricity generated by the turbines in a nuclear power plant is sent to the electrical utility grid, from which power is drawn to operate the electrical facilities in the plant, including the coolant pumps. If external electric power is lost, diesel generators automatically come on to supply electric power to the power plant. But, at least at Chernobyl, it took a full minute for the diesel generators to start up and reach their full pumping power. Power during that minute was to be supplied by the residual angular momentum (see Chapter 2) remaining in the turbogenerator as it spun down.[b]

In the Soviet nuclear power program, it was fairly common to test the ability of the turbine to provide backup electric power as it spun down. However, such tests were always performed with the reactor scrammed by insertion of the SCRAM rods, and with the emergency core cooling system (ECCS) and backup diesel generators available.

---

[b] Medvedev, or his translator, uses the term "the residual inert force" instead of angular momentum. "Residual inert force" is not a term used in physics in English. The phenomenon at play is the angular momentum of the spinning turbine. The term "inertia" would also be correct, but less precise.

On this occasion, however, the test was to be performed with the SCRAM rods withdrawn, the ECCS chained and locked shut, and the backup diesel generators blocked.

At Chernobyl Unit 4, this experiment had been performed three times previously with the safety systems operational; the experiments had been concluded safely but with unsatisfactory results for extracting power from the residual angular momentum. The system voltage of the turbogenerator dropped prematurely because the generator's magnetic field decayed as the rotation slowed, and between successive tests the system was modified to try to keep the magnetic field from decaying. In those three tests, however, the reactor was scrammed at the beginning of the experiment.

Medvedev does not say whose idea it was to perform the test with the safety systems disabled; however, he does say that "numerous" other nuclear power stations had refused to perform such a test because of the risks, but that the Chernobyl administration agreed. The test was to be performed as the reactor was being shut down for maintenance, but before the SCRAM rods had been inserted. The purpose of performing the test before the SCRAM rods had been inserted was to permit the test to be repeated if it failed the first time.

The ostensible reason for disabling the ECCS was that cold water flowing into the hot core could cause thermal shock. However, as Medvedev puts it by quoting a Russian proverb, "There's no point in crying over a man's hair when his head has been chopped off." Disabling the ECCS was a profoundly stupid thing to do. So was the decision to make the diesel generators unavailable. Medvedev says that the reason for blocking the diesel generators was to ensure a "pure experiment." I suppose you could say that they did.

13.3 SOME CONTRIBUTING FEATURES OF THE RBMK REACTOR

Medvedev's book suffers from its lack of any illustrations. Diagrams of the core and the power plant would have helped make his account easier to understand. However, the main points in his account are pretty clear. As noted above, the RBMK reactor has an inherent positive void coefficient of reactivity. To keep void reactivity insertions manageable, operators were required to ensure that at least 28 neutron absorber rods were fully inserted into the regions of the core with the highest "reactivity worth"—i.e., where additional neutrons give the greatest increases in core power. (Since the Chernobyl accident, remaining operational RBMK reactors must have 72 neutron absorber rods in the core.)

The concept of prompt supercriticality is explained in Section 6.1. Prompt super-criticality occurs when the reactivity ρ is greater than β, the delayed neutron fraction.[c] For thermal reactors in which the fissile fuel is U-235, β=0.0065. The 28 or more neutron absorber rods required to be in the Chernobyl core ensured that ρ remained below β when the coolant void fraction in the core increased. In general for thermal reactors, as long as the reactivity remains below β, the time between successive generations of neutrons is greater than about a quarter of a second, and control rod movements can keep up with power fluctuations. In the Chernobyl reactor, SCRAM rods could insert negative reactivity at a rate of 1β per second, which was enough to offset the positive reactivity insertion from voiding under normal conditions. Once a reactor becomes prompt supercritical, however, the neutron generation time shortens to milliseconds, and control rods cannot be inserted fast enough to prevent a runaway power transient.

The channels in the Chernobyl reactor into which the SCRAM rods were insert-ed were filled with coolant water during reactor operation. There were either 205 or 193 SCRAM rods in the withdrawn position above the core before the accident, depending on which account is correct. Insertion of the SCRAM rods displaced the water that was in the channels. This displacement effectively increased the coolant void fraction. The main section of the SCRAM rods, 5 m long, contained a neutron absorbing material, but the top and bottom sections of 1 m each were hollow, with graphite end plugs. Insertion of the all the SCRAM rods at once produced an initial positive reactivity insertion of 0.5β while the graphite and hollow sections of the rods moved into the core. By itself, this reactivity insertion would not be enough to create prompt supercriticality, but in the Chernobyl accident it wasn't by itself.

13.4 THE ACCIDENT SEQUENCE

The experiment was supposed to be performed with the reactor thermal power between 700 and 1000 MW. The test had been scheduled for 2 p.m. on 25 April, and accordingly a gradual reduction of power was begun at 1:06 a.m. The power level had reached 1700 MWt when another power station in the region unexpectedly shut

---

[c] The explanation given by Medvedev is incorrect. He says that reactor control depends on keeping the proportion of "slow" neutrons below 0.5β. I presume that by "slow" neutrons he means thermal neutrons. If so, he is wrong; the key is to keep the reactivity less than β. If he ac-tually means delayed neutrons, the delayed neutron fraction is fixed at β by definition. Perhaps Medvedev's meaning was garbled in translation.

down, and the Kiev grid controller asked the Chernobyl crew to delay further power reductions.

At the lower power level, fission product poisons (most importantly Xe-135—see Section 6.3.3) began to accumulate. When the grid controller told the Chernobyl crew that they could proceed with further power reductions, it was 11:04 p.m. The power had been reduced to 700 MW by five seconds after midnight on 26 April. However, the built-up fission-product poisons caused the power to continue to fall. Complicating matters was the shift change that took place at midnight. The night crew had not been briefed extensively on the experiment, which was to have been completed long before they arrived.

Besides the power reduction caused by the fission-product poisoning, the senior reactor control engineer on the new shift, Leonid Toptunov, made an error when the power reached 500 MWt and inserted the control rods too far. These combined effects reduced the power to 30 MWt. The test could not be completed unless the power was increased again to the range between 700 and 1000 MWt. This required withdrawing control rods. Toptunov refused to do so, but the deputy chief engineer in charge of the experiment, Anatoly Dyatlov, insisted.

By 1 a.m., the power had been raised to 200 MWt, but, according to Toptunov's deathbed testimony, only 18 control rods remained in the core instead of the required 28. The Soviet government's report to the International Atomic Energy Agency (IAEA) stated that the remaining number of rods was only between 6 and 8. In either case, the reactor's operating procedures required an immediate shutdown of the reactor. However, Dyatlov insisted on going ahead with the experiment.

Each reactor in the Chernobyl plant had two turbogenerators, and the ones attached to Unit No. 4 were numbered 7 and 8. The number 7 turbogenerator had already been shut down, and at 1:23:04 a.m., the number 8 turbogenerator was shut down to initiate the experiment. As this turbogenerator spun down, it ceased to remove energy from the coolant, so the coolant in the main circulation pumps began to steam up and the coolant flow rate through the core began to decline. This reduced flow rate caused the coolant to spend more time in the core and thus to boil more. Because of the positive void coefficient of the reactor, this boiling caused the reactor power to increase slowly.

Toptunov noticed the power rise and told the shift foreman, Aleksandr Akimov, that they had to SCRAM the reactor. At 1:23:40 a.m., Akimov pushed the SCRAM buttons.[d]

---

[d] Akimov actually pressed two buttons. The first, called the MPA button, should have been connected to the ECCS, the emergency feedwater pumps, and the standby diesel generators.

The voiding in the core had already produced a positive reactivity insertion of more than 0.5β. Insertion of the SCRAM rods added an additional 0.5β, making the reactor prompt supercritical. To lower the SCRAM rods mechanically would have taken 18 seconds, so the prompt supercritical transient would have lasted far too long in any case. However, the rapidly increasing power caused the channels for the SCRAM rods to overheat and warp, and the SCRAM rods became stuck after moving only about one-third of the way into the core, locking in the prompt supercritical state. Akimov switched off the electric current to the SCRAM rod servo-drives so that the rods could fall by gravity, but he was too late. The rods remained stuck before the absorber portions had fully entered the core.

From the moment the SCRAM was initiated, the fate of the reactor was sealed. But the explosion that destroyed the reactor took another 20 seconds to develop.

There were actually at least three explosions. The first was a steam explosion that blew apart the reactor's pressure relief valves (which were not designed to cope with pressures of the magnitude of those in the accident—approximately 300 atmospheres) and the lower water and upper steamwater lines. Destruction of these components led to even more loss of coolant from the core and further positive reactivity from the additional voiding. The next two explosions were hydrogen explosions. As the temperature in the core increased, the steam reacted with the zirconium cladding of the fuel rods and formed hydrogen and zirconium oxide. The hydrogen accumulated in the core and also passed into a watertight compartment beneath the reactor vault, mixing with the air there.

Hydrogen and oxygen don't like to be separated, so when the temperature of the mixture got high enough, they recombined explosively. The explosions vaporized about 50 tons of fuel and sent it more than 40,000 feet into the atmosphere, and they blew out another 70 tons of fuel and 700 tons of radioactive graphite, which landed in the vicinity of the Chernobyl site. Despite heroic efforts by firefighters, the remaining 80 tons of fuel and 800 tons of graphite burned completely during the next few days. The fire could not be extinguished because it was not sustained by the heat of combustion but by the decay heat from radioactive fission products.

When he was informed of the explosions shortly after they occurred, Viktor Petrovich Bryukhanov, the Chernobyl plant director, refused to believe that the

---

However, all of these emergency protection systems had been bypassed, and the MPA signal went only to the secondary electrical circuits. When pressing the MPA button did nothing, Akimov pressed the other emergency button, called the AZ button, which activated the SCRAM rods and also inserted all of the withdrawn control rods.

reactor had been destroyed. He directed Anatoly Andreyevich Sitnikov, the deputy chief operational engineer of construction phase 1, to go to the reactor personally and take a look. Sitnikov toured the reactor unit and climbed up to the roof of the chemical water treatment plant building to see the damage, thus receiving a fatal dose of radiation. Despite Sitnikov's reports of the devastation to them, Bryukhanov and Fomin refused to believe him and ordered firefighters to continue trying to extinguish the blaze. Nevertheless, Bryukhanov requested permission from headquarters in Moscow to evacuate Pripyat. Boris Yevdokimovich Shcherbina, the deputy chairman of the Council of Ministers, refused permission, saying "Don't start a panic! There must be no evacuation until the government commission gets there!"

That commission had been hastily assembled after news of the accident had reached Moscow, and it departed from Moscow at 9 a.m. on the morning of the accident. A second group followed at 4 p.m. These groups comprised high-ranking officials and technical experts. Members of the first group observed the devastation by helicopter by mid-afternoon, finally being forced to admit that the reactor had been destroyed. Nevertheless, Shcherbina, who arrived in Pripyat at 9 p.m. with the second group from Moscow, continued to stall on the obviously necessary evacuation. Some people from Pripyat had departed on their own judgment, but the official evacuation didn't begin until the next morning. Meanwhile, almost all the children in Pripyat had gone to school on the 26th, exposing themselves unnecessarily to radiation all day. The evacuation took three days to complete, taking the citizens away on a huge fleet of buses assembled for the occasion. All the livestock and pets in Pripyat were hunted down and shot to prevent them from spreading radioactive contamination.

13.5 THE AFTERMATH

The Chernobyl reactor explosions released four hundred times more radioactive fallout than had been released by the atomic bomb that was dropped on Hiroshima. About 60% of this landed in Belarus, but western Russia, Ukraine, and eastern, western, and northern Europe were also contaminated. About 336,000 people in Ukraine, Russia, and Belarus were eventually evacuated and relocated. Radioactive rain fell as far away as Ireland. The immediate area around the plant (the Chernobyl Exclusion Zone) remains uninhabited, but most of the affected areas are now considered safe for occupation.

With the huge amount of radioactive fallout released and the large human population exposed, it is remarkable how limited the impact of the accident on human health has been. An assessment was published in 2005 by the Chernobyl Forum, led by the IAEA and the World Health Organisation (WHO) of the United Nations.[4] This report attributes 47 deaths among plant workers and firefighters to acute radiation sickness caused by the accident. Also, among about 5000 cases of thyroid cancer believed to have been caused by iodine-131 released by the accident, nine cases, in children, had been fatal by the time the report was issued. Thyroid cancer is a highly treatable cancer, however, and most cases are cured. The report projected that about 4000 people will eventually die from various cancers caused by the Chernobyl accident. This impact will be superimposed on the approximately 100,000 cancers that would probably occur even without the Chernobyl accident. Rumors of large numbers of deaths and diseases from Chernobyl are regarded by the report as highly exaggerated, resulting from a human tendency to blame the Chernobyl accident for any death or disease in the affected area. The objective epidemiological studies cited in the report give the much smaller numbers noted above.

It should be noted that the predictions of cancer deaths from these radiation exposures are based on the Linear No-Threshold Hypothesis, which has recently been proven to be excessively conservative, as mentioned in Chapter 4.

The Chernobyl Exclusion Zone, a region of 488.7 km^2 (190.9 square miles, equivalent to a square 13.8 miles on a side) around the reactor site, has become a haven for wildlife, and several rare species have either moved into it on their own or been reintroduced. However, some adverse health effects have been noted, particularly among barn swallows, which return to the site in a depleted state of vigor after a long migration.

After the accident, the three remaining operating reactors at the Chernobyl site were shut down permanently, and the two units under construction were cancelled. Seven other RBMK reactors under construction or planned in the Soviet Union were cancelled, and one other operational reactor was shut down permanently. However, as of March 2014, eleven RBMKs are still running, all of them in Russia.[5] To prevent another event like the Chernobyl accident, design features and operating procedures in those plants have been modified. The most important change was the redesign of the SCRAM rods to eliminate the graphite tips and hollow sections. If this design flaw had not existed, the partial SCRAM might have shut the reactor down. The number of manual control rods was increased from 30 to 45, and 80 additional absorber rods are permanently placed in the core to reduce positive void reactivity at low power.[e] To compensate for the extra neutron absorbers and the control rod modifications, the

---

[e] As noted above, Medvedev says the total number is now 72. The figure of 80 additional rods comes from Reference 1.

fuel was enriched to 2.4% U-235. The time for mechanical insertion of the SCRAM rods was reduced from 18 to 12 seconds. Finally, precautions have been taken to prevent unauthorized access to the emergency safety systems that were bypassed for the experiment that went wrong.

The Chernobyl accident caused an immediate and long-lasting antipathy to nuclear power worldwide. Some European nations, such as Germany, legislated bans against new nuclear power plants and formulated plans to shut down their existing plants. However, economic realities and the limitations of natural resources have intruded into the consciousness of people and their governments, and such policies are now being reconsidered. Other reactor types do not share the instability of the RBMK reactors, and, as shown in Chapter 8, other energy resources, when examined closely, are not superior to nuclear power in safety, reliability, and environmental impact.

CHAPTER 14

THE FUKUSHIMA DAIICHI ACCIDENTS

On Friday, 11 March 2011, at 0546 UTC[a] (2:46 p.m. Japan Standard Time), the 2011 Tohoku-oki Earthquake (also called the Great East Japan Earthquake and the 311 Earthquake) struck at a point 70 km (43 mi) east of the Oshika Peninsula of the Tōhoku region of Japan at an undersea depth of 32 km (20 mi).[1, b] The earthquake unleashed a powerful tsunami. The combined effects of the earthquake and the tsunami killed 15,854 people and injured another 26,992 as of 12 March 2012, at which time another 3155 remained missing. The World Bank estimated the total economic loss at US$235 billion, the greatest in history from a natural event. The earthquake and tsunami also caused severe damage to three of the six nuclear reactors in the Fukushima Daiichi Nuclear Power Station, with far-reaching local and global consequences.

As I write this, more than eight years have passed since the event, but knowledge about the accident and its consequences is still being gathered, and conditions in and around the power plant complex are still changing. It will be many years before the story comes to a close, but in this chapter I summarize what is known now and what is expected in the future, both at Fukushima Daiichi and for the global nuclear industry.

Most of what I know about the Fukushima Daiichi accidents was gleaned from References 2 and 3. To avoid peppering the text with references to these sources, I

---

[a] UTC means Universal Coordinated Time. It is also known as Greenwich Mean Time (GMT), and in aviation it is called Zulu time, denoted by the letter Z as in 0546Z. Zulu is the word used for the letter Z in the international phonetic alphabet.

[b] The point within the Earth at which an earthquake fault begins to rupture is called the hypocenter. The point on the Earth's surface above the hypocenter is called the epicenter.

will not flag specific assertions taken from them again. Unreferenced assertions in the following text may be assumed to be derived from these sources.

14.1 THE EARTHQUAKE AND TSUNAMI

The Tohoku-oki earthquake was a type of event called a megathrust earthquake, which is an extremely large displacement at a thrust fault, where two tectonic plates push against each other and one is forced underneath the other.[4]

There are several ways to measure the intensity of an earthquake. The Richter scale[5] was developed in the 1930s and is based on the amplitude of waves measured by seismographs. This scale was supplanted in the 1970s by the moment magnitude scale,[6] which is based on the energy released in the earthquake. The moment magnitude is given by

$$M_w = (log_{10}M_o - 9.1)/1.5,$$

based on the seismic moment $M_o = \mu A D$, where μ is the shear rigidity of the rock around the fault, A is the area of the rupture, and D is the average relative displacement of the two sides of the fault. In the formula above for the moment magnitude, M_o must be expressed in joules. When it is expressed in other energy units, the constants in the formula will be different. The moment magnitude scale is still referred to as the Richter scale in the press. By the moment magnitude measure, the Tohoku-oki earthquake was rated at 9.0, which makes it the fourth largest earthquake recorded since 1900, when seismographic data-keeping began.[7]

Yet another measure is the value of the displacement at the fault. Displacements at the Tohoku-oki fault were as much as 60-80 m, the largest recorded for any earthquake in history.[7]

The last measure we shall consider is the seismic moment rate function. This is the rate of change of the seismic moment M_o with time, and it is roughly proportional to the power (energy per unit time) released in the earthquake. The seismic moment rate function for the Tohoku-oki earthquake was over 10^{21} N-m/s, more than 20% greater than any other earthquake ever recorded.[7]

The most relevant measure of the ability of an earthquake to do damage at a specific location is the ground acceleration at that location. Structures on the ground have to move with the ground if they are not to be broken loose from their foundations, and that means that the force on them is proportional to the ground acceleration (i.e.,

$F=ma$, cf. Chapter 2). The maximum ground accelerations at the Fukushima Daiichi site were 0.45g, 0.56g, 0.52g, 0.45g, 0.56g, and 0.46g for reactors 1-6, respectively. The design-basis accidents for the six reactors ranged from 0.42g to 0.46g.

By any measure, the Tohoku-oki earthquake was one of the most powerful earthquakes ever recorded, and by some measures it was the most powerful.

The earthquake unleashed a huge tsunami,[8] which reached the Fukushima Daiichi Nuclear Power Station about 50 minutes after the earthquake. *Tsunami* is a Japanese word meaning "harbor wave," which is a good description because tsunamis have long wavelengths and do not cause abrupt disturbances to ships at sea; instead, they cause problems when they arrive on shore. They are not breaking waves like ordinary surf; they begin like a rising tide, and are therefore also called tidal waves, even though they are not related to tides, which are caused by the moon's gravitational force on the water in the oceans. Tsunamis are caused by the displacement of large volumes of water, often by undersea earthquakes but also by landslides, undersea volcanoes, meteorite impacts and even glacial calvings. The connection between tsunamis and undersea earthquakes was first suggested by Thucydides in 426 B.C.

Large tsunamis have been the greatest natural disasters in history. There were 28 tsunamis between 1993 and 2012,[9] including the great Indian Ocean tsunami in 2004 that killed over 230,000 people in 14 countries. The tsunami launched by the Tohoku-oki earthquake reached estimated heights up to 40.5 m (133 ft), at Miyako in Tōhoku's Iwate prefecture, and traveled as much as 10 km (6 mi) inland. The greatest previously recorded height in Japan was 38.2 m in 1896.[1] At the Fukushima Daiichi reactor site, the height of the tsunami was 13-15 m (43-49 ft); the seawall that was in place to block tsunamis was only 5.7 m (19 ft) high.

14.2 THE REACTORS IN THE LINE OF FIRE

The Fukushima Daiichi Nuclear Power Station, which I abbreviate henceforth as Fukushima Daiichi, is one of five such complexes on the northeast coast of Honshu, Japan's largest island. A second complex is also located within the Fukushima prefecture; this one is called the Fukushima Daini Nuclear Power Station. The Japanese word *daiichi* (pronounced *dye-ee-chee*) means number one, or first,[10] and *daini* (pronounced *dye-nee*) means number two, or second. All five of these complexes were affected to some degree by the earthquake, but only Fukushima Daiichi suffered significant damage. Three of these complexes have more than one reactor, distinguished by unit number.

Recall from Chapter 2 and Section 7.1 that a heat engine, such as the coolant in a nuclear reactor cooling system, cannot convert all the heat supplied by a heat source, such as the reactor core, into work. Some heat must be rejected into a heat sink. Since these Japanese complexes are all located on the seacoast, the ocean is used as the heat sink. As shown in Figure 14.1,[11] when the reactor is operating, pumps supply water from the heat sink (a "river" in the figure) to a condenser unit where reactor cooling water from the steam turbine is cooled and condensed into liquid before being pumped by the primary coolant pumps back into the reactor pressure vessel. When the reactor is shut down, the turbine loop is bypassed and the cooling water is pumped through a residual heat removal system, which also dumps heat into the heat sink. Even if the reactor's primary cooling system is intact, the residual heat removal pumps must continue to operate in order to remove decay heat to its ultimate destination. Water from the heat sink must also be pumped to cool backup diesel generators. Electricity is required to operate all of these pumps.

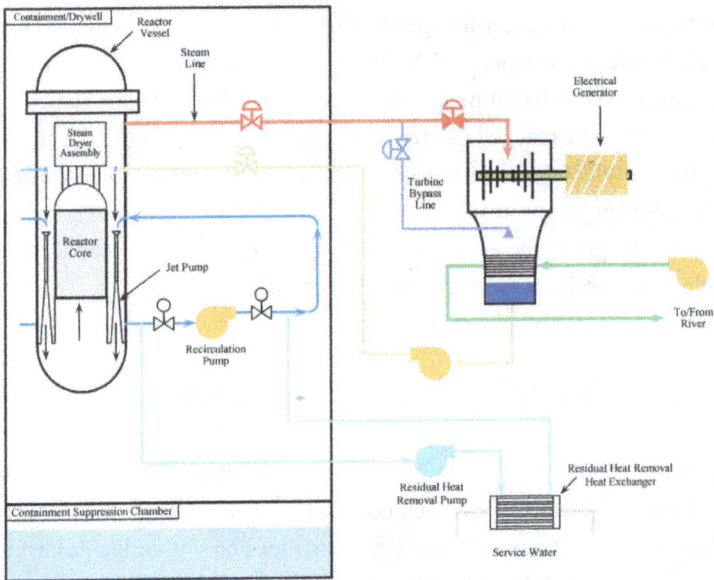

Figure 14.1—Schematic diagram of a typical BWR, showing the heat exchanger in the condenser and the residual heat removal system.[11] This figure was taken from a work of the U. S. government and is in the public domain under the terms of Title 17, Chapter 1, Section 105 of the U.S. Code.

All of the reactors in these complexes are General Electric BWRs. The Fukushima Daiichi reactors are the oldest. Their construction began between July 1967 (Unit 1) and October 1973 (Unit 6), and they began commercial operation between March 1971 (Unit 1) and October 1979 (Unit 6).[12] All the units at Fukushima Daiichi were built by the Japanese construction company Kajima;[13] they are owned and operated by Tokyo Electric Power Company (TEPCO). As noted in Section 11.1, Generation II BWRs progressed through six evolutionary designs, denoted as BWR/1 through BWR/6, and three containment designs, denoted as Mark I, Mark II, and Mark III. Fukushima Daiichi Unit I is a BWR/3 reactor; all the other reactors in these power stations are BWR/4 or BWR/5 reactors. Fukushima Daiichi Units 1-5 have Mark I containment systems, with toroidal wetwells as shown in Figure 11.7; all but one of the other units in the five nuclear power stations have Mark II containment systems. The Mark II containment has its wetwell below the drywell. The three containment systems are shown in Figure 14.2.

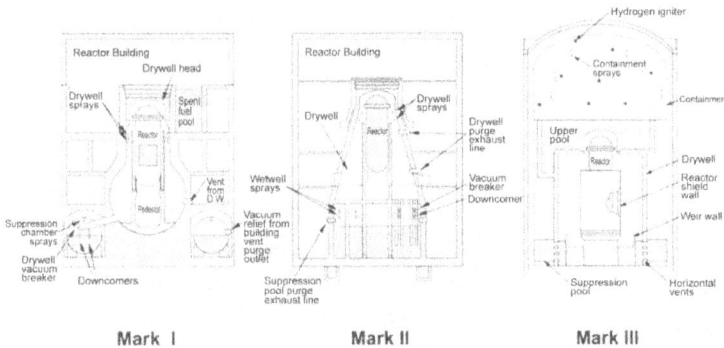

Figure 14.2—The three containment designs in Generation II BWRs. From Reference 3; reproduced by courtesy of the American Nuclear Society

The reactors have emergency core cooling systems as described in Section 11.1, but since they are Generation II reactors, which do not have passive safety, electric power is needed to operate their safety systems. Power from the operating reactors' turbogenerators was lost upon automatic reactor shutdown (SCRAM), and external power was lost in some cases because of the earthquake. Backup emergency diesel generators came on-line when external power was lost, but the tsunami disabled the diesel generators at Fukushima Daiichi and some of the other power stations. The reactors also have backup DC batteries to provide power for a short time if all other sources of electricity are lost. More detail about each of the power stations is given below.

Each of these reactor systems has a spent fuel pool in the reactor building outside the containment structure. Fuel is placed in the spent fuel pool during refueling shut-downs as well as after final removal from the core to wait as it cools enough to be moved to external storage or disposal. The fuel in the spent fuel pool generates heat by radioactive decay just as the post-shutdown core does, and it needs to be immersed in the pool to be kept cool; the water in the pool must be circulated through a heat removal system to keep it from boiling away, and makeup water must be provided to replace water that evaporates. The power plant complex also has a common spent fuel pool, to which fuel from any of the reactors may be sent.

The northernmost of the five power stations is the Higashidori Nuclear Power Station, near the north end of Honshu. Higashidori has only one reactor, a BWR/5 with Mark II containment, rated at 1100 MWe. At the time of the earthquake, this reactor was in a cold shutdown state for periodic inspection, and all of the fuel had been removed and placed in the spent fuel pool. One of the two backup emergency diesel generators was offline for inspection, but the other was available. The earthquake caused the loss of the off-site power supply, but the available diesel generator came on and provided cooling for the spent fuel pool. The integrity of the spent fuel pool was not compromised, and the reactor continued in its state of cold shutdown.

The next reactor complex, going south from Higashidori, is the Onagawa Nuclear Power Station. Onagawa has three reactors; Unit 1 is a BWR/4 with Mark I containment, rated at 524 MWe, and Units 2 and 3 are BWR/5s with Mark II containment, rated at 825 MWe. Units 1 and 3 were operating, and Unit 2 was going through startup procedures. Onagawa is the complex closest to the earthquake epicenter, but one of five offsite power lines survived the earthquake and some offsite power remained available. Unit 1 tripped from high seismic acceleration and the two emergency diesel generators started automatically, successfully maintaining electric power to the emergency core cooling systems. Cold shutdown was achieved the next day. In Unit 2, startup operations were automatically shut off because of high seismic acceleration, and the three emergency diesel generators started automatically, but they were kept in standby condition because offsite power had not been lost. The tsunami flooded and disabled two of the diesel generators, but the third remained in standby and cold shutdown was achieved within the day using offsite power. Unit 3 tripped from high seismic acceleration, and offsite power was lost when the tsunami arrived. However, cold shutdown was achieved the next day; it is unclear to me whether that was due to the operation of the emergency diesel generators or use of the single remaining offsite power line into the entire nuclear power station. The Onagawa Nuclear Power Station is set on a level 14 m above sea level, which enabled it to avoid most of the damage from the tsunami that affected Fukushima Daiichi. In fact, the gym at the Onagawa station served for three months as a shelter for residents in the nearby area whose homes were destroyed by the earthquake and tsunami.[14]

Next in line from north to south is Fukushima Daiichi. The Fukushima Daiichi Nuclear Power Station has six reactors. Unit 1 is a BWR/3 with Mark I containment, rated at 460 MWe. It was operating at full power at the time of the earthquake. Units 2-5 are BWR/4s with Mark I containment, rated at 784 MWe.[15] Units 2 and 3 were operating at full power when the earthquake struck, but Unit 4 was in cold shutdown with the fuel removed to the spent fuel storage pool, and Unit 5 was in cold shutdown but the fuel was in the reactor. Unit 6 is a BWR/5 with Mark II containment, rated at 1000 MWe. It, too, was in cold shutdown with fuel in the core. Units 5 and 6 are at a higher elevation than Units 1-4, so the tsunami did less damage there. Units 1-3 shut down automatically when they registered high seismic acceleration. The earthquake cut off the offsite power to all six units, and the emergency diesel generators came on. Backup power from the diesel generators was operating the coolant pumps until the tsunami arrived, but the seawater retaining wall, which was only 5.7 m high, did little to restrain the 15 m tsunami. The wave flooded the electrical switchgear and the cooling systems for all the emergency diesel generators except one generator at Unit 6. Unlike the others, which were water-cooled, that generator is air-cooled, and it was not dependent on the flooded cooling systems; also, its higher location helped protect it. All the rest of the diesel generators were disabled. When the tsunami struck, Units 5 and 6 temporarily lost the cooling provided by their residual heat removal systems, and the radioactive decay heat in their cores started to raise their core temperatures, but backup diesel power from the remaining generator at Unit 6 was used to restore operation of the residual heat removal systems in both Units 5 and 6. These units were returned to cold shutdown by the next day. The course of events in Units 1-4 after loss of backup diesel power is discussed at length below.

The next nuclear power station, less than 100 miles south of Fukushima Daiichi, is Fukushima Daini. Fukushima Daini has four BWR/5 reactors with Mark II containment, each rated at 1100 MWe. They were all operating at the time of the earthquake, and they all shut down automatically upon high seismic acceleration. One of three offsite power transmission lines remained connected. The tsunami flooded the seawater pumps that supplied seawater to the condensers in Units 1, 2, and 4, rendering them inoperable and depriving the reactors of their heat sinks. In unit 3, one of the pumps remained in service, and the reactor was brought to cold shutdown the day after the earthquake. In Units 1, 2, and 4, efforts by Fukushima Daini operating staff that Reference 3 terms "extraordinary" succeeded in replacing damaged pump motors and electrical cables, restoring decay heat removal. Units 1 and 2 were brought to cold shutdown on 14 March, and Unit 4 was brought to cold shutdown on 15 March.

The southernmost of the five nuclear power stations is Tokai Daini, which formerly had two reactor units, but Tokai Daini Unit 1 was closed in 1998. Unit 2 is a BWR/5 with Mark II containment, rated at 1100 MWe. It was operating at the time of the earthquake, but it shut down automatically when the earthquake caused a large vibration signal in the turbine system. Offsite power was lost, and the three emergency diesel generators started automatically. The tsunami disabled one of the diesel generators, but the other two continued to operate. Offsite power was restored on 13 March, and the reactor was brought to cold shutdown on 15 March.

14.3 THE ACCIDENT SEQUENCES AT FUKUSHIMA DAIICHI UNITS 1-4

14.3.1 Unit 1

Fukushima Daiichi Unit 1 was shut down automatically because of high seismic acceleration. Unit 1 has an isolation condenser system, as described in Section 11.1.2, for removing heat from a post-accident primary cooling system. This system at Fukushima Daiichi has two "trains," and it was working prior to the tsunami, so well, in fact, that one train of the system was shut down and the other was operated intermittently. However, power is needed to operate the valves and other components in the system, and after the tsunami struck at about 3:30 p.m. Japan Standard Time on 11 March, the system had only limited effect, if any. The tsunami disabled all the instrumentation needed to monitor and control the reactor safety systems, and it also shorted out the DC batteries. None of the emergency core cooling systems, except possibly partial function of the isolation condenser system, was available without a source of electric power. Therefore, the core began to heat up, and the primary cooling system pressure would have increased if the primary cooling system pressure boundary had been intact. However, by 2:45 a.m. on 12 March, the pressure in the reactor pressure vessel was found to be low. Some damage must have caused the loss of pressure, but it is uncertain whether this damage was a leak in the primary cooling system piping caused by the earthquake, safety relief valves stuck open because of the earthquake, or a breach in the pressure vessel from the overheated core, which had probably already partially melted by that time. However, workers who had been in the reactor building at the time of the earthquake reported that some piping had burst in the reactor building as a result of the violent shaking induced by the earthquake. The pressure in the containment vessel around the reactor pressure vessel had already risen beyond its design pressure limit by 12:45 a.m. on 12 March. This was an indication that the primary cooling system had been breached and hydrogen from the steam-zirconium reaction (cf. Section 11.1.1.1) had been released as the water level in the reactor pressure vessel fell below the top of the core.

The reduced pressure in the reactor pressure vessel enabled operators to inject water into the pressure vessel using fire engines starting at 5:46 a.m. on 12 March, but they were unable to fill the pressure vessel past the core midplane, indicating a breach in the primary cooling system at that level. Attempts were begun to vent the containment vessel to reduce its pressure, but high radiation levels in the reactor building frustrated these efforts. High radiation levels in the reactor building were probably due either to failures in the containment boundary caused by the excessive pressure in the containment vessel or to the opening of a vent rupture disk; the breached primary cooling system created a path for radionuclides to escape from the reactor to the space in the containment vessel around the reactor pressure vessel, and these radionuclides went on into the reactor building through the leaks in the containment boundary or the rupture disk. By about 2:30 p.m. on 12 March, the pressure in the containment vessel had decreased even without action from the crew, confirming that the containment boundary was leaking somewhere. Hydrogen built up in the reactor building, and at 3:36 p.m. on 12 March a hydrogen explosion occurred in the upper part of the reactor building, outside of the containment vessel. This explosion blew out the sides of the upper part of the reactor building, exposing the spent fuel pool to the atmosphere. Subsequently, the roof collapsed, covering the floor and some machinery with debris. Some debris may have fallen into the spent fuel pool.

At 8:20 p.m. on 12 March, at the order of the Japanese government, TEPCO began injecting seawater into the primary cooling system with fire trucks. This measure allowed more water to be injected, but it had been resisted because seawater would have ruined the core if it had not already been damaged by the zirconium-steam reaction and/or melting. Also, the salt in seawater may lead to the formation of more mobile forms of uranium, allowing greater dispersal of uranium into the environment. Seawater injection continued until 25 March, when freshwater injection resumed. Electric power began to be restored to Unit 1 by 24 March, and over the next few months the reactor was gradually brought to a state of cold shutdown (by 21 August, all temperature sensors registered less than 100 °C). Cleanup of the contaminated air and water in the reactor building and containment volume began on 28 March. On 28 October, a cover over the damaged reactor building was completed.

TEPCO staff estimates that the core began to melt within six hours of the earthquake, and that some of the molten core fell onto the bottom of the reactor pressure vessel within 16 hours. This molten fuel melted its way through the pressure vessel and fell onto the concrete floor of the containment vessel. TEPCO estimates that no more than 70 cm of this concrete have been eroded, out of the 7.6 m thickness of the floor.

Loss of electric power disabled the cooling system for the spent fuel pool, but there is enough water in the spent fuel pools to last for a period from several days to a couple of weeks, depending on how much fuel is in a pool and how long it has been cooling. Starting on March 31, a concrete pumping truck was used to pump water into the Unit 1 spent fuel pool. No damage occurred to the spent fuel in the hydrogen explosion, and, according to a report submitted by the Japanese government to the IAEA, water was restored to the spent fuel pool while the fuel was still fully immersed in water.[16] Therefore, no radionuclides were released from the Unit 1 spent fuel pool.

Radionuclides were released from the breached containment structure into the atmosphere, and seawater passing through the core to cool it was vented into the ocean, carrying radionuclides with it. Assessments of total radionuclide releases from the accidents are given in Section 14.4 below.

14.3.2 Unit 2

High seismic acceleration triggered an automatic SCRAM in Unit 2 at the time of the earthquake, and the two diesel generators came on to replace lost offsite power. Until the tsunami arrived, core cooling was provided primarily by the high-pressure coolant injection system and the residual heat removal system. The tsunami disabled all the diesel generators and the backup DC batteries, but the reactor core isolation cooling system (RCIC) was activated manually at 3:39 p.m., about nine minutes after the tsunami. The RCIC should not generally be expected to operate for more than about 8 hours, but it operated until 1:25 p.m. on 14 March, about 70 hours after it began operation.

At 11:53 a.m. on 14 March, a hydrogen explosion occurred in Unit 3, which blew the upper section of the Unit 3 reactor building apart, as had happened earlier in Unit 1. The force of this explosion damaged the walls in the Unit 2 reactor building.

The RCIC stopped working because of falling water level in the reactor, and operators began to reduce the primary cooling system pressure in order to be able to inject water with fire trucks. However, this pressure reduction required the operation of safety relief valves by the use of electricity or pressurized nitrogen, and the depressurization process was difficult. During efforts to depressurize the system, the core was fully exposed twice, for a total of about 6.5 hours, and substantial core melting and hydrogen production resulted. Injection of water into the reactor pressure vessel was not established until 4:11 a.m. on 15 March. The water level in the reactor pressure vessel could not be raised above core midplane, evidently because of a breach at that level in the primary cooling system.

As explained in Section 11.1.1.2, the safety relief valves vent steam into the pressure suppression pool in the toroidal wetwell when the pressure in the reactor pressure vessel exceeds a set limit. (They are forced open, without needing electricity or pressurized nitrogen, by the difference between the pressure in the reactor pressure vessel and that in the wetwell.) The wetwell is contained within the containment vessel and is at the same pressure. The pressure in the containment vessel did not rise as much as it should have from the deposition of hot steam into the wetwell from the reactor pressure vessel. This anomaly indicates that the containment boundary had been damaged and was leaking. At 6:14 a.m. on 15 March, a sound originally thought to be a hydrogen explosion was heard in the vicinity of the pressure suppression pool. Later reviews cast uncertainty on the nature of the sound, but in any case the pressure in the containment vessel dropped abruptly to near-atmospheric pressure and remained there. Whatever the source of the sound, it did not damage the fuel in the spent fuel pool, nor did the damage to the reactor building caused by the hydrogen explosion in Unit 3.

Seawater was used for core cooling from 15 March until 26 March. On 26 March, partial offsite electric power was restored, and fresh water was used to cool the core from that time. On 27 March, the temperature at the bottom of the pressure vessel fell below the boiling point, although it subsequently rose again. On 14 September this temperature was 114.4 °C, but by 12 February 2012 it was 78.3 °C (water boils at 100 °C at standard sea level atmospheric pressure).

Over the months after the accident, conditions in Unit 2 gradually stabilized, and the water injection techniques available for reactor cooling improved. As of 2 July 2011, the cooling water was routed through the site's waste water treatment plant. The spent fuel pool was cooled by seawater from 20 March until 29 March, at which time freshwater injection began. On 31 May, a circulatory cooling system for the Unit 2 spent fuel pool was put in service. Since September 2011, the temperature in the Unit 2 spent fuel pool has remained below 35 °C.

14.3.3 Unit 3

Unit 3 was also shut down automatically from high seismic acceleration. In Unit 3, the DC batteries were not disabled, so electric power was available until the batteries became fully discharged after about 15 hours. This power was used to operate the high-pressure coolant injection (HPCI) system. The operation of the HPCI system reduced the pressure in the reactor pressure vessel, so that the RCIC system, which uses high-pressure steam, only operated for about 20 hours instead of 70 as in Unit 2. After the HPCI system ran out of power and the RCIC system stopped working, the buildup of heat in the reactor pressure vessel caused pressure there to rise again. In order to inject water at low pressure using fire trucks, operators needed to reduce the pressure, but the same issues with electric power or pressurized nitrogen for the safety relief valves that affected Unit 2 delayed the start of depressurization for 7 hours, during which time the top three-fourths of the core became exposed and significant melting and hydrogen production occurred. Fire engines began injecting fresh water (containing boron to prevent recriticality) at about 9:25 a.m. on 13 March; this injection was switched to seawater at 1:12 p.m. The supply of seawater in the reserve pool from which it was drawn ran out at 1:10 a.m. on 14 March, and the seawater injection was interrupted until 3:20 a.m. The water level in the reactor pressure vessel never recovered beyond the level of the core midplane, probably because either the reactor pressure vessel or piping in the primary cooling system had been breached. Most of the water pumped into the pressure vessel leaked out of it, but after 2 July 2011 water pumped through the reactor has been treated at the site's water treatment plant. On 11 January 2012, 300 m^3 of water contaminated with radionuclides was found in two underground tunnels, one of which is near Unit 3 (the other tunnel, containing less water, is near Unit 1).

Hydrogen from the steam-zirconium reaction on the exposed fuel rods escaped into the containment vessel. It also entered the reactor building outside the containment vessel, indicating a breach in the containment boundary. As in Unit 2, the pressure in the containment vessel rose more slowly than expected from the amount of heat being deposited in the pressure suppression pool, which also indicated a breached containment boundary. The hydrogen in the reactor building exploded at 11:15 a.m. on 14 March. This was a much more violent explosion than the one in Unit 1, not only blowing the upper section of the reactor building apart, but also, as noted above, flinging debris through the walls of the Unit 2 reactor building. The blast was felt as far as 40 km away. Nevertheless, although damage to the Unit 3 spent fuel pool was feared at first, it appears that little if any damage actually occurred in the spent fuel. Beginning on 17 March, makeup water was supplied to the spent fuel pool by water cannons and by dropping water from helicopters. On 27 March, this task was taken over by a concrete pump. Supply of water to the spent fuel pool by the normal spent fuel pool piping resumed in late April. In June, the water in the spent fuel pool was found to be highly alkaline (with a pH value[c, 17] of 11.2); on 25-26 June, 90 tons of water containing boric acid were pumped into the spent fuel pool to reduce its alkalinity and to prevent criticality from occurring if the racks holding the spent fuel should collapse and allow the fuel assemblies to fall together. Since September 2011, the water in the spent fuel pool has remained below 35 °C.

14.3.4 Unit 4

Because the neutron flux in a reactor core is not uniform, the core is loaded so that fresher fuel is located in regions of lower neutron flux and older fuel is located in regions of higher neutron flux. This loading arrangement has the effect of evening out the power distribution in the reactor. In order to achieve such an arrangement, crews must remove all the fuel from the core during refueling outages. When the earthquake and tsunami struck, Unit 4 had been defueled, all of the fuel from the reactor was in the spent fuel pool, and it had been placed there more recently than the fuel from Units 1-3 had been placed in their spent fuel pools. Therefore, the reactor itself posed no threats, but the spent fuel in Unit 4 was generating more decay heat than the spent fuel in the other reactors' spent fuel pools.

Early in the morning of 15 March, an explosion occurred in Unit 4 that severely damaged the roof and walls of the Unit 4 reactor building and the refueling floor containment structure. This explosion was originally assumed to have been caused by the zirconium-steam reaction in the Unit 4 spent fuel pool after some of the fuel had been uncovered by the boiling away of the water in the pool. Water was sprayed intermittently into the spent fuel pool from trucks beginning on 20 March, but the water temperature in the spent fuel pool on 24 March was reported to be 100 °C, the boiling point of water at sea level. On 25 March a more reliable supply of water into the spent fuel pool was initiated using concrete pumps.

---

[c] pH is a measure of the acidity or alkalinity of a solution based on the hydrogen ion concentration. In aqueous solutions, pH ranges from 0 to 14, with 7 being a neutral solution. The pH scale is logarithmic (base 10), so a change of one unit in pH corresponds to a tenfold change in hydrogen ion concentration.

On 30 April, an observation was made of the Unit 4 spent fuel pool using video cameras, and the spent fuel racks seemed to be essentially intact. There was no hydrogen released from the Unit 4 spent fuel pool. It was later determined that the probable source of the hydrogen that caused the explosion in Unit 4 on 15 March was actually Unit 3; a pathway from Unit 3 to the refueling floor in Unit 4 was present via piping shared by the two units.

By September, a new system had been installed to cool the Unit 4 spent fuel pool, and the temperature of the water in the pool had been reduced to less than 40 °C.

Because the damage caused by the explosion had weakened the structure supporting the Unit 4 spent fuel pool, there was concern that further seismic disturbances could cause the structure to collapse, possibly causing the fuel to collect in a critical configuration. To strengthen the damaged structure, steel support pillars were installed between 31 May and 20 June.

14.4 RADIONUCLIDE RELEASES

Open pathways between the cores and the atmosphere allowed volatile radionuclides to escape into the air. Leaks and deliberate releases of water carried radionuclides into the ocean. Most of the released radioactivity, both into the air and into the ocean, consisted of iodine-131 (^{131}I), cesium-134 (^{134}Cs), and cesium-137 (^{137}Cs).

Pure iodine[18] is a solid at standard temperature and pressure, but it melts at 113.7 °C and boils at 184.3 °C. It sublimates, or passes into vapor directly from the solid state, at standard temperature and pressure, so it easily becomes airborne. It is a member of the halogen group like fluorine and chlorine, but it is less reactive chemically than the other halogens, so it can occur in the elemental state under everyday conditions. Iodine-131 beta-decays to stable xenon-131 with a half-life of 8 days. (One-half percent of the beta decays stop at an excited state of ^{131}Xe, which decays by internal transition, emitting a low-energy gamma ray, with a half-life of 12 more days.[19]) As I write this, it has been more than 365 half-lives of ^{131}I since the accidents, and the fraction of ^{131}I remaining is less than 10^{-99} of the original (that's the smallest number my scientific calculator can display). There is essentially no ^{131}I left from the accident. For a short period after the accident, ^{131}I posed a health risk to residents of the area around Fukushima Daiichi, but evacuation of the immediate area began promptly, and exposure to ^{131}I was limited.

Cesium (or caesium)[20] is an alkali metal that is a soft solid at standard temperature and pressure, but it melts at 28.44 °C (83.19 °F), which is barely above room temperature. However, it does not normally occur in the elemental state, as it is highly pyrophoric (it burns spontaneously in air) and it reacts explosively with water. The normal oxide of cesium, Cs_2O, vaporizes at 250 °C, so it can easily be dispersed into the atmosphere in the conditions present in a nuclear accident, and it will fall out of the air over an area that depends on weather conditions. Cesium-134 beta-decays to stable barium-134 with a half-life of 2 years, so it will be present in substantial fractions of the original released quantity for a few decades. Cesium-137 beta-decays to stable barium-137 with a half-life of 30 years, so the ^{137}Cs released by the accidents will not disappear from the environment for centuries. Note that a radionuclide decays in 10 half-lives to one thousandth of its original activity (i.e., 0.5^{10}), and it decays in 20 half-lives to one millionth of its original activity (0.5^{20}).

Water supplied by fire trucks, concrete pumping trucks, and helicopters to cool the cores and spent fuel pools of the stricken reactors flowed through breaches in the primary cooling systems and containment vessels into the reactor buildings. Rainwater joined this accumulation through openings created by the hydrogen explosions. Some of this water stayed in the buildings, some of it was transferred into tanks, and some of it leaked or was deliberately released (because there was no more room in which to store it) out into the ocean. Water sprayed onto the cores picked up some radionuclides and delivered a substantial amount of contamination into the ocean. In June of 2011, a water cleanup system began to operate, and two water cleanup systems were deployed to remove radionuclides from the water in the tanks and reactor buildings. Some radionuclides that fell out of the air onto land around the reactor site were carried into the ocean by rainwater runoff following the accident.

An estimate released by TEPCO on 24 May 2012 stated that the total releases of the three principal radionuclides by 30 September 2011 were 511,000 TBq of ^{131}I, 13,500 TBq of ^{134}Cs, and 13,600 TBq of ^{137}Cs. A TBq, or terabecquerel, is 10^{12} Bq. How much contamination do these numbers imply? Recall from Chapter 4 that a becquerel is one radioactive disintegration per second, and that a curie is 3.7×10^{10} Bq. So the releases of the two cesium isotopes amount to about 365,000 Ci, which is a lot of radioactive contamination. The release of ^{131}I was nearly two orders of magnitude higher, but fortunately its rapid decay made it only a short-term problem. The total released radioactivity was about 11% of the radioactivity released by Chernobyl. However, in the Fukushima Daiichi accidents, only these three radionuclides were released in significant quantity, whereas in the Chernobyl accident the entire core, with its full range of fission products, activation products, and unused fuel, either went up in smoke or was blown out onto the surrounding area.

One Bq is a very small amount of radioactivity, so releases of radionuclides measured in Bq are expressed in very large numbers. The sheer magnitude of these numbers can be intimidating; however, a more realistic perspective on their meaning can be gained by comparing the resulting concentrations of contaminants to regulatory limits.

As shown in Table 4.6, the concentration limits permitted by the U.S. Nuclear Regulatory Commission for ^{137}Cs are 0.2 pCi/L in air and 1000 pCi/L in water. (The same limits also apply to ^{131}I.) Japan has its own limits, but comparison of contamination with USNRC limits gives a useful perspective for American readers. A picocurie (pCi) is 10^{-12} Ci. A little arithmetic shows that the total amount of ^{137}Cs released would require 3.65×10^{11} m^3 of water, or a cube of water 7.14 km on a side, to be diluted to the USNRC limits. The ocean is much bigger than that. The same amount dispersed in air would require a cube of air 122 km on a side to be diluted to the USNRC limits. That's a lot more air, but it's still small compared to the size of the atmosphere. Serious contamination of both the air and the water was confined to the area near the accident site.

Soil in the surrounding area was contaminated by the fallout from the radionuclides dispersed in the air, and soil contamination leads to contamination of drinking water and agricultural products. Elevated levels of ^{131}I and ^{137}Cs in tap water were reported in 13 Japanese prefectures, including Tokyo, for about a month after the accident, but readings after that were near zero. The highest reading in Tokyo for ^{131}I was about 40 Bq/kg, which is slightly over the NRC limit of 1000 pCi/L, on 26 March 2011; the highest level in Tokyo for ^{137}Cs was about 3 Bq/kg, well under the limit, on 24 March 2011. The highest concentration of ^{131}I in tap water anywhere in Japan was in Tochigi Prefecture on 24 March, at about 115 Bq/kg; the highest concentration of ^{137}Cs was in Ibaraki Prefecture on 21 March, at about 18 Bq/kg.

Radioactive contamination was found in agricultural products from the Fukushima Prefecture and several prefectures in the surrounding area. Restrictions were established on the shipping of agricultural products from these prefectures,[21] but by 2017 those products had been deemed safe and the restrictions had been lifted.[22] Removal and decontamination of soil containing radioactive cesium is ongoing; as of 2018 over 700 million cubic feet of soil had been stripped and stored in plastic bags. After decontamination, almost all of this soil can be reused.[23] Decontaminated areas become inhabitable and usable again much sooner than the natural decay of radioactive contamination would allow. The long-term cleanup goal is to reduce dose rates received by people occupying the decontaminated areas to 0.1 rem per year above natural background, which is about the same dose delivered to a medical patient receiving a full-body CT scan (see Chapter 10).

14.5 SHORT-TERM AND LONG-TERM HEALTH CONSEQUENCES

As the Fukushima Daiichi accidents were unfolding, news media sensationalized the events. From all the panic the accidents caused, one might expect people to be dying from radiation sickness all over Japan like mosquitoes flying through an insecticide fogger. But this has not been the case.

Two workers drowned in the turbine building of Fukushima Daiichi Unit 4, and one worker died by being crushed in a crane operating console at Fukushima Daini.[24] Workers in the post-accident cleanup efforts have had to wear full-body protection gear, which gets very hot in summer weather; by 19 July 2011, 33 workers had suffered heat stroke and two workers in their 60s had died of heart failure. Two other workers died by October 2011 of causes unrelated to the accidents.

The standard occupational dose limit in Japan for workers in facilities where sources of radiation are present is 50 mSv/yr and 100 mSv over any five-year period.[d] The usual emergency dose rate limit is 100 mSv/yr, but it was raised to 250 mSv/yr for this accident. The maximum external dose received by any worker was 199 mSv, the maximum internal dose received by any worker was 590 mSv, and the maximum total dose received by any worker was 670 mSv. At least six workers at Fukushima Daiichi have exceeded the emergency legal limits for radiation exposure, and 408 workers have received radiation exposure above 50 mSv/yr. These radiation exposures increase the workers' long-term probability of developing cancer. Based on the conservative Linear No-Threshold (LNT) Hypothesis (see Chapter 4), the lifetime cancer rate of an exposed typical person is estimated to increase by 10%/Sv. (For actual individuals, this varies; for example, a 60-year-old worker is more likely to die of other causes before developing cancer than a 30-year-old worker is.) As mentioned for predictions of health effects from the Chernobyl accident, as discussed in Chapter 13, the LNT hypothesis is excessively conservative. The only short-term health effects from radiation among the workers were radiation burns sustained on the legs of two workers who were wading in contaminated water without waterproof wading equipment.

No member of the public has yet suffered any discernible effects from radiation exposure from the accidents. Exposure of the public to radiation can be categorized four ways: radiation from contaminated air around an individual, radiation from inhaled contaminated air, radiation from ingesting contaminated food and water,

---

[d] Recall from Chapter 4 that 1 sievert (Sv) is 100 rem. Then the standard limit of 50 mSv/yr is equal to 5 rem/yr as in the U.S. The prefix m implies thousandths.

and radiation from contamination on the surface of the ground. At the time when Reference 3 was published, reliable estimates were not yet available for any except the last of these categories; radiation doses from cesium-contaminated soil would be about 16.6 mSv per MBq/m^2 of combined ^{134}Cs and ^{137}Cs during the first year after the accidents. This figure was calculated with the assumption that the average person spends half of each day indoors, where exposure is reduced. Analysis of soil samples indicates that about 874 km^2 of land was contaminated with a cesium concentration greater than 0.6 MBq/m^2, which corresponds to a dose of 1 rem in the first year.

Residents within 3 km of the reactor site were given a mandatory evacuation order at 9:30 p.m. on 11 March, and the evacuation was extended to a 10-km radius on the next day and to a 20-km radius shortly thereafter. On 25 March, residents living within 30 km of the reactor site were urged to leave. From the evacuation zone, 160,000 people were displaced from their homes. Therefore, most of the people living in the zone of highest contamination were not there long enough to receive significant radiation doses. However, some of the 874 km^2 area of highest contamination lies outside of the 20-km evacuation zone. While no member of the public has yet been confirmed to have suffered adverse effects from exposure to radiation from the accident, the trauma of displacement was fatal to about 1000 people as of October 2012.[25]

Several survey programs have been initiated to monitor exposure levels in people near the Fukushima Daiichi site. By July 2011, more than 210,000 residents had been screened for contamination by governmental and academic institutions. In June and July 2011, an internal dose survey was performed on 122 residents of communities in Fukushima Prefecture, and follow-up surveys were performed on 109 of them. Of the 109, 52 had up to 3100 Bq of ^{134}Cs, and 32 had up to 3800 Bq of ^{137}Cs, and 26 were contaminated with both. The total dose of radiation from this contamination did not exceed 100 mrem for any of these individuals. At the time of publication of Reference 3, long-term monitoring plans were being laid by Fukushima Prefecture and by the United Nations.

The most probable long-term health effect of the radiation exposure sustained by the public is development of cancer. The very low doses absorbed suggest that the cancer rate from the Fukushima Daiichi accidents will be very low. A study performed at Rensselaer Polytechnic in June 2011 predicts that the cancer rate will be only 0.001% above the natural cancer rate; this increase amounts to about a hundred cases of cancer. The largest impact of the accidents on people's lives is the disruption caused by the relocation of residents in the area around the accident site.

14.6 THE CONTRIBUTING INSTITUTIONAL FAILURES

There are parallels between the characteristics of the Soviet nuclear institutions that contributed to the Chernobyl accident as outlined in Chapter 13 and some characteristics of Japanese culture and its nuclear institutions. In the Soviet bureaucracy, maintaining the appearance of success was so important that problems were kept hidden and solutions were never even sought. In Japan, appearances are also valued over substance, and there is a strong cultural tradition not to question authority. As Amory Lovins put it, Japan's "rigid bureaucratic structures, reluctance to send bad news upwards, need to save face, weak development of policy alternatives, eagerness to preserve nuclear power's public acceptance, and politically fragile government, along with TEPCO's very hierarchical management culture, also contributed to the way the accident unfolded."[26]

There is also a career path called "descent from heaven," whereby regulators move into executive positions within the industry they regulated; regulators who cause problems for their industries are not likely to be hired.

Whatever the impact of such cultural factors, TEPCO had a history of violating safety regulations, falsifying records, and ignoring warnings based on risk analysis dating back to 1967. In particular, a TEPCO in-house study in 2007 predicted the possibility of a tsunami over 10 m high at the Fukushima Daiichi site. This study was ignored by the TEPCO management. Furthermore, poor communication between TEPCO and the Japanese government, and between different governmental agencies, along with inadequate disaster response plans, made the accident worse than it could have been.

After the accident, the Japanese government formed the Fukushima Nuclear Accident Independent Investigation Commission (NAIIC), led by Kiyoshi Kurokawa, former president of the Science Council of Japan and a science advisor to the Japanese Cabinet.[27] He said, "It was a profoundly man-made disaster—that could and should have been foreseen and prevented." The NAIIC report stated, "Across the board, the commission found ignorance and arrogance unforgivable for anyone or any organization that deals with nuclear power."

14.7 THE IMPLICATIONS OF FUKUSHIMA DAIICHI FOR THE FUTURE OF NUCLEAR POWER

The Fukushima Daiichi accidents triggered a strong worldwide response. The reactions of the world's nations have been mixed. Reference 28 is an interactive map showing the status of nuclear programs worldwide; each country's reaction is categorized by color, and in some cases placing the cursor over the country brings up further information.

The strongest negative reactions have been expressed by Germany, Switzerland, Italy, and Belgium. These countries have all decided to phase out their nuclear power systems. Germany has 17 nuclear power reactors, which supplied 17.7% of the nation's electricity in 2011. On 30 May 2011, Germany decided to phase out all of its nuclear power plants by 2022; six reactors that had been shut down for testing and two that had been shut down for several years because of technical difficulties will not be restarted.[29] However, as of March 2014, nine of Germany's nuclear power reactors were still operating;[30] doing without 18% of Germany's electricity may have been seen as more disruptive than the politicians had imagined. Switzerland has five nuclear reactors, which produce 40% of its electricity. It had plans to replace these reactors with five modern plants as their existing reactors reached the ends of their lives, but these plans were abandoned in May 2011. The existing reactors will be phased out by 2034.[31] Italy has no operating reactors, but in 2008 plans had been laid to build some. On 11-12 June 2011, Italian voters voted to cancel these plans.[32] Belgium has seven reactors, but in November 2011, the country decided to phase them out by 2025, depending on its ability to find adequate replacement power.

Most of the world's countries that have nuclear power have decided to review the safety of their nuclear plants and emergency response procedures in the light of lessons learned from the Fukushima Daiichi accidents. These countries are the United States, Canada, Russia, France, Spain, the United Kingdom, the Netherlands, Sweden, Finland, the Czech Republic, Slovakia, Ukraine, Hungary, Romania, and Bulgaria. In addition, China, India, Japan, and Mexico are reviewing their safety standards and debating the future of their nuclear power programs.

Particularly, in Japan a strong anti-nuclear public sentiment has arisen, with 80% of the public now allegedly against nuclear power.[33] But, prior to the accident, nuclear power accounted for 30% of Japan's electrical generating capacity. By May 2012, all of Japan's nuclear power plants had been shut down, and power rationing and voluntary cutbacks of electricity use were necessary. As of 2017, 11 of Japan's nuclear power plants had been restarted. In order to meet its commitments under the Paris climate accord,[34] Japan will need to generate 20-22% of its electricity by nuclear power. Accordingly, 26 restart applications have been submitted; plans call for 12 reactors to come back on-line by 2025 and six more by 2030.[35]

In the United States, the Nuclear Regulatory Commission established a task force to review NRC regulations and determine whether U.S. nuclear plants could continue to operate safely.[36] On 12 July 2012, the task force reported that plant operations and licensing activities could continue without imminent risk, but made a dozen recommendations for enhanced safety. Subsequently, additional recommendations have been made and the recommendations have been prioritized. The following recommendations, which I have paraphrased, are notable:

- Review seismic and flooding hazards and requirements for resistance to them.
- Review requirements for dealing with prolonged loss of offsite power (station blackout).
- Establish mitigating strategies for beyond-design-basis accidents.
- Provide the capability to make up water lost from spent-fuel storage tanks.
- Provide measures to control and mitigate hydrogen released into containment and other buildings.

The NRC has established the Japan Lessons Learned Project Directorate, a permanent working group of 20 full-time NRC staff members, to focus exclusively on formulating NRC requirements in response to the lessons learned from the Fukushima Daiichi accidents.

In the January 2013 issue, *Physics Today* published the following update:[37]

Acting on a set of recommendations developed by its Japan Near-Term Task Force, the NRC instructed reactor operators to deploy additional generators and other equipment, both onsite and at nearby offsite locations, to ensure that backup power, coolant systems, and other vital equipment will remain operational if power from the grid is lost for days or even weeks in a disaster.

In addition, the 21 US boiling-water reactors with containment systems similar to the Fukushima BWRs must be equipped by 2016 with containment vents that can cope with the increased pressure and temperature of steam generated early in an accident and can withstand possible fires and small explosions if the vents are used to release hydrogen later in an accident. The upgraded vents must be capable of operation even if the reactor loses all electrical power or if other hazardous conditions exist. The NRC also mandated that thermometers and water-level gauges be installed in the spent-fuel storage pools of all plants.

A few countries—Pakistan, Brazil, Argentina, Iran, and South Africa—have nuclear power plants but have not announced any safety reviews or other actions to apply lessons learned from the accidents. Some countries, such as Vietnam, that currently have no nuclear reactors but have plans to build them, are determined to proceed with their plans.

The degree to which natural human shortcomings undermine safety regulations and designed safety features has to make a nuclear power advocate like me pause and reconsider the proposition that nuclear power can be safe. I believe that in the United States, our cultural willingness to challenge authority, and our legal protections against "whistle-blowers," can help mitigate those human shortcomings. In some other countries, cultural traditions and governmental opacity seem to work against safety. I believe that nuclear power can be truly safe only when nuclear reactors are made idiot-proof, or passively safe. Some passively safe designs are described above in Chapter 11. Others are discussed in some of the following chapters. Nevertheless, it should be kept in mind that, although the economic and social costs of the Fukushima Daiichi accidents have been severe, nobody has died from the radiation released in the accidents, and the additional cancer deaths predicted for the long term are such a small increase over background cancer rates that they will never be discernible. Furthermore, these additional cancer deaths were predicted on the basis of the Linear No-Threshold Hypothesis, which has no basis in empirical evidence and is almost certainly excessively conservative; as explained in Chapter 4, it would be extremely difficult to perform experiments to determine the effects of extremely low radiation doses, and such experiments have never been done. The Linear No-Threshold Hypothesis is applied only because there is also no experimental basis for choosing a threshold. So I remain a nuclear power advocate, but I would like to see passively safe reactors eventually replace Generation II LWRs.

CHAPTER 15

HEAVY-WATER REACTORS

As noted in Chapter 11, ordinary water is slightly poisonous to thermal neutrons. As a consequence, it cannot be used as a moderator for reactors fueled with natural uranium: The production of neutrons from U-235 fission in natural uranium is too small to overcome the absorption of neutrons by the water. Therefore, the fuel in LWRs is enriched in U-235 above the concentration in natural uranium.

However, the neutron absorption cross section in deuterium is very low, so if the moderator is D_2O, natural uranium can be used for reactor fuel. It is much easier to enrich water in deuterium than to enrich uranium in U-235, so reactors moderated with D_2O, called heavy-water reactors (HWRs), are a viable alternative to LWRs. The Canadian nuclear power industry is based on their heavy-water reactor design, the CANDU (Canadian Deuterium-Uranium) reactor.[1, 2]

Although they are technically pressurized-water reactors, CANDU reactors are unique in almost every respect. Figure 15.1 illustrates the principal features of a CANDU reactor.[3]

The core is contained in a vessel called the calandria, which is basically a low-pressure tank for the heavy-water moderator. The calandria is about 7.6 m in diameter and 4 m long. The fuel is contained in bundles in tubes called pressure tubes. Each fuel bundle is about 50 cm long, and there are several fuel bundles loaded end-to-end in each pressure tube. There are about 380 pressure tubes in the calandria. The fuel bundles can be loaded or removed at the ends of the pressure tubes without shutting the reactor down; they are introduced at one end and removed at the other. This capability allows the fuel to be consumed more completely than if it were stationary as in an LWR.

Each pressure tube is contained in a tube called a calandria tube, from which it is separated by spacers and a gas annulus. This separation minimizes heat transfer from the pressure tube to the moderator in the calandria, so that the moderator can be maintained at low temperature and pressure. Nevertheless, enough heat is deposited in the moderator by gamma-ray heating that the moderator must be circulated through its own heat-removal loop, which is not shown in the figure.

The figure shows only one primary coolant loop, but actually there are two, which pass coolant through the core in opposite directions in adjacent pressure tubes. This approach permits the average reactor temperature to be more uniform, so that temperature-dependent nuclear reaction cross section variations are minimized and the neutron flux is more uniform. Refueling is accomplished in the direction of coolant flow, so that the direction of movement of the fuel bundles through the core is also in opposite directions in the two primary systems. Then fresh fuel in one tube is located next to depleted fuel in adjacent tubes. This technique also minimizes flux-shape variations and permits more uniform power generation throughout the reactor.

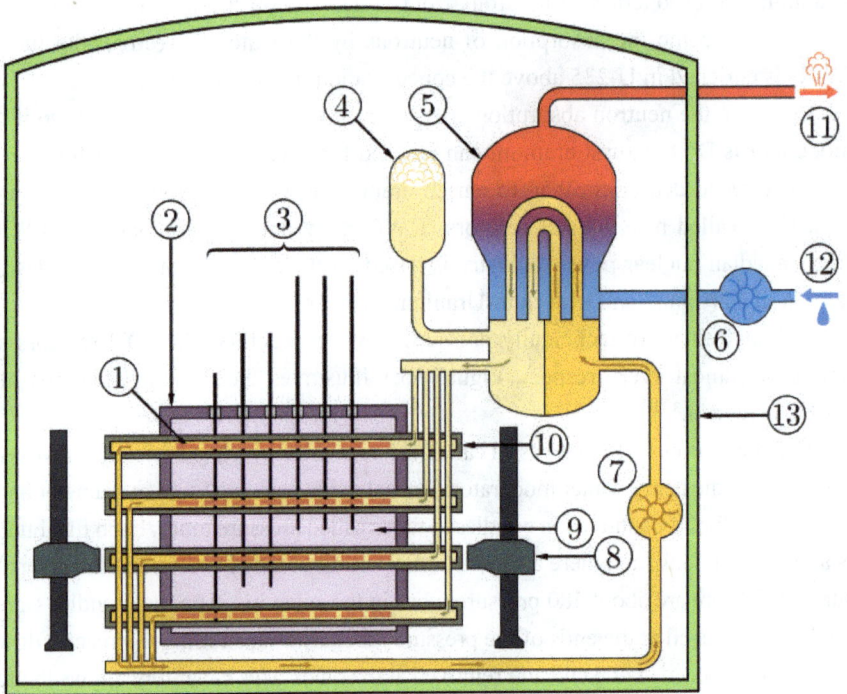

#	English	Français
1	Nuclear fuel rod	Combustible nucléaire
2	Calandria	Cuve
3	Control Rods	Barres de contrôle
4	Pressurizer	Préssurisateur
5	Steam generator	Générateur de vapeur
6	Light water condensate pump (secondary cooling loop)	Pompe de l'eau légère condensée (boucle de refroidissement secondaire)
7	Heavy water pump (primary cooling loop)	Pompe de l'eau lourde (boucle de refroidissement primaire)
8	Nuclear fuel loading machine	Machine de chargement du combustible nucléaire
9	Heavy water (moderator)	Eau lourde (modérateur)
10	Pressure tubes	Tubes de force
11	Steam	Vapeur
12	Water condensate	Eau condensée
13	Reactor containment building	Enceinte du réacteur nucléaire

Figure 15.1—Schematic layout of CANDU reactor. This file was taken from the Wikimedia Commons and is in the public domain.[3]

The coolant in the primary coolant loops is also heavy water. Heat from the primary loops is exchanged in the steam generators with light water in the secondary loops. It is easier to prevent coolant losses at lower pressure, so the pressure and temperature in the primary loops are limited to 10 MPa (1450 psi) and 310°C (590 °F), respectively. The temperature in the secondary loops is limited accordingly, so that the efficiency of the CANDU reactor is lower than that in an LWR—about 29%.

The large mass of cool moderating heavy water in the calandria serves as a heat sink in case of an accident, providing inherent passive safety. The large proportion of U-238 in the natural uranium provides a strong negative temperature coefficient of reactivity by the Doppler effect (see Section 5.4), which ensures shutdown of the nuclear chain reaction in a significant temperature excursion.

Besides using the CANDU design for all of its power reactors, Canada has sold CANDU reactors to several other countries, including South Korea, China, India,

Argentina, Romania, and Pakistan. As of March 2019, India has two CANDU reactors and 16 reactors of its own CANDU-derivative design, with six more under construction.[4]

As discussed in Chapters 21 and 23, plutonium suitable for use in nuclear weapons is made in nuclear reactors from U-238 that does not reside in the reactor for a long time. In ordinary "batch-loaded" reactors like LWRs, in which the fuel remains in place for a year or more at a time, the plutonium accumulated in fuel assemblies is not suitable for nuclear weapons, because it accumulates too much ^{240}Pu and ^{242}Pu, which undergo too much spontaneous fission. However, in the continuously fueled CANDU design, it is possible to move fuel through the reactor quickly and produce weapons-grade plutonium. Because India has CANDU and CANDU-derivative reactors, it was suspected that it used those reactors for production of their nuclear weapons. However, it now appears that it uses a dedicated weapons-production reactor called Dhruva, based on a different reactor design altogether, derived from the Canadian NRX research reactor.[5] The decision of a nation to pursue or not to pursue nuclear weapons is made in accordance with its perception of its national interest, and most nations perceive nuclear weapons to be contrary to their interest. Nations that want nuclear weapons will obtain them, regardless of what kind of reactors they may have.

CHAPTER 16

GAS-COOLED REACTORS

Current interest in gas-cooled reactors is due to the ability of helium-cooled reactors to operate at high temperatures, as explained below. Reactors designed to exploit this ability are called high-temperature gas-cooled reactors (HTGRs) or very-high-temperature reactors (VHTRs). The density of gas coolants is so low that neutron absorption by the coolant is not important, so in principle a variety of gases could be used to cool reactor cores. However, the two gases that have been used in practice are carbon dioxide and helium. The British Magnox[1] and AGR (Advanced Gas-cooled Reactor)[2] use carbon dioxide, but most other gas-cooled reactors use helium. One of the advantages of helium is that it is chemically inert, whereas carbon dioxide causes corrosion in piping and steam generator components at high temperatures.[3]

Helium-cooled reactors have been operated in many parts of the world. In the United States, the Peach Bottom Unit 1 reactor near Harrisburg, Pennsylvania, was a prototype helium-cooled reactor that operated from 1966 to 1974.[4] The Fort St. Vrain reactor near Platteville, Colorado, operated from 1977 until 1992.[5] After that, it was converted into a natural gas power plant because it suffered numerous recurring problems and never operated reliably. These reactors were based on the prismatic fuel design, described in detail in Section 16.4. An alternative design, the pebble-bed reactor (PBR), was developed in Germany. Their AVR (Arbeitsgemeinschaft Versuchsreaktor, meaning a test reactor developed by a consortium) reactor operated successfully from 1967 to 1988.[6] Their pebble-bed Thorium High-Temperature Reactor (THTR) operated from 1983-1989.[7] A newer experimental PBR, the HTR-10, has been operating in China since 2003.[8] In Japan, an experimental prismatic design, the HTTR, began operation in 1998.[9] It was shut down for maintenance in February 2011, and did not restart after the Fukushima Daiichi accident. A safety review prior to restart was initiated in 2014.[10] As of 2017, the HTTR was considered operable, but it evidently was still not operating.[11]

Currently, efforts to develop both prismatic and pebble-bed designs as Generation IV reactors are being pursued throughout the world. The HTR-10[12] was intended to spearhead the development of a commercial PBR venture in China; in 2005, authorities authorized construction of a two-module pebble-bed demonstration plant called the HTR-PM, which will produce 210 MWe at Rongcheng in Shangdong Province. The first unit of HTR-PM has been completed, and as of February 2019 it was being commissioned, with connection to the electric grid expected in late 2019.[13] PBMR (Pty.), Ltd. of South Africa was developing PBR technology for domestic commercialization and international sales, but the worldwide economic downturn of 2008 and the resulting lack of a committed customer led them to suspend further work on the project.[14] A DOE-funded program was begun to design and construct a prototype HTGR called the Next-Generation Nuclear Plant (NGNP) at the Idaho National Laboratory (INL). The NGNP program was led by the INL, but in 2009 a cost-sharing entity called the NGNP Industry Alliance, including major reactor vendors and potential end users, was established to represent the interests and views of its members.[15] The responsibility to decide among designs was allocated to the Alliance. Parallel studies were conducted for both pebble-bed and prismatic NGNP designs. A consortium led by Westinghouse, which included PBMR and the Shaw Group, was developing the pebble-bed design, and Areva and General Atomics were separately developing prismatic designs. In March 2010, the DOE awarded a $40 million dollar contract to Westinghouse and General Atomics for further development.[16] But in May 2010, the partners in the pebble-bed consortium were unable to agree on their future path, and Westinghouse withdrew from the consortium.[17] Their withdrawal was one of the factors that led South Africa to close the PBMR program, including its participation in the NGNP.[18] A decision among the designs was made by the Alliance in February 2012 in favor of an Areva prismatic design[19] based on their Antares reactor concept.[20] The Alliance stated that neither the prismatic configuration nor the pebble-bed configuration had any clear technical superiority, and that the decision was based on capital cost and "the business case for reactor design development and licensing."[19] The NGNP was scheduled to become operational in about 2021, but because of disputes between the DOE and the Alliance over cost-sharing issues, the program was suspended in 2013.[21] However, the Alliance was quickly resurrected[22] and is currently conducting reactor design studies and experimental investigations of fuel configurations.[15]

General Atomics, in San Diego, California, is developing its EM^2, a unique helium-cooled reactor that uses silicon carbide composite fuel elements, fuel rods, and core structures. Since it is a "small modular reactor," it is discussed in Chapter 19.

Generation IV reactors are advanced designs (not based on conventional water-cooled configurations) intended to meet eight technology goals divided into categories of sustainability, safety and reliability, and economics.[23] The sustainability goals require improved resource utilization, reduced waste production, and low suitability for weapons proliferation. The safety and reliability goals do not specify numerical performance criteria, but they require very low probability of core damage, excellent safety and reliability, and a complete lack of any need for offsite response. The economic goals require a clear advantage in life-cycle cost over other energy sources and a level of financial risk comparable to that in other energy sources.

Although the safety goals do not specify performance criteria, it is implicit that the safety of Generation IV systems must be superior to that of Generation III or III+ systems. In practice, that goal implies that Generation IV systems will feature complete passive safety: In no event can a core melt occur even with a complete loss of coolant.

HTGRs can achieve passive safety by a combination of fuel properties and design. First, the fuel can withstand very high temperatures. The components of the fuel and their melting points are usually uranium dioxide (2800 °C),[24] graphite (3675 °C),[25] and silicon carbide (2730 °C).[26] However, fuel damage occurs at temperatures lower than these melting points, and the fuel temperature is limited in accident scenarios to 1600 °C. This is still a much higher temperature than the temperature limit in LWRs, which is based on the zirconium-water reaction (1200 °C). Even so, the attainment of passive safety depends on core designs with low power density and short heat transfer paths from the core to the surrounding heat sink (which in practice is the Earth).

The prismatic and pebble-bed HTGR concepts are described in detail in Sections 16.3 and 16.4. However, the advantage of high-temperature operation that both concepts have is discussed first in the next section.

16.1 HTGRS AND PROCESS HEAT

Most nuclear reactors produce electricity. In that role, they can replace coal-fired electric power plants, which is certainly a good thing for the environment and public health. (Environmental and health effects of all available energy resources are discussed in Chapter 8.) However, electricity production accounts for only about 38% of our total energy consumption (2017 figures).[27] About 29% of total U.S. energy consumption goes into transportation;[27] almost all of this is supplied as gasoline or other petroleum derivatives. The remaining energy consumption is devoted to space heating and industrial processes, and it comes from a variety of sources.

Not only is petroleum a limited resource that will eventually become more and more expensive as supplies are depleted, but much of the petroleum resource is controlled by nations whose interests are different from, if not outright hostile to, those of the United States and western Europe. Energy independence is vital to America's future national security. Furthermore, petroleum, like all fossil fuels, emits carbon dioxide when burned, and increasing atmospheric carbon dioxide is almost certainly contributing to global climate change. There are several approaches to replacing petroleum for transportation fuels. One is to shift to purely electric vehicles (as opposed to hybrids, which use only slightly less gasoline), and another is to use hydrogen fuel in place of petroleum.

Electric vehicles are the subject of intense current research and development, but for now they are limited in practicality. The Tesla cars have ranges of 250-300 miles, and their batteries can be charged rapidly (in about an hour) only at "Supercharger" stations;[28] on household 220-volt systems, they can take several hours.[29] This may be acceptable for trundling around town, but it doesn't work for long trips. The Chevrolet Volt can go only 53 miles on its battery; then it reverts to a gasoline engine.[30] The Tesla Semi, a planned long-distance truck tractor, will allegedly travel up to 500 miles on a charge,[31] but its charging time will be longer. However practical electric vehicles become, the electricity has to come from some energy source, which in the short term will usually be coal.

Hydrogen can be used in either hydrogen fuel cells[32] or internal-combustion engines.[33, 34] In either case, hydrogen is combined with oxygen taken from the surrounding air to produce water. In the internal combustion engine, some air pollution may be produced by oxidation of atmospheric nitrogen, but that can be mitigated by catalytic converters as in modern gasoline-powered cars. However, hydrogen does not exist in a free state on earth. It is chemically very reactive, and it is almost all bound up in chemical compounds, mostly water and biomass. It is also present in hydrocarbons, such as petroleum and natural gas. Most industrially produced hydrogen comes from methane, the major constituent of natural gas.[33, 35]

Hydrogen production costs energy. Because it is bound in chemical compounds, hydrogen production requires the breaking of the chemical bonds that bind it into these compounds, and that requires energy from some external source. Because no process is completely reversible, the energy required to produce hydrogen is greater than the energy obtained from burning it—in practice, a lot greater.

The most abundant hydrogen resource is water. Hydrogen is obtained from water by electrolysis[36, 37] or thermochemical water splitting.[38] In electrolysis, an electric current is passed through the water, separating it into hydrogen and oxygen. Ideally, the electric current comes from a non-polluting source such as windmills or nuclear reactors. In thermochemical water splitting, iodine and sulfuric acid are used as catalysts at a temperature up to 900 °C.[38] This temperature can be achieved with heat supplied by an HTGR. Efficient electrolysis also requires high temperatures.[36] At 800-850 °C, electrolysis is 40% more efficient than at room temperature.[39] To supply all the U.S. highway transportation energy requirement as hydrogen (based on 1997 figures; more recent figures would give more current, but qualitatively similar, results) would require 241 GW of electricity, which could be supplied by 241 nuclear power plants of 1000 MWe or 640,000 windmills of 1.5 MW, covering 71,000 square miles.[37] For perspective, consider that the area of the state of Washington is 71,303 square miles.[40]

There are many drawbacks to hydrogen as a transportation fuel. First, storage of hydrogen in an automobile's fuel tank is difficult. Three approaches are being studied: storage as a compressed gas, storage in liquid form, and storage as a hydride of a porous metal. None of these techniques is currently able to store enough hydrogen for a satisfactory range (such as the 370-mile range of the typical gasoline-powered car) in a practical way.[41] The energy density of hydrogen per unit mass is 2.6 times better than that of gasoline, but hydrogen has very low mass density even in liquid form; the energy density per unit volume of liquid hydrogen is 4.5 times lower than that of gasoline.[42] Therefore, for the same fuel tank volume, the mileage and range obtainable from liquid hydrogen fuel are only about 22% of those available from gasoline. This implies a range of only about 80 miles for the typical car. Attaining reasonable range would require much larger fuel tanks, which would add weight and compromise the utility of vehicles. However, if the environmental effects of using petroleum are seen to be intolerable, hydrogen fuel still may emerge as the best choice.

Some industrial processes also require high temperatures. Heat from HTGRs can be used for such processes in facilities sited near the reactors.

The NGNP program in Idaho was established to build a prototype HTGR for both electricity and process heat production.[43] Specific examples of NGNP design features and goals are used below to illustrate HTGR concepts, as there are no commercial HTGR power plants currently in operation to use as examples. (After China's HTR-PM plant comes on line, this will change. But detailed information on HTR-PM may not be as readily available as information on the NGNP.) In the next section, the fuel design common to both types of HTGR is discussed, and in the following sections the two types are discussed in detail.

16.2 HTGR FUEL DESIGN

All fuel for currently planned HTGRs adheres to the TRISO (tri-structural iso-tropic) fuel design.[44] (The word *isotropic* refers to materials whose properties are the same in all directions.) TRISO fuel consists of layers made of four different materials: a fuel kernel in the center, two different types of carbon, and silicon carbide. The dimensions and densities of the fuel materials are subject to refinement in actual HTGR designs, but the values given in Table 16.1, which apply to an NGNP design from 2003, are representative.[45] I make use of 2003 design versions because I was heavily involved in that design effort and know more about it than I do about more recent versions. Qualitatively, the 2003 designs are still illustrative.

Table 16.1—NGNP TRISO particle fuel parameters (2003), from Reference 45

Parameter	Value
Kernel composition	UCO
Kernel diameter	350 microns*
Kernel density	≥ 10.5 g/cm³
Buffer thickness	100 microns
Buffer density	1 g/cm³
IPyC thickness	40 microns
IPyC density	1.9 g/cm³
SiC thickness	35 microns
SiC density	3.2 g/cm³
OPyC thickness	40 microns
OPyC density	1.9 g/cm³
Total particle diameter	780 microns (0.78 mm)

*A micron, or μm, is 10^{-6} m, equal to 10^{-3} mm.

The central fuel kernel has usually been made of uranium dioxide (UO_2), although it can be made of other materials, including uranium oxycarbide (UCO) or, as in the THTR, a mixture of uranium and thorium oxides. For the NGNP, the fuel was to be UCO. The diameter of the kernel design for the 2003 NGNP is 0.35 mm.

The carbon and silicon carbide layers are successively vapor-deposited. The first carbon layer (the buffer) is porous, to provide void space for the containment of fission-product gases that escape from the kernel. The density of the buffer is about 1 g/cm³. The thickness of the buffer is 0.1 mm.

The second carbon layer is a strong and dense material called pyrolytic carbon. Since it is one of two such layers, it is called the inner pyrolytic carbon layer (IPyC). The final carbon layer is the outer pyrolytic carbon layer (OPyC). These layers are both 0.04 mm thick, with a density of 1.9 g/cm³.

Between these pyrolytic carbon layers is a shell of silicon carbide (SiC) 0.035 mm thick, with a density of 3.2 g/cm³. Silicon carbide is a very strong material—strong enough that the silicon carbide shell acts as a tiny containment vessel for the individual TRISO fuel particle. The strength of the SiC shell allows HTGRs to be built without containment buildings. The TRISO fuel particle is illustrated in Figure 16.1.[45]

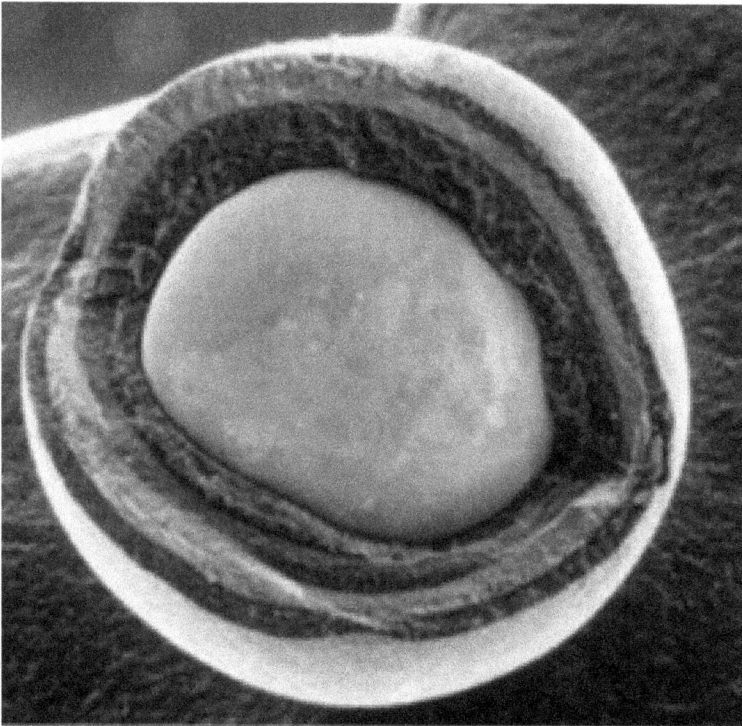

Figure 16.1—A photograph of the TRISO fuel particle with a portion of the coatings cut away. From Reference 45, an externally released report of an agency of the U.S. government; this figure is in the public domain under the terms of Title 17, Chapter 1, Section 105 of the U.S. Code.

Despite the strength of the SiC shell, the performance of the TRISO particles in early HTGRs was mixed. One of the problems that led to the conversion of the Fort St. Vrain reactor to natural gas was excessive failure rates in its American-made TRISO fuel. In contrast, the fuel in the German AVR proved very reliable. The difference was not understood until the INL began an investigation of HTGRs in collaboration with the Massachusetts Institute of Technology. INL scientists discovered differences in the manufacturing process that accounted for the greater reliability of the German fuel.[46] The South African company that was pursuing HTGR development, PBMR (Pty) Limited, licensed the German manufacturing process, while the Chinese have bought the German equipment and licensed their process for the HTR-10. Modern TRISO fuel is very reliable.

In both the prismatic and pebble-bed configurations, TRISO fuel particles are embedded in a graphite matrix. In the prismatic HTGR, the graphite matrix takes the form of cylindrical compacts about 5 cm long and 1.25 cm in diameter. In the pebble-bed HTGR, the matrix is a sphere about 5 cm in diameter, which is surrounded by a graphite layer without fuel about 0.5 cm thick. There are approximately 14,500 TRISO particles in a PBR fuel sphere.[47]

16.3 PRISMATIC HTGR CONFIGURATIONS

In the prismatic HTGR design, the cylindrical fuel compacts are stacked in blind holes (i.e., holes that do not extend all the way through) bored in hexagonal prismatic blocks as shown in Figure 16.2, which shows the whole hierarchy of fuel components.[45] The figure also shows the scales of the various components. In the version of the NGNP design proposed by Areva, the prismatic blocks are 79.4 cm high and 36.0 cm across the flats.[48] There are 1020 blocks arranged in three rings to form an annular core. The annular configuration provides a very short heat conduction path out of the core for decay heat removal in a loss-of-coolant accident, which enables passive safety.

In that same design, there are 210 fuel holes and 3126 fuel compacts in a typical block, for an average of about 15 compacts per hole. This varies a little bit in different portions of a fuel block. There are also 108 coolant holes bored all the way through the block, lining up with the coolant holes in the blocks above and below it. Some blocks have holes for control rods in place of some of the fuel holes.

The annular core and other pressure vessel internals are shown in Figure 16.3, taken from the 2003 design study.[45] As shown in the figure, hexagonal blocks fill in an area like a honeycomb. Inside and outside the annular core are reflector regions composed only of graphite. Hexagonal fuel blocks are stacked about 10 deep, with graphite axial reflector regions above and below the active core.

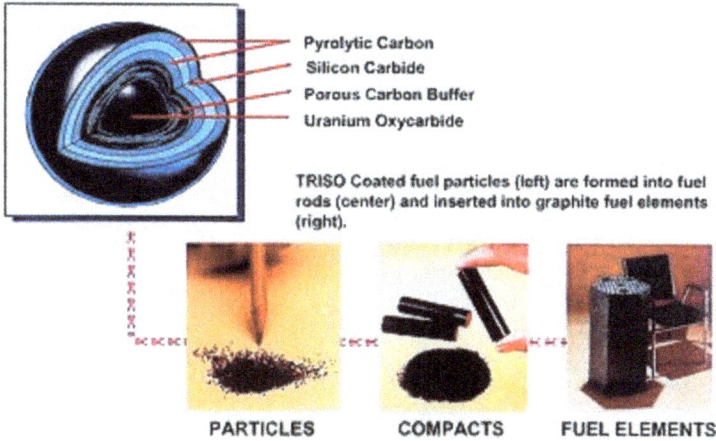

Pyrolytic Carbon
Silicon Carbide
Porous Carbon Buffer
Uranium Oxycarbide

TRISO Coated fuel particles (left) are formed into fuel rods (center) and inserted into graphite fuel elements (right).

PARTICLES   COMPACTS   FUEL ELEMENTS

Figure 16.2—Hierarchy of fuel components in a prismatic HTGR. From Reference 45, an externally released report of an agency of the U.S. government; this figure is in the public domain under the terms of Title 17, Chapter 1, Section 105 of the U.S. Code.

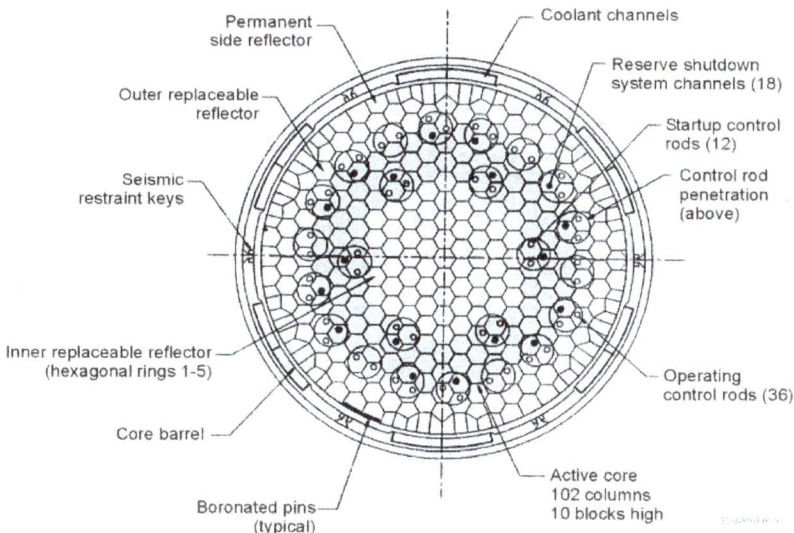

Figure 16.3—Arrangement of prismatic blocks in an HTGR. From Reference 45, an externally released report of an agency of the U.S. government; this figure is in the public domain under the terms of Title 17, Chapter 1, Section 105 of the U.S. Code.

Figure 16.4 shows a cutaway view of the preconceptual design of the NGNP reactor vessel and internals from Reference 45.

In this conception, the inlet and outlet ducts for the helium coolant are concentric, carrying helium between the reactor vessel and the primary cooling system (PCS) and intermediate heat exchanger (IHX) vessels. The PCS converts some of the heat

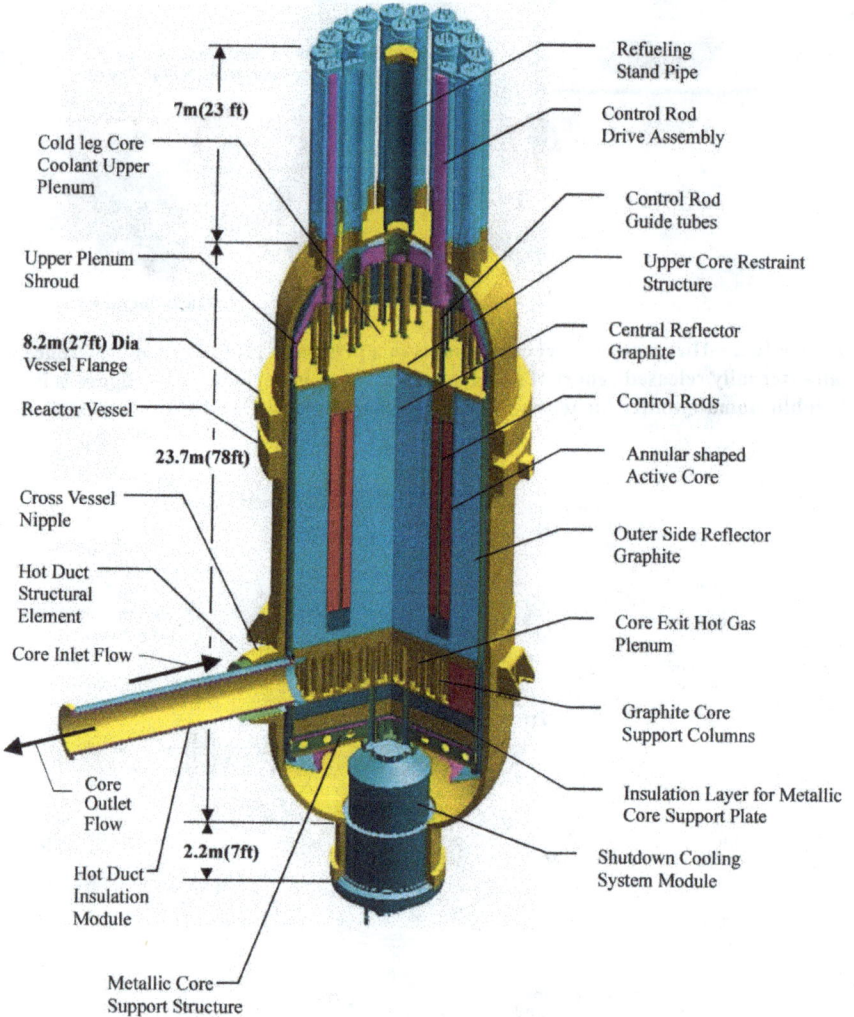

Figure 16.4—Prismatic HTGR pressure vessel and internals. From Reference 45, an externally released report of an agency of the U.S. government; this figure is in the public domain under the terms of Title 17, Chapter 1, Section 105 of the U.S. Code.

supplied by the reactor to electricity. In this General Atomics design, the PCS uses a gas turbine in the PCS vessel, as shown in Figure 16.5, to avoid an intermediate heat exchange step that would reduce efficiency. However, a steam generator connected to a steam turbine could also be used, or, as in the Areva design, a gas turbine could be used in a secondary coolant loop. In the IHX vessel, heat is exchanged with a secondary loop to provide process heat for hydrogen generation.

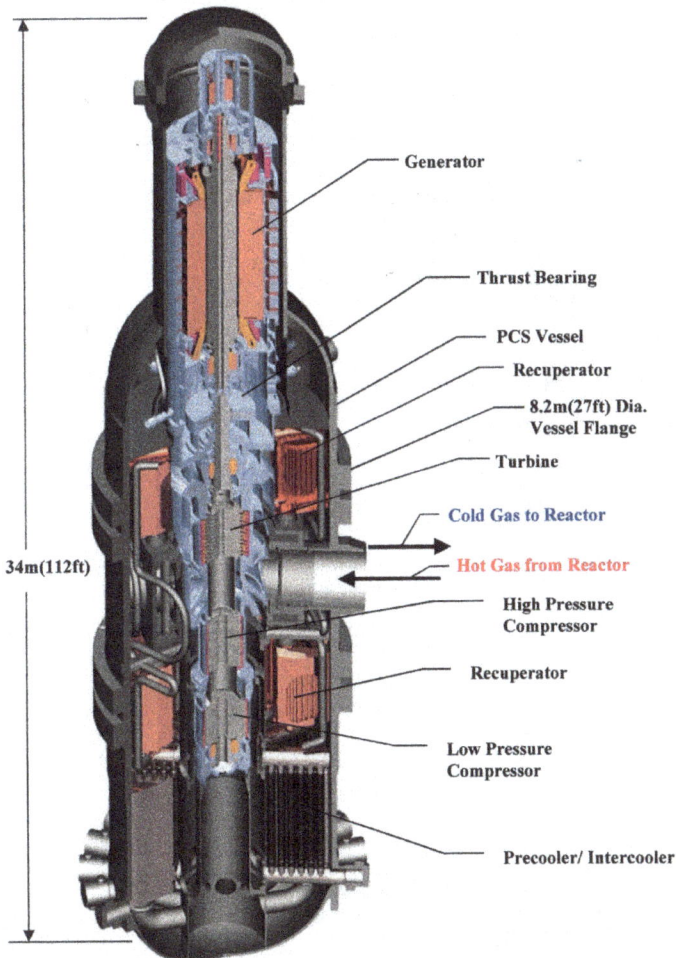

Figure 16.5—General Atomics NGNP PCS. From Reference 45, an externally released report of an agency of the U.S. government; this figure is in the public domain under the terms of Title 17, Chapter 1, Section 105 of the U.S. Code.

The helium flows upwards through the outer reflector to the upper plenum, then downwards through the core. The flow of cool helium in the outer reflector reduces the temperature of the pressure vessel by about 100 °C, enhancing the performance of the pressure vessel steel. The control rods are inserted from the top.

The shutdown cooling system module shown at the bottom of the reactor vessel is a water-based system that removes decay heat from the core when the PCS is shut down or in an accident. But the shutdown cooling system is not necessary to prevent fuel damage in a loss-of-coolant accident. The cavity around the reactor is passively cooled by natural convection of the air in the cavity.

Table 16.2 shows design parameters and performance goals for the two prismatic NGNP designs as of November 2007.

Table 16.2—Proposed prismatic configurations for the NGNP in November 2007[49]

Parameter	Areva Design	General Atomics Design
Thermal power (MWt)	565	550-600
Pressure vessel height (m)	25	31
Pressure vessel outside diameter (m)	7.5	8.2
Pressure vessel thickness (cm)	15	26.1
Core outlet temperature (°C)	900	Up to 950
Core inlet temperature (°C)	500	490
Coolant pressure (MPa)	7-9	7-9
Cycle configuration	Indirect gas turbine	Direct gas turbine
	Parallel hydrogen	Parallel hydrogen
Secondary fluid	He and Nitrogen	He
Hydrogen plant power	10% of reactor power	5 MW: HTE*
		60 MW: S-I*

*HTE: high-temperature electrolysis

S-I: sulfur-iodine thermochemical water-splitting process

16.4 PEBBLE-BED HTGR CONFIGURATIONS

Essentially, a pebble-bed reactor is a vat of pebbles that are dropped in at the top and withdrawn from the bottom. The pebbles are very rugged and can be dropped several meters onto a hard surface without damage. Pebbles withdrawn at the bottom are checked for damage and fuel consumption ("burnup"), and if they are intact and the fuel in them has not been consumed beyond a prescribed limit, they are reintroduced at the top. Pebbles may typically be expected to make about 10 trips through the core before being taken to a spent-fuel repository or recycled.

The continuous circulation of fuel permits the reactor to be operated continuously, without interruptions for refueling. The CANDU reactor shares this property, which potentially offers substantial increases in the capacity factor (the fraction of time during which the reactor is operating). This property is one of the advantages of PBRs.

In the AVR, THTR, other early conceptual PBR power plant designs, and in the HTR-10, the vat was just a cylindrical volume. However, for passive safety, the core in the NGNP pebble-bed design is a tall, thin annulus as in the prismatic HTGR designs.

Figure 16.6 shows a simplified conceptual sketch of the pebble-bed reactor with the core as a simple vat.

There are about 450,000 pebbles in the core region, with graphite-block reflectors surrounding the core on all sides. Control rods and shutdown rods are inserted into channels in the side reflector region. An emergency secondary shutdown system is composed of B_4C spheres that are dropped into channels in the central reflector (boron has a very high absorption cross section for thermal neutrons). In the THTR, control rods were inserted into the core, simply pushing pebbles aside as they moved downwards, but this approach did too much damage to pebbles.

Coolant flows through the core in the interstitial passages around the pebbles, and through the reflectors in bored coolant channels. One of the disadvantages of most PBRs is that the coolant flow path through the pebble bed is long, convoluted, and irregular, which leads to high pressure losses through the core and requires high pumping power, reducing the overall efficiency of the plant. Although the NGNP does not apply it, a concept has been introduced in which the coolant would flow radially through the core, so that the flow path would be relatively short.[50] Then the pressure losses through the core would be no greater than in prismatic HTGRs.

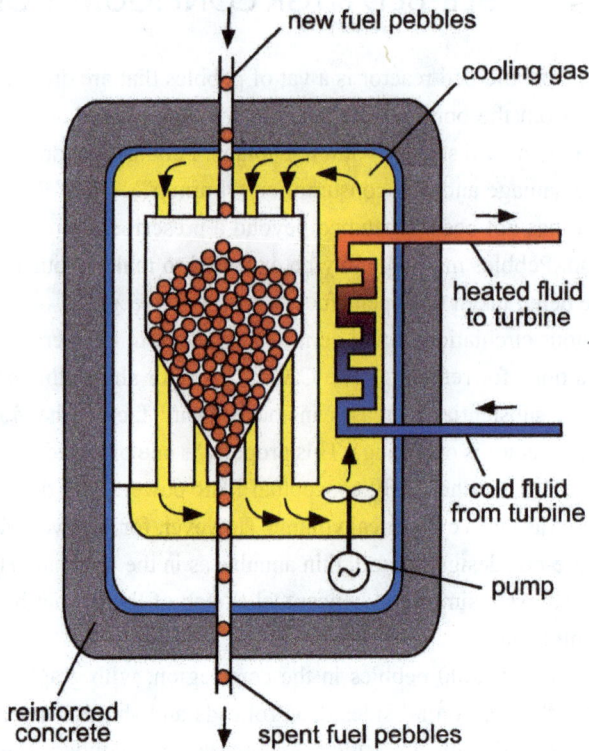

Figure 16.6—Pebble-bed reactor concept with annular core. This picture was taken from Wikimedia Commons and has been released to the public domain.[51]

Table 16.3 shows a comparison of parameters in a PBR design and a prismatic design aimed purely for hydrogen production. This comparison was made by PBMR (Pty) Ltd., the South African company that was developing PBRs. PBMR (Pty) Ltd. was a member of the Westinghouse consortium and the principal source of PBR technical expertise in the consortium. The table shows several aspects of the differences between PBRs and prismatic HTGRs that are generally true.

The power level of the PBR is lower. It is more difficult to achieve high power in a PBR than in a prismatic HTGR, although it can be done by making the core sufficiently tall and thin.[45] Prismatic HTGRs are subject to axial neutronic instabilities, which make it more difficult to make their cores taller. With axial coolant flow, to keep the core pressure drop reasonably low one must limit the core power density in the PBR, although the power density in a pebble can be higher than in a compact because the coolant is closer to the fuel in the PBR. Differences shown in the table for fuel enrichment and burnup limits

**Table 16.3—Comparison of PBR and prismatic
HTGR parameters for hydrogen production**

Parameter	PBMR PHP (Pebble bed)	H2-MHR (Prismatic)
Inner/outer active core diameter (m)	2.0/3.7	2.96/4.83
Active core effective height (m)	11.0	7.93
Fueled region power density (W/cm³)	16.9	32
Fuel element power density (W/cm³)	9.8	8.3
Core power density (W/cm³)	6.0	6.6
Core inlet/outlet coolant temp. (°C)	350/950	590/950
Normal operation max. fuel temp. (°C)	~1150*	1250-1350
Off-normal max. fuel temperature (°C)	~1670	<1600
Reactor power rating (MWt)	500	600
Primary He coolant inlet pressure (MPa)	9.0	7.1
Primary He flow rate (kg/s)	160	320
Core pressure drop (KPa)	202	58
Fuel composition	UO_2	UCO
Fuel enrichment (%)	5.0 startup, 9.6 equilibrium	19.8 fissile, 14.5 avg with fertile
Fuel burnup limit (GWd/MT U)	90	120

*The symbol ~ means approximately.

are due to the selection of UO_2 in the PBR versus UCO in the prismatic reactor in these particular examples. However, the burnup is not uniform in the prismatic block, so not all fuel compacts will reach their burnup limit. In the PBR, each pebble is recirculated until it reaches its burnup limit. Therefore, fuel utilization is actually better in the PBR. Both versions of the HTGR use fuel much better than LWRs, in which burnup limits of 50 GWd/MT (gigawatt-days per metric ton of uranium) are currently being sought.

Table 16.4 shows parameters for the pebble-bed version of the NGNP corresponding to those of Table 16.2 for prismatic reactors.

In the Westinghouse design, electricity is generated by a standard Rankine steam cycle. There are two intermediate heat exchangers (IHXs) that transfer heat from the primary helium coolant to a secondary helium loop. IHX A operates at temperatures up to 900 °C, which degrade the heat exchanger materials; thus, IHX A is replaceable.

IHX B, operating at temperatures below 710 °C, will last the whole 60-year design lifetime of the plant. As in PWRs, the use of a secondary loop isolates everything downstream from the primary system from any radioactive contamination that may occur in the primary system.

Table 16.4—Proposed pebble-bed configuration for the NGNP in November 2007

Parameter	Value
Thermal power (MWt)	500
Pressure vessel height (m)	30
Pressure vessel outside diameter (m)	6.8
Core outlet temperature (°C)	950
Core inlet temperature (°C)	400
Coolant pressure (MPa)	9
Cycle configuration	Indirect for electricity and hydrogen
Secondary fluid	He
Hydrogen plant power	10% of reactor power

16.5 HTGR SAFETY

In an HTGR, the helium coolant has very low density, as it is a gas. Also, the neutron absorption cross section of helium is extremely small. Therefore, the reactivity feedback coefficient from changes in the coolant density is negligible. Reactivity feedback in HTGRs comes primarily from the Doppler effect in the uranium-238 constituent of the fuel. This is a negative temperature feedback effect, which stabilizes the reactor in normal operation and shuts the reactor down in large temperature excursions as in a loss-of-coolant accident.

Because some coolant remains in the core even if it is completely depressurized, the condition in an HTGR analogous to a LOCA in an LWR is called a low-pressure conduction cooldown event (LPCC). This is the design-basis accident (DBA) in an HTGR. As explained in Chapter 11, a reactor must be able to sustain a DBA without loss of the systems, structures, or components necessary to ensure public health and safety. Since the SiC layer in the TRISO particle is the containment structure for radionuclides, an HTGR must sustain its DBA without fuel damage. Currently, a conservative limit of

1600 °C is imposed on the fuel particles. As noted above, this is accomplished in an HTGR by making the heat transfer path out of the core sufficiently short, usually by making the core a tall annulus with narrow radial thickness. Also, HTGRs operate at much lower power density (the amount of heat generated by a unit volume of the core) than LWRs—about 6 W/cm³ versus 54 W/cm³ in a BWR and 100 W/cm³ in a PWR; this means that the decay heat is more spread-out in an HTGR, so that the heating of any unit volume is less intense. The high heat capacity of the graphite materials—i.e., they absorb a lot of heat for a relatively small temperature increase—is also an important factor contributing to the integrity of the core during an LPCC.

In the event of fuel failures, radionuclides are contained by the graphite matrix of the fuel pebbles or compacts, the reactor pressure vessel, and the reactor building. All the NGNP designs call for putting the reactor below grade, so that the soil around the reactor not only provides the final heat sink for decay heat but also retards the escape of radionuclides into the atmosphere. Still, the primary defense against the escape of radionuclides is the prevention of fuel damage by natural heat transport processes. Figure 16.7 shows the results of a heat-transfer study in an early 300-MWt version of the NGNP.[45] The three curves represent three different locations in the core; the highest curve is the hottest location in the core. The fuel temperature remains well below 1600 °C. Analogous detailed studies will be presented for any HTGR design as part of the licensing process.

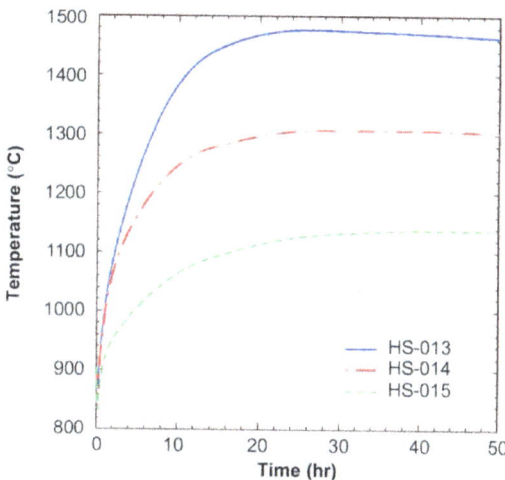

Figure 16.7—Fuel temperature versus time in an LPCC event in a 300-MWt pebble-bed version of the NGNP. From Reference 45, an externally released report of an agency of the U.S. government; this figure is in the public domain under the terms of Title 17, Chapter 1, Section 105 of the U.S. Code.

Another consequence of depressurization in an HTGR is entry of air into the reactor vessel. Oxygen in the presence of sufficiently hot graphite can cause oxidation. Graphite does not "burn" in air, in the sense that the heat released by oxidation causes self-sustaining combustion.[a] However, the heat released by radioactive decay can cause sustained oxidation, which resembles combustion. Such sustained oxidation occurred in the Chernobyl accident, which is discussed in Chapter 13. For all practical purposes, such combustion is a fire.

Because a Chernobyl-type fire in a graphite-moderated HTGR must be made impossible, it is appropriate to recall why it did happen in the Chernobyl reactor, and why it cannot happen in an HTGR. The Chernobyl reactor was graphite-moderated and water-cooled, a combination that produces a positive coolant void coefficient of reactivity in some circumstances. HTGRs are helium-cooled, and they have essentially no coolant void coefficient of reactivity. All the reactivity feedback processes in HTGRs are negative. Furthermore, the safety rods in the Chernobyl reactor were tipped with graphite and a hollow segment, which introduced positive reactivity as they were being inserted; this was a serious design flaw. This introduction of positive reactivity into a condition in which the reactivity was already positive briefly created prompt supercriticality, which caused a steam explosion and hydrogen explosions. The lid was blown off the reactor vessel, one-fourth of the core was blown out into the atmosphere, and the top of the remaining core was exposed to air. The remaining core then oxidized fiercely.

None of these conditions exists in an HTGR. In an HTGR LPCC, the coolant ducts leading to the pressure vessel are presumed to suffer a double-ended guillotine break (i.e., a section of the ducts is removed so that no impediment exists to the immediate blowdown of the coolant in the pressure vessel). The vessel depressurizes within a few seconds, but the vessel itself is not ruptured. There is no plausible process by which a vessel rupture could occur. The Doppler effect will shut the reactor down, and there is no steam to cause a steam explosion. Then the air does not flow quickly into the vessel; it seeps in first by molecular diffusion, then, after many hours, it begins to be sucked in by natural convection. When natural convection begins, oxygen in the air oxidizes the graphite in the lower end of the bottom reflector, but it is depleted by the oxidation process before it reaches the core and does not cause oxidation in the core. Before the oxidation front reaches the core, the reactor has

---

[a] This claim was made and supported by experimental observations reported in a short course I attended on gas-cooled reactors in San Diego, CA, in approximately 2000. The course was sponsored by the U.S. Department of Energy and General Atomics, Inc.

cooled sufficiently by conduction, thermal radiation, and natural convection that the oxidation process ceases. This event sequence was predicted analytically for a PBR by the INL[52] and shown experimentally in Germany and Japan.[53, 54] The results of the INL calculations are shown in Figure 16.8. The natural convection and the oxidation of the lower reflector begin at 214 hours. The rate of cooling in the lower part of the core increases at that time because of the natural convective cooling.

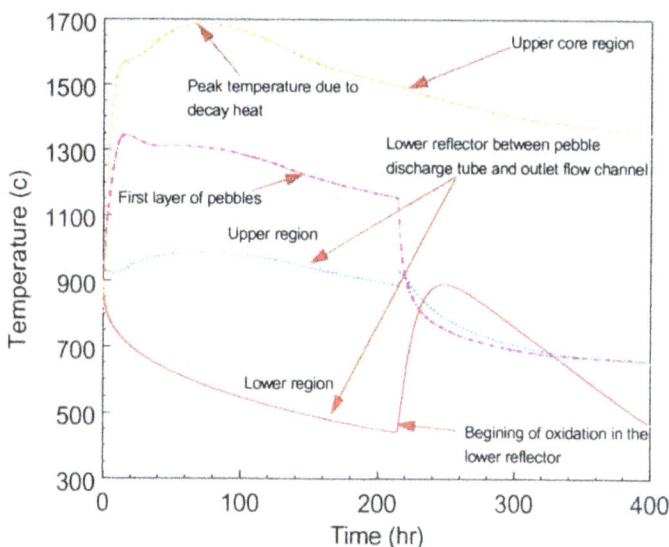

Figure 16.8—Temperature history in a PBR during an air ingress event. From Reference 52, an externally released report of an agency of the U.S. government; this figure is in the public domain under the terms of Title 17, Chapter 1, Section 105 of the U.S. Code.

A very impressive safety demonstration was performed in China's HTR-10.[12] The reactor's coolant flow was shut off while the reactor continued to run. The reactor temperature increased, but the neutron chain reaction was shut down by the Doppler effect without operator intervention (see Section 5.4),[b] and a balance was reached be-

---

[b] In Section 5.4, the role of the Doppler effect as a stabilizing process is described: If a power fluctuation produces a temperature increase, Doppler broadening of neutron absorption resonances causes increased neutron absorption and reduces the power. The opposite happens if a power fluctuation produces a temperature drop. However, in these scenarios, coolant flow is assumed to be uninterrupted. In the HTR-10 experiment, coolant flow was shut off, so the temperature continued to rise and the Doppler effect continued to suppress the power until it

tween radioactive decay heating and heat dissipation by thermal conduction while the peak core temperature remained below the 1600 °C temperature limit. Even though the HTR-10 core is not annular, this benign passive shutdown was possible because the HTR-10 power density is low, its core has a high surface-to-volume ratio, and the reactor size and power are small. These qualities are preserved in the HTR-PM demonstration plant. The electric power output of 210 MWe will be provided by two PBR modules of 250 MWt each. The small power rating of the reactors was determined by the requirement for passive safety and the desire to avoid using the annular configuration of the South African designs and the pebble-bed version of the NGNP.

Generation IV HTGRs, like all Generation IV reactors, will provide new levels of safety even beyond the extremely high safety offered by Generation III and III+ reactors. They will essentially be immune to large releases of radionuclides into the environment.

---

ultimately shut the neutron chain reaction down.

CHAPTER 17
BREEDER REACTORS

Uranium-235, a fissile isotope, constitutes only 0.7% of natural uranium. Almost all the rest is U-238, which is not fissile. (Recall that fissile nuclides can undergo fission when they absorb thermal neutrons.) However, U-238 is said to be fertile, which means that it can be converted into a fissile nuclide by the parasitic absorption of a neutron—i.e., absorption that does not induce a fission reaction. The reaction is

$$^{238}U\ (n,\gamma)\ ^{239}U \rightarrow \beta + {}^{239}Np; \ ^{239}Np \rightarrow \beta + {}^{239}Pu.$$

The product of the first beta-decay reaction is neptunium-239, and this decay reaction has a half-life of 23.45 minutes. The product of the second decay is plutonium-239, which is fissile, and this reaction has a half-life of 2.3565 days. The half-life of plutonium-239 is 24,000 years.

In thermal reactors, some plutonium-239 is produced. But more can be produced in reactors with a higher-energy neutron spectrum. The number of new neutrons produced in a fission reaction is dependent on the energy of the neutron that induces the reaction. Two commonly used parameters measure the efficiency of neutron production. One of them is the average number of neutrons produced per fission reaction, denoted by the lower-case Greek letter nu (υ), and the other is the average number of neutrons produced by fission per neutron absorbed in fuel, denoted by the lower-case Greek letter eta (η). The latter parameter accounts for the effect of parasitic absorption in fuel. Figure 17.1 shows η as a function of neutron energy for several fissile nuclides.

For criticality, η must be at least equal to 1.[a] If $\eta > 1$, some neutrons may be left

---

[a] The condition $\eta = 1$ is necessary, but not sufficient, for criticality. In Chapter 6 we defined criticality as the condition in which exactly one neutron emitted from each fission reaction, on average, goes on to induce a subsequent fission. Since η is the average number of neutrons

over to convert a fertile material to a fissile fuel, if the extra neutrons aren't lost by leakage into neutron-absorbing shielding materials around the core and reflector or absorbed by nonfuel materials in the core. If $\eta > 2$, it may be possible to produce more fissile fuel by conversion of fertile material than is consumed by fission. Reactors in which more fissile fuel is produced than consumed are called breeder reactors. Breeder reactors offer the possibility of using the entire uranium resource instead of only 0.7% of it.

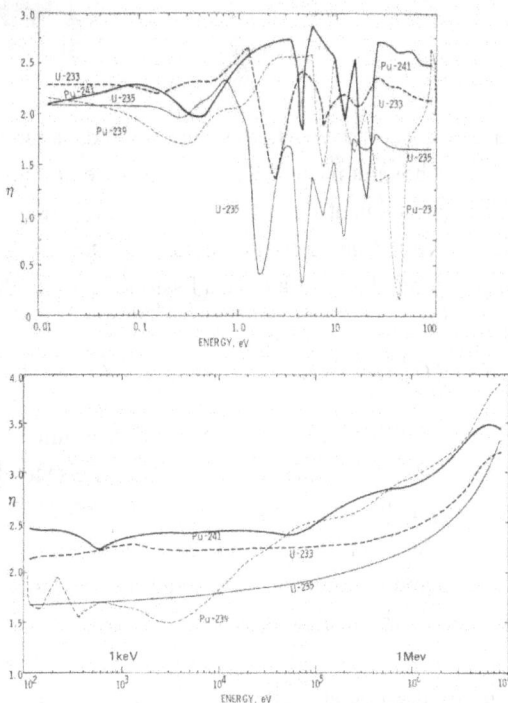

Figure 17.1—η versus neutron energy for several fissile nuclides. From "Assessment of Thorium Fuel Cycles in Pressurized Water Reactors," N. L. Shapiro, J. R. Rec, and R.A. Matzie, EPRI-NP-359, Electric Power Research Institute, Inc. (EPRI), February 1977. Copyright owned by EPRI. Figure reproduced courtesy of EPRI.

---

emitted by fission per neutron absorbed *by fuel*, we can easily understand the situation where $\eta > 1$ but most of the neutrons are absorbed by control rods, so that even though more than one neutron is emitted by fission for each neutron absorbed by fuel, not enough neutrons are absorbed by fuel for the reactor to remain critical. For a given neutron energy, the parameter η is a property of the nuclide, whereas criticality depends on the configuration and composition of the whole reactor.

For all of the nuclides shown in Figure 17.1, η is considerably greater at energies above 1 MeV than it is for thermal neutrons, which have energy in the range of 0.025-0.09 eV. In particular, for U-235, $\eta \leq 2.1$ (except in a narrow band at 1 eV) unless the neutron energy is nearly 1 MeV, and then it increases to above 2.5 for neutron energies greater than about 2 MeV. Use of uranium as fuel in a breeder reactor requires a high population of neutrons with energy above 1 MeV. Pu-239 is considerably better than U-235 as a source of neutrons for breeding. At first look, a reactor that breeds more Pu-239 than it consumes may seem like some sort of perpetual motion machine, but it is not. The Pu-239 all comes from U-238, which eventually has to be replaced in order for the reactor to continue breeding.

High-energy neutrons are fast neutrons. Thermal neutrons travel at an average of 2200 m/s at room temperature, and an average of up to 4200 m/s in high-temperature reactors. Neutrons are usually emitted from fission reactions with energies of 2 to 3 MeV, which works out to as much as 2.4×10^7 m/s. A reactor in which fission is predominantly induced by fast neutrons is called a fast reactor, and a fast reactor that breeds is called a fast breeder reactor. (The term "fast breeder reactor" does not mean that the reactor breeds fissile fuel rapidly.) A predominantly high-energy neutron spectrum is also called a "hard" spectrum, in contrast to a spectrum with a high population of thermal neutrons, which is called a "soft" spectrum.

The purpose of breeder reactors is to produce fuel for thermal reactors. Except for the Pu-239 that needs to return to the reactor to start a new fuel cycle, the Pu-239 bred in the fast breeder reactor is taken from it and used as fuel in thermal reactors after reprocessing. Reprocessing is presently not done in the U.S., and the U.S. has no operating breeder reactors, but France does reprocess fuel, and the French have had a breeder reactor development program. Reprocessing is discussed in Chapter 21.

So that the neutron energy spectrum can remain hard, fast reactors have no moderator. This allows the core to be more compact, so fast breeder reactors can operate at higher power density. For reactors of high power density, a coolant is needed that has high heat capacity and high thermal conductivity, such as a liquid metal. Most fast breeder reactors have used liquid sodium as a coolant. Fast breeder reactors cooled by liquid metals are called liquid-metal fast breeder reactors (LMFBRs).[1]

17.1 HISTORY OF BREEDER REACTORS

Liquid-metal fast reactors have been used from almost the beginning of the nuclear era. In 1952, the Experimental Breeder Reactor I (EBR-I), on what is now the site of the Idaho National Laboratory, provided the city of Arco, Idaho, with the world's first nuclear-generated electricity.[2] Two years later the Soviet PTGR became the first commercial LMFBR.[3]

A second LMFBR in Idaho, EBR-II, operated from 1965 to 1995. It generated 19 MWe for local electrical utilities, and it served as a test bed for the LMFBR fuel cycle and materials development.[4]

The second nuclear submarine in the United States Navy, the U.S.S. Seawolf (SSN-575), was initially powered by a liquid-sodium-cooled reactor that supplied heat to a superheated steam cycle.[5] (Superheated steam is steam at a temperature above the saturation temperature for the pressure at which the steam is maintained. Recall from the discussion of the Carnot Cycle in Chapter 7 that water boiled at constant pressure remains at constant temperature until it completely vaporizes. If, after the water vaporizes completely, more heat is added at constant pressure, the temperature will increase and the steam is called superheated.) Seawolf SSN-575 was one of four U.S. Navy ships to bear that name; the fourth is a nuclear-powered attack submarine currently in service. Seawolf SSN-575 was launched in 1957; she was returned to the shipyard at the end of 1958 for replacement of her power plant by a standard submarine PWR because of problems with the superheated steam system.

Unit 1 of the Enrico Fermi Nuclear Generating Station near Detroit, Michigan, was a commercial LMFBR of 94 MWe that operated from 1957 to 1972. On 5 October 1966 Fermi 1 suffered a partial fuel meltdown when a loose piece of zirconium blocked coolant flow in several fuel channels, but the reactor was repaired and resumed operation. It was finally shut down on December 31, 1975. No radioactive materials were released from the reactor vessel by the accident.[6]

Many other countries have built sodium-cooled reactors, including the United Kingdom, the Soviet Union,[7] India,[8] Japan,[9] Germany,[10] and France. The French effort has been the most productive. The French LMFBR program began with the experimental Rapsodie reactor, which began operation in 1967, and continued with the Phénix prototype reactor, which began operation in 1973.[11] Phénix operated at 233 MWe until 1998, when it was shut down for modifications until 2003. It resumed operation at 140 MWe until March 2009, and continued operation until October 2009 as a research reactor, addressing problems in waste disposal.[12]

The largest fast reactor yet built was the French Superphénix, which was rated at 1200 MWe but never operated at that power level.[13] For most of its operational life, the reactor produced between zero and 33% of design power. Construction of Superphénix began in 1974, and the plant operated from 1985 to 1998. Superphénix was plagued from the start by both political opposition and operating problems.

Opposition to all things nuclear was gaining political clout during the mid-1970s; in July 1977, a march of 60,000 people to protest Superphénix was broken up by French riot control police with the death of one protester and the serious injury of over one hundred others. On the night of January 18, 1982, opponents stooped to terrorism by launching five rocket-propelled grenades at the unfinished containment building. Two grenades narrowly missed the reactor vessel, which had not yet been loaded with fuel. Operation was impeded by corrosion and leaks in the liquid-sodium cooling system. In 1990 the reactor building suffered structural damage from heavy snowfall. By December 1996 the problems with the cooling system had been fixed, and the reactor reached 90% of its design power. However, when a political coalition dominated by ministers from the Green Party took power in 1997, the government initiated steps to close the plant permanently. Special focus on Superphénix by antinuclear groups was probably motivated both by the hazards associated with the sodium coolant and by its purpose of producing plutonium, which is erroneously regarded as a substance of unprecedented danger (see Section 4.8).

Although some sodium-cooled fast reactors, such as EBR-II, have operated successfully for many years, concerns over the safety of sodium coolant have led to the consideration of other kinds of breeder reactors.

For uranium-233, η is sufficiently greater than 2 for thermal neutrons that breeding might be practical in a thermal reactor using water as a coolant and thorium-232 as the fertile material by applying the reaction

$$^{232}\text{Th} \, (n,\gamma) \, ^{233}\text{Th} \rightarrow \beta + \, ^{233}\text{U}.$$

Thorium is an abundant resource, and other efforts have been devoted over the years to applying this fuel cycle (e.g., the German THTR pebble-bed reactor mentioned in Chapter 16). The Shippingport reactor in Pennsylvania, which began operation in 1957 as the first commercial nuclear power plant in the U.S. (generating 60 MWe), was converted to an experimental light-water breeder reactor in 1977. The experiment ended in 1982. Subsequently, the reactor was decommissioned and dismantled, and the site was returned to unrestricted use.[14]

Lead is a highly toxic material, but it does not burn and its boiling point is very high (1740 °C,[15] compared to 883 °C for sodium).[16] Consequently, molten lead or a lead-bismuth eutectic mixture[b, 17] can be used in place of liquid sodium. The lead-

---

[b] The melting point of a solid mixture depends on the proportions of the different constituents of the mixture. For a mixture of certain specific proportions, the melting point is a minimum.

cooled fast reactor (LFR)[18] is a Generation IV reactor concept that has been tested on Soviet submarines, but solidification of the coolant has been a problem. Further development is needed for LFRs, as with all Generation IV concepts.

Finally, breeding is possible in molten salt reactors (MSRs), either fast reactors cooled by molten salts such as sodium chloride or in thermal reactors cooled by molten salts such as a fluorine-lithium-beryllium ("FLIBE") eutectic mixture.[19] The light lithium and beryllium nuclei in FLIBE serve as moderators. In thermal molten-salt breeder reactors, the thorium-uranium fuel cycle must be used. In some MSR designs, the fuel is contained in conventional fuel rods, but in others, the fuel is composed of uranium tetrafluoride and dissolved in the coolant. Then the possibility of an accident involving a core meltdown disappears, as the fuel is already melted. This concept has been tested on a small scale at the Oak Ridge National Laboratory (ORNL), first in a series of experiments in the 1950s aimed at developing a nuclear-powered aircraft[20] (definitely a bad idea, for reasons that should be obvious), and next in the Molten-Salt Reactor Experiment (MSRE), which operated from 1965 to 1969.[21] Both the molten-fuel and the solid-fuel versions of MSRs are being explored as Generation-IV designs, as they have many advantages over conventional reactors, including a high degree of passive safety. But they also will require a great deal of development before they can be deployed on a commercial scale.

Improved versions of sodium-cooled LMFBRs are also being explored as Generation IV concepts. Because of the longer development history of sodium-cooled LMFBRs, the time required to bring Generation IV sodium-cooled LMFBRs into production may be less than that required for other breeder reactor concepts.

17.2 DESIGN OF LMFBRS

Breeder reactors may take many forms, as summarized above, but since most actual breeder reactors have been LMFBRs, this section focuses on them. Two distinct types of LMFBRs have been built: pool-type and loop-type reactors. Both types are illustrated in Figure 17.2.[22]

A mixture with those proportions is called a eutectic mixture. The melting temperature of a eutectic mixture is called the eutectic temperature. Molten mixtures which do not have the composition of a eutectic mixture comprise both liquid material and solid crystals at temperatures between the eutectic temperature and the melting point of that particular mixture. In a eutectic mixture, all the constituents melt at the same time.

As shown in the figure, in both types of LMFBR heat passes from the primary sodium loop to a secondary sodium loop, and then to a water loop that goes to a steam turbine to generate electricity. Ideally, the primary coolant would have a very low neutron absorption cross section, but none of the possible coolants have that property. Thus, the primary sodium becomes activated, and the secondary sodium loop isolates the activated sodium from the water coolant. Regardless of the activation, it is important to keep water and sodium separated, as they can react explosively upon contact.[23]

In the pool-type LMFBR, the reactor vessel is a tank, and the primary sodium coolant fills the tank to a level near the top. The core, the primary coolant pumps, and the primary heat exchanger are immersed in the tank, and the primary coolant pumps suck coolant from the tank into a core inlet plenum within a flow baffle that directs the coolant flow into the core. Another set of piping takes heated coolant from the core outlet plenum into the primary heat exchanger. In the loop-type LMFBR, the reactor vessel fits around the core more closely, and the primary heat exchanger is external to the reactor vessel as in most reactor designs.

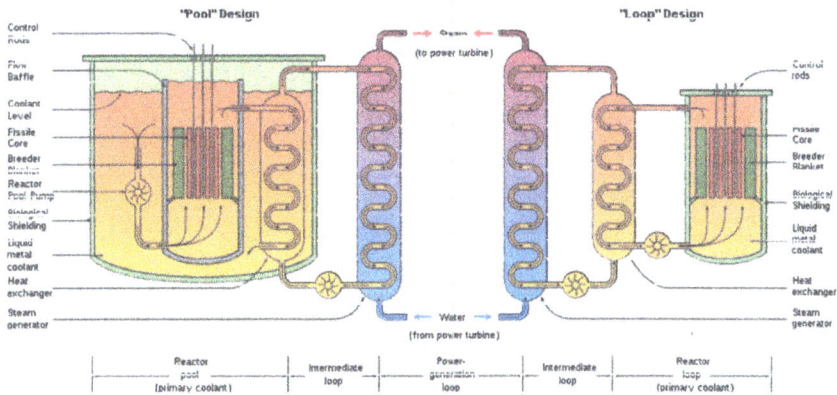

Figure 17.2 –Schematic of pool- and loop-type LMFBRs. This file was created by Graevemoore and put on the Wikipedia Commons. It may be reproduced freely under the terms of the Creative Commons Attribution ShareAlike 3.0 license, q.v. (http://creativecommons.org/licenses/by-sa/3.0/)

It is convenient to separate the active core, where most of the fission occurs, from the region where most of the breeding occurs. This region surrounds the active core, so the term "blanket" is a natural choice for its name. Figure 17.2 shows blanket regions (in green) around the cores of both versions of the reactor.

Because of sodium's high heat capacity and thermal conductivity, not much sodium is needed to carry away the heat generated in the fuel rods. The fuel assemblies in an LMFBR comprise thin stainless-steel-clad fuel rods around which wires are wrapped helically, as shown in Figure 17.3. The wires around the fuel rods provide separation between adjacent rods to create coolant flow paths between rods. The rods are bundled in hexagonal fuel assemblies. The fuel in the rods is formed in pellets as in LWRs, but the diameter of the pellets is smaller than in LWRs.

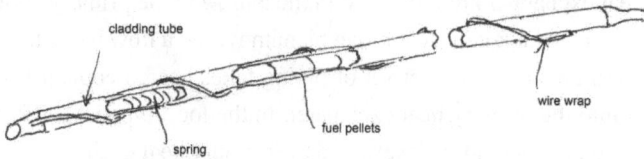

Figure 17.3—Fuel rod design in a typical LMFBR design.

Since breeder reactor technology has never been widely deployed, there is no typical LMFBR design for which to present design parameters. Table 17.1 shows design parameters for Superphénix.

Table 17.1—Design parameters for Superphénix[24]

Parameter	Value
Thermal power	3000 MW
Electric power	1200 MW
Primary coolant pressure	Atmospheric
Core inlet temperature	395 °C
Core outlet temperature	545 °C
Reactor vessel inside dimensions	21 m diameter x 19.5 m height
Reactor vessel thickness	2.5 cm
Active core height	1 m
Active core diameter	3.55 m
Reactor configuration	Pool-type

17.3 SAFETY OF LMFBRS

In addition to the hazard in sodium-cooled reactors from the potential contact of sodium and water in the heat exchanger between the secondary sodium loop and the steam loop, there are reactivity feedback concerns in sodium-cooled reactors. Even though sodium is not a strong moderator, it does slow neutrons down when they collide with it, and therefore local sodium boiling would harden the neutron energy spectrum. This effect would increase both η in the fuel isotope (U-235 or Pu-239) and the fission rate in U-238, giving an increase in reactivity—a positive void reactivity coefficient. On the other hand, a harder neutron energy spectrum suffers more leakage (a small effect in a large reactor) and incurs greater absorption through the Doppler effect; both of these phenomena give negative void reactivity. The overall behavior of the reactivity depends on the details of the reactor configuration. The issue is moot if the reactor is designed so that sodium boiling can never happen. Void reactivity feedback is much smaller in lead-cooled or lead-bismuth-cooled reactors, because both lead and bismuth nuclei are very massive, and they moderate neutrons very slowly.

In both the pool-type and loop-type LMFBRs, and also in molten-salt-cooled reactors, the pressure is essentially atmospheric, so depressurization accidents will not occur. These coolants have high heat capacity and thermal conductivity, and the radioactive decay heat can be carried away by natural convection if power to the primary coolant pumps is lost. These features are significant safety advantages for breeder reactors. The MSR with dissolved fuel takes these advantages even further by eliminating the possibility of a core meltdown accident.

Whatever type of breeder reactor is built in the future will have to meet Generation IV safety standards, which require a substantial improvement over the safety of today's LWRs.

CHAPTER 18

RESEARCH REACTORS

Reactors built to perform research have taken many forms over the years. Some of them have been prototypes for specific power reactor designs, and many of these have already been mentioned in previous chapters. Others have been designed specifically to study reactor physics or the behavior of materials in high-radiation environments. This chapter is not a comprehensive review of such reactors; instead, it gives examples of a few important reactors or reactor types.

18.1 TRIGA REACTORS

The TRIGA reactor[1] (for Training, Research, Isotopes, General Atomics) is a swimming-pool reactor. In this type of reactor, a small core is immersed in an open tank of water, so that the core can be seen from above. Figure 18.1 shows an example of a TRIGA reactor. TRIGA reactors are made by General Atomics for universities, government and industrial laboratories, medical centers for training of students, basic research, nondestructive testing of materials, medical radioisotope production, and treatment of tumors. General Atomics has sold 66 TRIGA reactors in 24 countries as of 2008. They range in steady-state thermal power from 0.1 to 16 MW, although some can be pulsed up to 22,000 MW for very short time intervals.[2] The relatively low steady-state power allows the core to be cooled by natural convection. The pressure in the tank is just the hydrostatic pressure of the water. The core is composed of plates of uranium-zirconium hydride, which has a large negative temperature coefficient of reactivity. These features provide such a high level of safety that TRIGA reactors are often located on college campuses. As it was put by Frederic de Hoffman, head of General Atomics at the time TRIGA reactors were developed, they can be trusted "even in the hands of a young graduate student."

Figure 18.1—A TRIGA reactor (this figure was obtained from the Wikimedia Commons and is in the public domain).[3]

The blue glow emanating from the core is Cherenkov radiation.[4] In general, Cherenkov radiation is electromagnetic radiation emitted when a charged particle travels through a non-conducting medium at a higher speed than the speed of light in that medium. As explained in Section 3.4, the speed of light c is the speed limit for any object in any reference frame. However, c is the speed of light in a vacuum. In materials, the speed of light is less than c. In water, the speed of light is only $0.75c$. Therefore, it is possible for high-energy particles passing through materials to exceed the speed of light in those materials, and when they do so in electrically insulating materials, they emit Cherenkov radiation.

18.2 ZERO-POWER REACTORS

Reactors operated at such low power that they need minimal shielding and no coolant are called zero-power reactors, even though the power at which they operate is not truly zero. They are used to make physics measurements, for example measurements of neutron flux distribution to validate reactor physics computer codes. Examples of zero-power reactors are the VENUS reactor at the Belgian

Nuclear Research Center,[5] the PROTEUS reactor at the Paul Scherrer Institute in Switzerland,[6] and the Zero-power Physics Reactor, originally named the Zero-power Plutonium Reactor, (ZPPR) at the Idaho National Laboratory (INL),[7] and the ATR Critical Facility (ATRC), also at the INL.

The VENUS reactor and the PROTEUS reactor can be reconfigured readily to simulate a wide variety of reactors. Both have been used for code validation studies in the International Reactor Physics Experiment Evaluation Project,[8] a program aimed at collecting a wide variety of pedigreed experimental data for use in confirming the accuracy of reactor physics computer codes. Figure 18.2 illustrates the VENUS reactor. Figure 18.2 (a) shows the exposed core during a loading operation, and Figure 18.2 (b) shows the hall in which the reactor resides. Figure 18.3 shows the PROTEUS reactor as it was configured for an experiment on a gas-cooled fast reactor concept.[9]

ZPPR was used from 1969 to 1992 to study LMFBR physics. The reactor, illustrated in Figure 18.4, is called a split-table reactor. It consists of two halves, which are separated for loading and brought together to achieve criticality. Even after the reactor had operated for many years, the radiation emitted by the core was so low that reloading could be done by hand, as shown in the photograph.

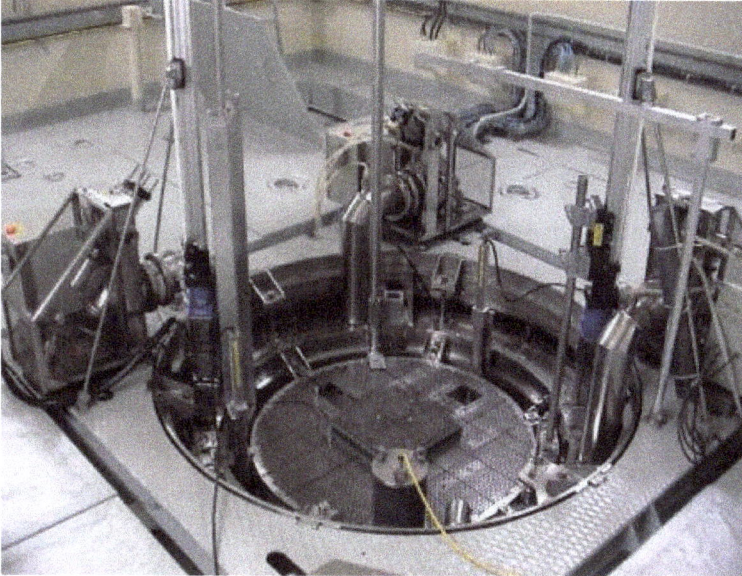

(a) **View into the reactor**

(b) The reactor hall

**Figure 18.2—The VENUS reactor and its setting. Copyright SCK•CEN
(Studiecentrum voor Kernenergie/Centre d'Etude de l'Energie
Nucléaire, also known as the Belgian Nuclear Research Centre, Brussels,
Belgium), reproduced with permission from SCK•CEN.**

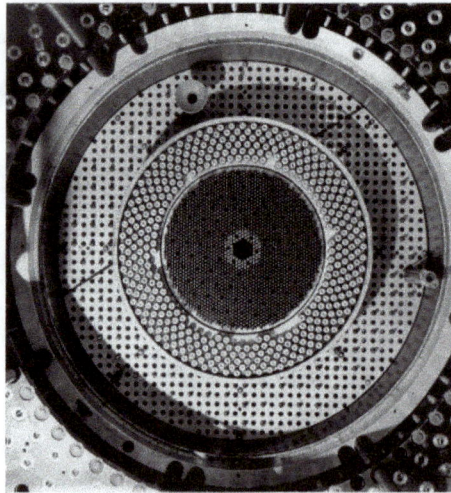

**Figure 18.3—The PROTEUS reactor configured for an experiment on a gas-cooled fast
reactor design. From Reference 9, reproduced by courtesy of the Paul Scherrer Institute.**

Figure 18.4—ZPPR , with the two halves separated for loading. Reproduced by courtesy of Argonne National Laboratory (ANL); ANL is managed and operated by UChicago Argonne, LLC, for the U. S. Department of Energy under Contract No. DE-AC02-06CH11357. The image was created by a contractor of the U.S. Government under said contract; accordingly, the U.S. Government retains for itself, and others acting on its behalf, a paid-up, nonexclusive, irrevocable worldwide license in said image to reproduce, prepare derivative works, distribute copies to the public, and perform publicly and display publicly, by or on behalf of the Government.

The ATRC is a zero-power mockup of the Advanced Test Reactor, which is discussed in the next section.

18.3 THE ADVANCED TEST REACTOR

The Advanced Test Reactor (ATR),[10, 11, 12] located at the INL, is a high-neutron-flux reactor designed primarily for accelerated testing of materials such as prototype nuclear fuel materials. The neutron flux is one or two orders of magnitude higher than in a normal power reactor, so that lifetime neutron exposure ("fluence") can be achieved in short time intervals. The reactor can operate at thermal power of up to 250 MW, and its test cycles are typically about seven weeks long. The ATR is also used to produce cobalt-60 for medical purposes and plutonium-238 for radioisotopic power sources used in spacecraft. Figure 18.5 is a photograph of the core taken from the upper plenum of the pressure vessel during reactor operation. As with the TRIGA reactor, the blue glow is Cherenkov radiation.

Figure 18.5—The ATR core in operation.[13] This photograph by Mike Howard was taken from the Wikimedia Commons and is licensed under the <u>Creative Commons Attribution ShareAlike 2.0</u> license.[14]

The core is constructed in a four-leaf-clover configuration. The leaves of the clover, called lobes, are loosely coupled neutronically, so that they can operate at significantly different power levels. Control of the lobes is obtained by turning beryllium drums containing hafnium neutron absorber material on portions of their circumferences; the drums are labeled outer shim control cylinders in Figure 18.6. The closer the hafnium is to a lobe, the more the flux in that lobe is suppressed. Fine flux adjustments are made by inserting hafnium "shim" rods into holes in the neck shim area (the blue cruciform region in Figure 18.6). The nine tubes extending into the lobes from above are called flux traps or in-pile tubes; they contain the largest test specimens that the ATR is capable of irradiating and expose specimens to the highest flux (up to 10^{15} neutrons/cm^2-s for thermal neutrons or up to 5×10^{14} neutrons/cm^2-s for fast neutrons). Some of the flux traps may contain smaller tubes. Five of the flux traps contain pressurized-water loops, enabling samples of fuel or structural materials to be tested for any proposed pressurized-water reactor design. There are many smaller specimen tubes in the core, in the beryllium reflector around the core, and in the irradiation tanks outside the pressure vessel; in Figure 18.6, these are the holes labeled B, H, and I positions and the squares in the outer irradiation tanks.

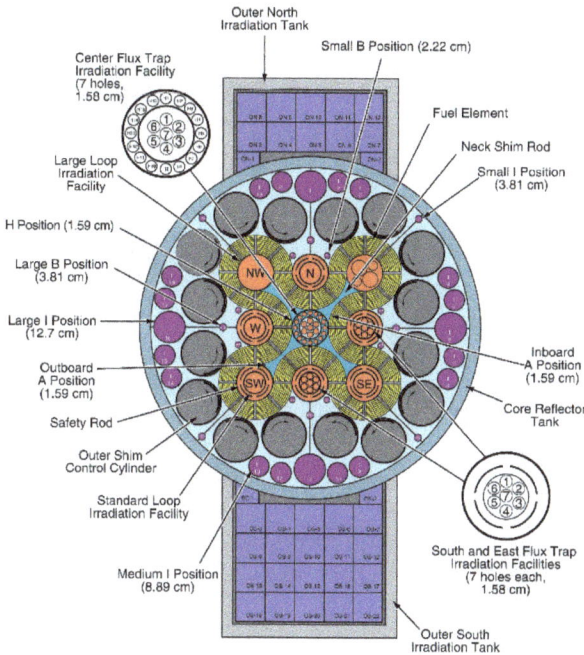

Figure 18.6—Layout of ATR test specimen positions. This figure was obtained from Wikimedia Commons; the author is Frances M. Marshall of the INL, and the figure is in the public domain.

The ATR is a small reactor, with a core volume of only 1.4 m^3. It operates at 60 °C and 390 psi. As pointed out in Reference 11, these conditions are similar to those in a home water heater. Thus, even though the reactor operates at high neutron flux, and thus at high power density, its operating environment is quite benign.

The four-leaf-clover configuration is formed by fuel assemblies of curved cross section. An ATR fuel assembly is shown in Figure 18.7. Each assembly subtends a 45° arc about its center of curvature. All the assemblies are geometrically identical, so to change the concavity of a sequence of assemblies, all one has to do is to rotate the assemblies 180° around their longitudinal axes. Each fuel assembly comprises 19 metal plates, which are formed as sandwiches of aluminum-uranium alloy surrounded by aluminum. The first and last plates are thicker for added strength. During the shutdown periods between operating cycles, fuel assemblies are shuffled around to meet the irradiation requirements for the next cycle, and the reactor always contains a mix of fresh fuel and fuel at varying degrees of burnup.

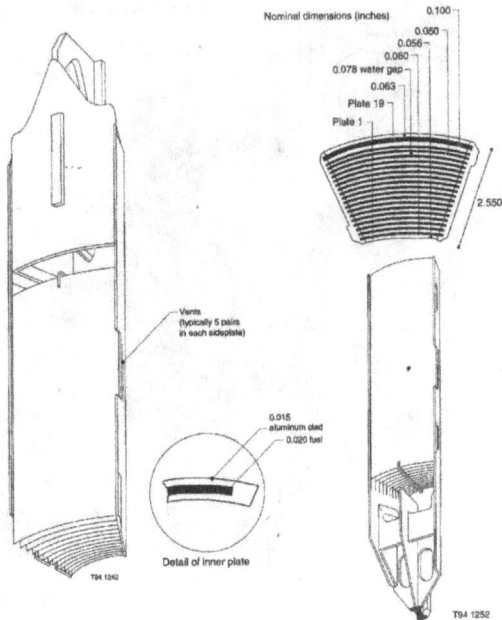

**Figure 18.7—An ATR fuel assembly. From Reference 12.
Reproduced by courtesy of Idaho National Laboratory.**

Because of the wide variety of experiments performed in the ATR, it is important to verify for each operating cycle that no experiment introduces unsafe conditions such as a positive void reactivity coefficient. Thus, before each operating cycle, INL reactor physicists perform a detailed Core Safety Assurance Package (CSAP). These packages are rigorously reviewed by independent reviewers. During part of my career at the INL, I performed some of the independent reviews of CSAP calculations.

The ATR is the most versatile test reactor in the world. For much of its history, the ATR was used primarily to test fuels used in the U.S. Navy's nuclear-powered ships, but now the reactor has been designated as a National Scientific User Facility, and projects with valid scientific or engineering goals are solicited. For information on the User Facility, go to the link of Reference 15; a user's guide in PDF format is available in Reference 16.

CHAPTER 19
SMALL MODULAR REACTORS

In the industrial countries where nuclear power was first developed, great demand for electric power exists, and large power plants have been easily absorbed into the electrical power grids. Large power plants offer economy of scale—costs per kilowatt decrease as the plant is made larger. Typical nuclear power plants in the United States generate about 1000 MW of electricity, about the same as a large coal-fired plant, and some Generation III+ reactors will generate up to 1700 MWe (see Chapter 11). However, in developing countries the electric infrastructure is limited, and it is more practical to add generating capacity in smaller increments so demand and supply can grow together. Furthermore, the high capital cost of large reactors may be prohibitive for smaller national economies. Also, smaller units are appropriate at remote locations even in industrialized countries. Finally, even at installations intended for large-scale power generation, there may be advantages to adding generating capacity in increments rather than all at once. Therefore, in recent years there has been a growing interest in reactors classified by the International Atomic Energy Agency (IAEA) as small (up to 300 MWe) or medium-sized (between 300 and 700 MWe).[1]

At the same time, the idea of modular reactors has gained momentum. Modular reactors are identical units that could be deployed in clusters at a single site or distributed among multiple locations. Production of identical units not only lowers manufacturing costs but also simplifies licensing.

Small reactors of many kinds have been deployed for many years. The U.S. Navy operates submarines and aircraft carriers with PWR power plants up to about 165 MWe.[2] Russia, India, France, and China operate nuclear-powered military vessels; Russia also operates civilian nuclear-powered icebreakers. The reactors in all of these vessels fall into the small-reactor classification.[3] Russia has operated four 62 MWt/11 MWe reactors in the Siberian Arctic for district heating and electricity production since 1976; these reactors are water-cooled and graphite-moderated like the RBMK reactors, but they have operated reliably.[1] However, most of the renewed interest in small reactors is focused on new modular designs with passive safety.

As summarized in Reference 1, small modular reactor (SMR) concepts span a variety of reactor types. Russia is developing several small PWRs based on their icebreaker reactors to be deployed on barges, along with some small BWR and PWR designs that can be built in factories and transported to sites where their power is needed, or even transported from one place to another. Argentina, South Korea, Japan, France, and China are developing small LWRs. In the United States, Westinghouse, NuScale, and Holtec were recently developing small-to-medium modular LWRs; as of early 2017, only the NuScale design was receiving U.S. government funding. All of these designs promise some degree of passive safety. The high-temperature gas-cooled power plant concepts discussed in Chapter 16 all call for modular reactors of small or medium-sized power output, although they will not be physically small. Companies in the United States, Russia, South Korea and Japan are working on small fast reactor designs, using either sodium or lead-bismuth eutectic as coolant. Many of these designs are in the early stages of conceptual development.

Two modern SMR designs are described below to illustrate the nature of SMR concepts. The first is the NuScale water-cooled design, and the second is the Gen4 Energy liquid-metal-cooled reactor.

19.1 THE NUSCALE REACTOR

The NuScale reactor design[4] began in 2000 as a collaboration of the U.S. Department of Energy (USDOE), Oregon State University (OSU), and the Idaho National Engineering and Environmental Laboratory (INEEL), as the Idaho National Laboratory was then known. One purpose of the collaboration was to produce a conceptual design for a small simplified LWR; the result of the effort was a report on a design called the multi-application small LWR (MASLWR).[5]

When the collaboration ended in 2003, OSU scientists continued to develop the SMR concept. In 2007, NuScale Power[6] was formed to pursue commercialization of their design.

The design layout of the NuScale SMR design is shown in Figure 19.1. The components in the red box, labeled "off-the-shelf," are the elements of a standard steam cycle. The remainder of the figure shows one reactor module; the power output of a module is 45 MWe, but as many as 12 modules can be housed in one reactor building, for a total power plant capacity of up to 540 MWe. The modules are placed below the ground surface for resistance against hurricanes, tornadoes, and aircraft impacts. A module is immersed in a stainless-steel-lined concrete pool (3) containing enough water to cool the core for 30 days in a post-accident scenario without replenishment.[7] Essentially, the NuScale reactor design is a PWR, but it is much different from a typical large PWR. The small size of the reactor permits enhanced safety in several ways, as explained below.

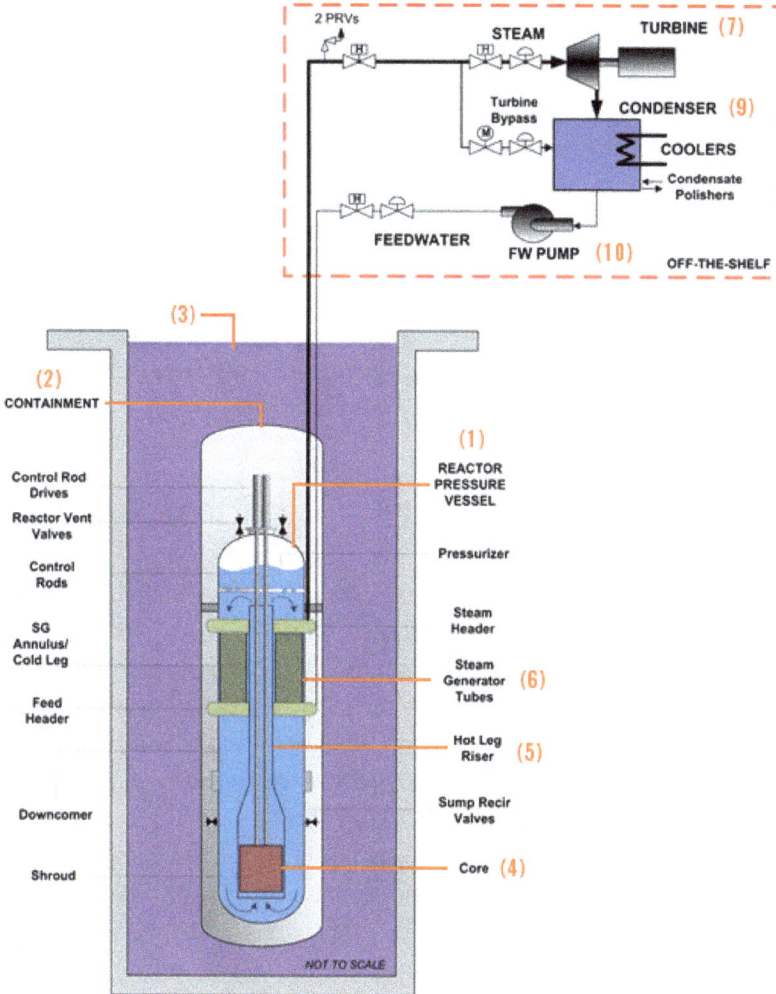

Figure 19.1—Conceptual layout of the NuScale SMR.
Reproduced courtesy of NuScale Power, LLC.[6]

First, there are no coolant pumps in the primary cooling system. The reactor pressure vessel (1) is a tank of water in which the core (4) is immersed; the coolant flow is driven upwards through the core by natural convection and channeled by the hot leg riser (5) through the steam generator (6), which is contained within the pressure vessel, and returns to the core inlet by the force of gravity after it becomes denser by giving its heat to the secondary loop in the steam generator.

The pressure vessel, 45 feet high and 9 feet in diameter, is housed in a stainless steel containment vessel (2), which is immersed in the pool (3). There are no pipes in the primary coolant system that could rupture in an earthquake, and the design is very robust against earthquake damage.

The small size of the containment vessel permits much greater strength—by a factor of 10— relative to internal pressure than in a conventional LWR. There is little or no oxygen in the space between the pressure vessel and the containment vessel, so that any hydrogen released into that space will have no potential for explosion. The core contains only 5% as much fuel as in a conventional LWR, so that residual decay heat can be transferred out of the reactor without overheating the core if the secondary coolant flow is lost. There is thus no need for emergency core cooling systems. Spent fuel is contained in an underground building, in a pool containing four times as much water per MW of thermal power as in conventional LWRs.

19.2 THE GEN4 ENERGY MODULE

Gen4 Energy, Inc., renamed from Hyperion Power Generation, Inc. in March 2012, was formed in 2007 to commercialize an SMR concept developed at Los Alamos National Laboratory.[8] It is aimed at three principal markets: remote communities, facilities for mining, refining, and producing minerals and petrochemicals, and government installations that need (mostly for security) to be independent of local electric power grids. The Gen4 Module (G4M) is designed to be assembled at the factory, shipped to the application site, operated for 10 years without refueling, and then returned to the factory. The power plant layout is shown in Figure 19.2.

The G4M reactor vessel,[9] like the NuScale module, is located below the ground for safety, and it is small both in physical size and power output. The diameter of the module is 1.5 m, which allows the module to be shipped fully assembled, and its power output is 25 MWe, appropriate for the small facilities for which it is intended. Figure 19.2 shows an installation that produces electric power, but it can also be used to provide heating for industrial processes and buildings.

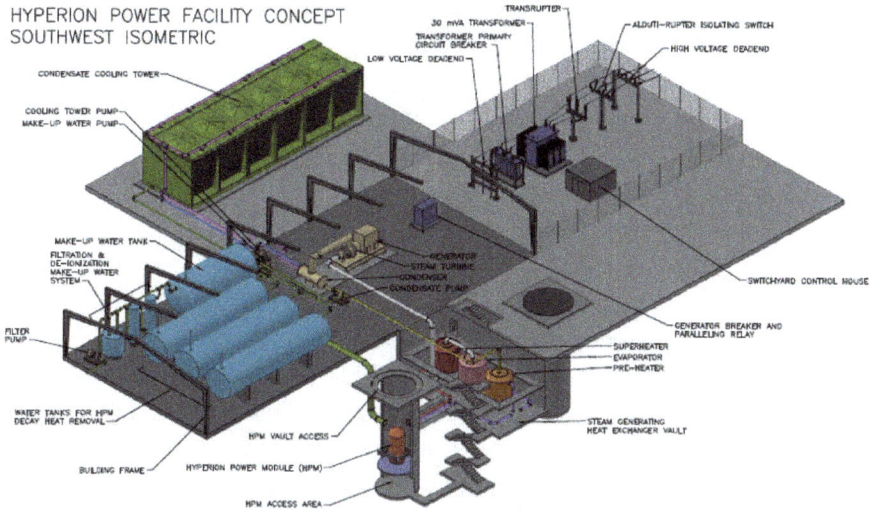

Figure 19.2—Power plant for Gen4 Module. This illustration was made when the company was called Hyperion Power Generation, Inc., and the modular reactor was called the Hyperion Power Module (HPM). It was taken via Wikimedia Commons[10] from work produced by the U.S. Nuclear Regulatory Commission and is in the public domain under the terms of Title 17, Chapter 1, Section 105 of the U.S. Code.

The reactor is cooled by a lead-bismuth eutectic (LBE) alloy (see Section 17.1). This coolant offers some significant advantages over water coolant. First, its boiling point at atmospheric pressure is 1670 °C,[11] which is far above the temperatures that would be reached in a worst-case accident scenario. Since LBE is liquid at operating temperatures at atmospheric pressure, the reactor vessel does not need to be able to contain high pressure. These qualities reduce the potential for releases of radionuclides. Second, LBE does not react chemically with air, water, fuel, metal, or concrete. The cladding in the fuel rods is stainless steel; the combination of stainless steel cladding and LBE coolant eliminates the possibility of the hydrogen explosions that have occurred in water-cooled reactors using Zircaloy cladding.

The fuel proposed for the G4M is uranium nitride.[12] This fuel has higher thermal conductivity than the oxide fuels used in most nuclear reactors, which enables the fuel to run at lower temperature than an oxide fuel would with the same heat output and coolant temperatures. The very high boiling point of LBE permits the coolant to operate at higher temperature than the water in LWRs; the core outlet temperature in the G4M is 500 °C. This high temperature will give higher thermodynamic efficiency to the power plant.

As in the NuScale reactor, coolant flow through the core is driven by natural circulation, so that no primary coolant pumps are needed.

In summary, the NuScale and Gen4 Energy designs offer advances in safety well beyond those in the latest large LWR designs.

19.3 OTHER SMR CONCEPTS

In addition to the well developed NuScale and Gen4 Energy designs, there are other SMR concepts worthy of mention.

First, General Atomics is developing a helium-cooled SMR, called the Energy Multiplier Module, or EM^2, which can use spent fuel from LWRs as fuel, thereby greatly reducing the amount of high-level waste that would eventually have to go to a repository.[13, 14] The design features silicon carbide materials that will allow higher operating temperatures for greater efficiency, and it will operate for 30 years without refueling.

Another effort in SMR development is being conducted by Terra Power,[15] a company founded by Microsoft's Bill Gates, among others. Terra Power was originally focused on developing a traveling wave reactor (TWR),[16] in which neutrons from fission in a layer of fissile fuel converts fertile fuel into fissile fuel in the adjacent layer, so that the conversion progresses along the vertical axis of the reactor over the lifetime of the core, which could be up to 60 years. This approach, in theory at least, could enable the full utilization of natural uranium without enrichment or reprocessing. Now, in addition to the TWR, Terra Power is exploring a molten salt reactor called the Molten Chloride Fast Reactor (MCFR) for both electricity and process heat generation.

CHAPTER 20
NUCLEAR FUSION

As a source of energy, nuclear fusion[1] has clear advantages over nuclear fission. It uses abundant, easily extracted resources for fuel. There is no possibility of a fusion-power accident that would release a large amount of radioactive material into the environment. Fusion produces much less nuclear waste, and most of what it does produce is short-lived. It has almost no relevance to nuclear weapons proliferation. But nuclear fusion has one major disadvantage: It is extremely difficult to sustain a controlled fusion reaction, and so far nobody has been able to do it. When I began my research career in 1975, I worked in fusion technology, and there was a joke in the fusion community: The time that was required to develop a working fusion reactor was a constant of nature. A working fusion reactor was always going to be about 25 years in the future.

However, in the ensuing years a great deal has been learned about plasma physics, which is the basic science of nuclear fusion, and great progress has been made in the required supporting technologies, such as superconducting magnets and high-power lasers. A prototype fusion reactor, ITER, which has been in planning for many years, is now scheduled to be built at the French research facility in Cadarache, and to begin operation in 2025. Site preparation and some equipment procurement activities have already started.[2] So now (2019) that "constant of nature" has shrunk to about 6 years.

This chapter first explains the difference between the fusion reaction and the fission reaction that makes controlled fusion so difficult, then presents the basics of plasma physics, and finally describes proposed fusion reactor designs, with a focus on ITER.

20.1 THE FUSION REACTION REVISITED

Fusion reactions are briefly discussed in Section 5.5. It is noted there that the binding energy per nucleon increases when heavy nuclei such as uranium undergo fission, in which they divide into lighter nuclei, and when light nuclei such as hydrogen undergo fusion, in which they combine into heavier nuclei. This behavior is illustrated in Figure 5.4. Let us review what that means.

As explained in Section 3.4, the rest mass of a nucleon in a nucleus depends on its state in the nucleus. For example, when a neutron is absorbed by a nucleus in an (n,γ) reaction, the energy carried off by the gamma ray comes by conversion of a small part of the masses of the nucleons in the new nucleus, including the absorbed neutron, into kinetic energy. When a uranium nucleus undergoes fission, the nucleons in the fission fragments give up mass to provide kinetic energy to the fission fragments, the fission neutrons, and the other radiation associated with fission. When two nuclei join in fusion, they give up mass to provide kinetic energy to the particles that are emitted from the reaction. (A fusion reaction always has at least two products; a reaction in which two nuclear particles join into only one particle cannot simultaneously satisfy the law of conservation of energy and the law of conservation of momentum.)

The direct impetus that generates the kinetic energy of the fission fragments in a fission reaction is the electrostatic repulsion between the positively charged fission product nuclei. When the fissile nucleus absorbs a neutron, some of the released binding energy is imparted to the nucleons in the compound nucleus as kinetic energy. The energized nucleons rattle around until one group of them gets far enough away from another group for the electrostatic repulsive force to overcome the strong nuclear force, which only acts over a very short distance. Then these groups, which are the fission fragments, zing away from each other, acquiring about 168 MeV of the total 200 MeV of energy released in the reaction.

In fusion, the whole process is reversed. In order for the fusing nuclei to get close enough to each other for the strong nuclear force to overcome their mutual electrostatic repulsion, they must start with sufficient kinetic energy. The easiest fusion reaction to achieve is the deuterium-tritium reaction,

$$D + T \rightarrow \alpha \ (3.5 \ \text{MeV}) + n \ (14.1 \ \text{MeV}).$$

(Recall that deuterium is the isotope of hydrogen that has one neutron in addition to the proton, and tritium is the isotope that has two neutrons in addition to the proton.) For this reaction, the deuteron and the triton must each initially have at least about 100 keV of kinetic energy, measured in the reference frame of their common center of mass. (This reference frame is the one that moves with a straight line drawn between the two particles, and the common center of mass is the point between them towards which they are converging. If you are moving with that reference frame, you see the line between the particles as stationary, and the particles move towards each other along that line.) This energy is much less than the energy acquired by fission fragments because the charges of fusing nuclei are much smaller than the charges of fission fragments—the fusing nuclei only have one proton each. This energy can be imparted by firing opposing beams of such nuclei at each other with accelerators, but not enough reactions occur to replace the energy required to create the beams. The practical way to impart such energy to fusion fuel nuclei is to heat the fuel to very high temperature. Then the thermal kinetic energy of fuel nuclei is what overcomes the electrostatic repulsive force. Fusion achieved this way is called thermonuclear fusion.[a]

It is not necessary for the fuel to be so hot that all the fuel nuclei have the required 100 keV of kinetic energy. Particles in a gas have a range of energies in a distribution called the Maxwell-Boltzmann distribution, illustrated in Figure 20.1. The location of the peak in this distribution depends on the temperature. If enough particles in the high-energy tail of the distribution have energies above 100 keV, then fusion reactions will occur with enough frequency to provide more energy than was required to heat up the fuel.

How hot does the fuel have to be? For the deuterium-tritium reaction, the temperature has to be something like 100 million kelvins. It is more convenient to express temperatures that large in terms of electron volts. The average kinetic energy in a Maxwell-Boltzmann distribution is $3kT/2$, where k is Boltzmann's constant, 1.38×10^{-23} J/K. It is customary to refer to the product kT as the temperature. When converted from kelvins to joules and then to eV (1 eV = 1.6×10^{-19} J), a temperature of 100 million K is 8625 eV, or 8.625 keV. The precision of this number is an artifact of applying precise conversion factors to a rough number. A more appropriate way to express the temperature requirement is to say that fusion temperatures are about 10 keV or more. Another handy equivalence is that a temperature of 1 eV is 11,600 K.

---

[a] Some interest persists in so-called "cold fusion," in which fusion reactions are alleged to occur at room temperature by some mechanism yet to be identified. Experiments claiming success in such processes have not been repeatable, and I think such processes are impossible, with one exception. A process called muon-catalyzed fusion definitely does occur at room temperatures, but it does not appear practical to apply this in a power reactor. This chapter deals only with thermonuclear reactor concepts.

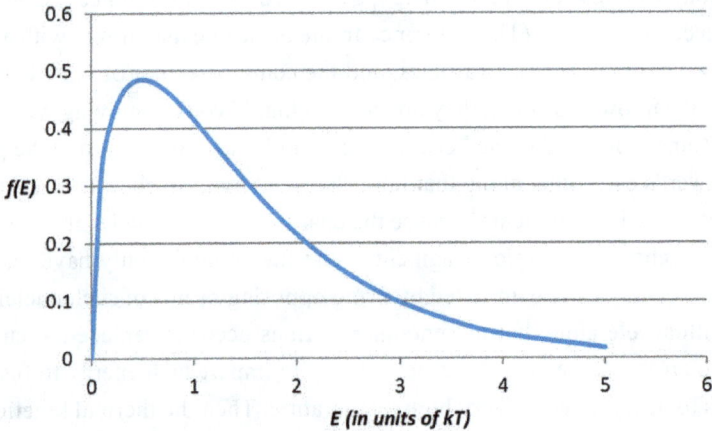

The vertical axis is labeled $f(E)$ with values from 0 to 0.6. The horizontal axis is labeled E (in units of kT) with values from 0 to 6.

Figure 20.1—The Maxwell-Boltzmann distribution function for particle kinetic energy. The product $f(E)dE$ is the probability that a molecule will have kinetic energy in a narrow energy band of width dE centered around E. The most probable energy is equal to $\frac{1}{2}kT$, and the average energy is equal to $3kT/2$.

Some simple rules of thumb have been developed to assess whether as much energy is released by fusion as was required to heat the fuel. This heat balance requirement is called "scientific breakeven." The simplest of these rules of thumb is the Lawson criterion, expressed as

$$n\tau > 10^{14} \text{sec/cm}^3 ,$$

which applies to the deuterium-tritium reaction (τ is the lower-case Greek letter tau). This criterion means that, at fusion temperatures, the product of the density n of the fuel in particles per cubic centimeter and the time τ in seconds for which that density is maintained must be at least equal to the indicated number. "Engineering breakeven," in which the heat delivered by fusion and then converted into electricity is sufficient to maintain the temperature of the fuel, is more difficult to achieve, and the criterion for it depends to some extent on the details of the whole plant design. A sufficient condition for engineering breakeven is simply $Q > 1$, where Q is the ratio of the power returned to the plasma by alpha particle heating to the power required to heat the plasma. (The alpha particles, being charged, interact with the plasma and deposit their energy in it, while the neutrons mostly just fly off into the pressure vessel wall.) Thus, a value of Q greater than 1 means that the plasma can heat itself, and no externally supplied heating from extracted electricity is needed.

A temperature of 100 million degrees is obviously well above the boiling point of anything, so one might think that the big problem in trying to contain fuel at fusion temperatures is to keep the fuel from vaporizing the container. But the real nature of the problem is more the other way around. If the fuel touches the container walls, it will lose heat into the walls so fast that the required temperature cannot be maintained. A fusion plasma is generally very rarefied—typical densities in fusion plasmas will be on the order of 10^{14} particles per cubic centimeter, compared to about 10^{19} particles per cubic centimeter in standard air—and there is not enough energy in the whole plasma to vaporize the containment vessel around it. Contact between the hot fuel and the walls may scorch the walls, but if the fusion reaction cannot be maintained, that problem is moot. So the challenge is to keep an extremely hot gas from touching the walls of the container in which it is kept, at least for a long enough time to satisfy breakeven criteria. The containment vessel's role in keeping the surrounding air out is just as important as its role in keeping the fuel in.

One approach to fusion is called inertial confinement, which is a misnomer. The fuel is really not confined from the walls at all, but only stays away from them for a very short but sufficient time. In this approach, a small pellet containing a mixture of deuterium and tritium is suspended in the middle of the container and irradiated on all sides by powerful laser beams. The surface layer of the pellet ablates away violently, and the reaction force exerted on the remaining pellet material by the escaping atoms compresses the pellet and heats it to fusion temperatures. The pellet stays together long enough for the breakeven criterion to be met, then explodes. The particles ejected from the reaction fly into the walls, where their kinetic energy is collected as heat; the radius of the containment vessel is made large enough that the particle flux into the walls is spread out and does not damage the walls.

Another way to confine fusion fuel is called magnetic confinement. This approach takes advantage of the nature of matter at extremely high temperatures. The three states of matter familiar to everyone are solid, liquid, and gas. At very high temperatures, a fundamentally different state exists, called plasma, a Greek word meaning something molded or fabricated.[3] On Earth, most plasmas are indeed fabricated, but plasma is probably the most common state of matter in the Universe, as stars and most interstellar gas[4] are made of it.[b]

---

[b] Actually, the most common type of matter in the Universe is believed to be so-called "dark matter," which has not yet been detected directly. It is believed to exist because the motion of galaxies does not proceed as it would if only the visible matter in them were present. The nature

At temperatures of only about 1 eV, collisions between molecules in a gas are so energetic that the molecular bonds are broken and some electrons begin to be knocked off their atoms. So the gas becomes ionized. At fusion temperatures, all the electrons are knocked free, and the gas becomes fully ionized. The electrons don't go anywhere—they are still present along with the nuclei—but they are no longer bound in atomic energy shells. Thus, the gas retains overall electric charge neutrality, but now the electrons and nuclei (ions) are free to respond separately to electric and magnetic fields. Groups of electrons or ions can bunch up in ways that create local net electric charge, and they can generate their own electric and magnetic fields in ways that affect other electrons and ions some distance away. Such behavior is called collective behavior, and overall charge neutrality coupled with local charge imbalance is called quasineutrality. A plasma is defined as a quasineutral gas of charged particles or charged and neutral particles that exhibits collective behavior. The fundamental science that explores the behavior of plasmas is plasma physics. Plasmas have many applications besides nuclear fusion, such as plasma television sets and astrophysics, but plasma physics is the basic science of fusion reactors. The aspects of plasma physics relevant to fusion reactors are addressed in the next section.

20.2 BASIC PLASMA PHYSICS

The collective response of plasmas to magnetic fields is the basis for magnetic confinement fusion. Most of the fascination of fusion plasma physics is related to magnetic confinement, so this section deals primarily with that. First, a brief discussion of electromagnetic theory is needed.

20.2.1 Elementary electromagnetic theory

As noted in Chapter 4, the basics of classical electromagnetic theory are contained in Maxwell's equations. First, we need to know the essential meaning of Maxwell's equations. In this chapter, I present them in the form of verbal descriptions. Their mathematical form is discussed in Appendix II for those who have the appropriate background, but all you need to understand is their physical significance, which I

---

of this dark matter has not yet been discovered. But plasma is probably the most abundant form of ordinary matter.

try to explain here in fairly simple language. Reference 5 is a very good elementary electromagnetic theory text.

Maxwell's equations tell us about the behavior of electric and magnetic fields. An electric field is a measure of the electrostatic force (both its magnitude and its direction) exerted on a stationary charge at any point in space by other charged particles. As you go from point to point in space, you can follow the direction of the force and construct curves, called electric field lines, connecting the points. A magnetic field is a measure of the force exerted at any point of space on a moving particle of charge in addition to the electrostatic force; this magnetic force depends both on the strength of the field and the velocity vector of the particle. One can also construct magnetic field lines by following the direction of the magnetic force from point to point.

Maxwell's first equation says that electric fields emanate from charged particles, as shown in Figure 20.2 for a single charged particle.

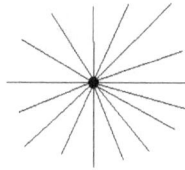

Figure 20.2—Electric field lines emanating from a single point charge

Electric charge has sign, so if a positive charge and an equal negative charge are placed next to each other, except very near them their combined electric field will be zero.

Maxwell's second equation says that there are no magnetic monopoles, or magnetic charges from which magnetic field lines emanate. That means that ultimately all magnetic field lines are closed curves.

Maxwell's third equation says that magnetic fields are created by the motion of charged particles. The left-hand side of the equation gives a property of the magnetic field called the "curl." This is a mathematical description of the circulatory nature of the field; this circulatory nature leads to the closure of field lines. There are two terms on the right-hand side of the equation. One arises from particle currents and the other arises from changes in the electric field over time. The form of this equation makes it appear that there are two different sources of magnetic fields, particle currents and changing electric fields, but they are actually two complementary ways of accounting for the motion of charged particles.[6] Maxwell's third equation also tells us that magnetic field lines induced by a single moving charged particle are perpendicular to the velocity of the moving charged particle. The magnetic field created by an electric current in a straight wire is illustrated in Figure 20.3. The sign of the electric current is defined so that a flow of positive charges is a positive current. Since a current in a wire arises from the flow of electrons, which are negatively charged, the current is in the direction opposite from the direction of the electron flow. The direction of the magnetic field follows a right-hand rule: If you align your right thumb with the current, the direction of the magnetic field is the direction in which your fingers curl around the wire. The symbol B is traditionally used for the magnetic field vector.

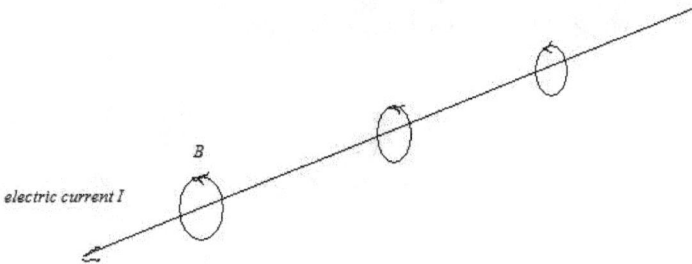

Figure 20.3—The magnetic field induced by a current in a straight wire

Maxwell's fourth equation says that a magnetic field changing in time induces an electric field. This principle is called Faraday's Law, named for the English scientist Michael Faraday (1791-1867), whom some historians of science consider the best experimentalist in the history of science.[7] This principle is applied ubiquitously in electric generators and transformers. One of the most memorable demonstrations I have ever seen was made in the beginning physics class I took as a sophomore at the University of Michigan from Professor Gabriel Weinreich, a wonderful teacher who first inspired my love of physics. Professor Weinreich held a toroidal (doughnut-shaped) fluorescent light bulb, unattached to anything, over his head. There was a nearby magnetic field generator that induced an oscillating magnetic field oriented vertically where he was standing. When he oriented the bulb horizontally, it lit up like a halo! The oscillating magnetic field drove an oscillating toroidal current in the bulb, which caused the bulb to light up. Figure 20.4 illustrates the basic principle.

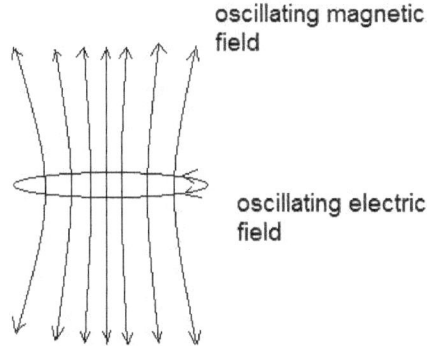

oscillating magnetic field

oscillating electric field

Figure 20.4—Electric field induced by a changing magnetic field

20.2.2 Motion of individual particles in magnetic fields

The force exerted on a charged particle by a magnetic field is perpendicular to both the particle's velocity and the magnetic field line at the particle's instantaneous location. (If the velocity is parallel to the field, the magnetic force is zero.) In Chapter 2, it was explained that if an object is subjected to a force of constant magnitude that is always perpendicular to its velocity, it will move in a circle. For the same reasons, if an object is subjected to a force of constant magnitude that is always perpendicular to one component of its velocity, it will move in a helical path. Therefore, a charged particle moving in a uniform magnetic field and subject to no other forces will move in a helical path: Parallel to the field its velocity is constant; perpendicular to the field its path is circular, and it moves on the circle at constant speed. The radius of the circular path is called the gyroradius or the Larmor radius, given by

$$R_L = \left| \frac{m v_\perp}{qB} \right| ,$$

where m is the mass of the particle, $v_\perp$ is the component of the velocity perpendicular to the magnetic field, q is the magnitude of the particle's charge, and B is the magnitude of the magnetic field $\mathbf{B}$. (The vertical bars denote absolute value, so that the gyroradius is a positive number whether the charge is positive or negative.)

Thus, the heavier and faster the particle, the larger its gyroradius; the higher its charge and the stronger the magnetic field, the smaller its gyroradius. The direction of the magnetic force is dependent on the charge of the particle. A positively charged particle gyrates in accordance with a left-hand rule: If you align your left thumb with the magnetic field, with your fingers curling around the field line, the particle gyrates in the direction in which your fingers are pointing. A negatively charged particle follows a right-hand rule.

Because the gyrating particle is charged, its motion creates its own magnetic field. Since it is moving in a helical path, it effectively creates a ring current, which induces a poloidal magnetic field as shown in Figure 20.5. (The term "poloidal" is explained in Section 20.2.4.) The induced field is opposite in direction from the externally applied field. The cumulative magnetic fields of large numbers of orbiting particles can modify the externally applied field drastically. This is an example of collective behavior.

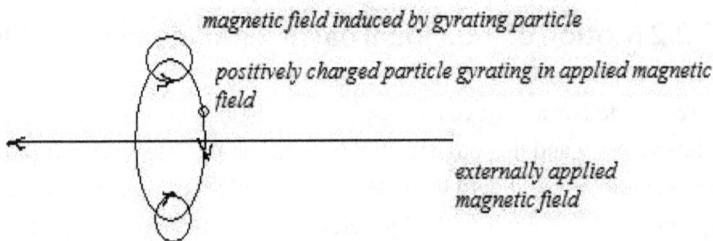

Figure 20.5—Magnetic field induced by a gyrating charged particle

The angular frequency of the circular motion is called the gyrofrequency, the Larmor frequency, or the cyclotron frequency. In units of radians per unit time, this is

$$\omega_c = \frac{|q|B}{m}.$$

(The radian is a natural unit of angular measure. By definition, there are 2π radians in a circle, so a radian is $360°/2\pi = 57.296°$. Thus, if the cyclotron frequency is expressed in radians per second, the number of times per second that the particle goes a full $360°$ around the field line is $\omega_c/2\pi$.) If there is an electric field in addition to a uniform magnetic field, any component of the electric field oriented parallel with the magnetic field will cause a charged particle to accelerate along the magnetic field. Any component of the electric field oriented perpendicularly to the magnetic field will cause the particle alternately to accelerate and decelerate as the particle moves in its orbit alternately with and against that component of the electric field. A positively charged particle will accelerate when it moves with the electric field and decelerate when it moves against the electric field; a negatively charged particle will behave oppositely. The varying speed will cause the gyroradius to vary, and the result is a drift of the center of gyration, or the guiding center, in a direction perpendicular to both the electric field and the magnetic field. This drift is called the "E cross B drift," or the ExB drift. The "cross product," or "vector product," denoted by the symbol x, is a mathematical rule for multiplying two vectors; I am not going to give the rule here, but I want to use the conventional notation because it reminds us that it is the crossed, or perpendicular, components of E and B that produce the drift. As shown in Figure 20.6, the direction of the ExB drift is the same for electrons and ions. The figure uses a common convention to show the direction of B: The black dot represents the head of an arrow coming towards the viewer; if B were pointing into the paper, it would be represented by a circle with an X in it, to represent the tail feathers of the arrow going away from the viewer.

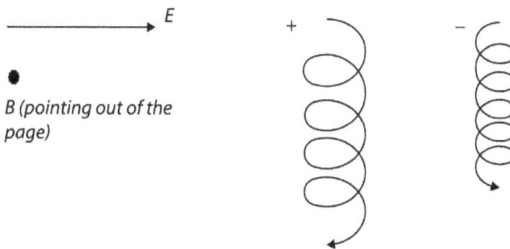

Figure 20.6—The ExB drift

Many other phenomena cause guiding center drifts. An important one is a gradient in the magnetic field. The varying magnetic field strength causes the gyroradius to vary, and the guiding center drifts in a direction perpendicular to both the magnetic field and the magnetic field gradient (the "grad-B drift"). In this case, however, the electrons and ions drift in opposite directions, so that the magnetic field gradient drives an electric current in the plasma.

Another important kind of motion occurs in the magnetic mirror effect. Consider a magnetic field in which the field lines are primarily in an axial direction but bunch together in some region along the axis. When the field lines bunch together the field is stronger. The magnetic force on a particle gyrating in the region of converging field lines has a component directed back into the region of the weaker field, as shown in Figure 20.7. It is easiest to see this for a particle whose guiding center lies on the axis, but it is true for any particle. A plasma contains particles moving in all directions, and almost all of them have some velocity component along the axis. If the velocity is mostly along the field, then the retarding magnetic force will not be sufficient to turn the particle around, but at some ratio of velocity components perpendicular to and parallel to the field, the retarding force will reflect the particle back into the region of the weaker field.

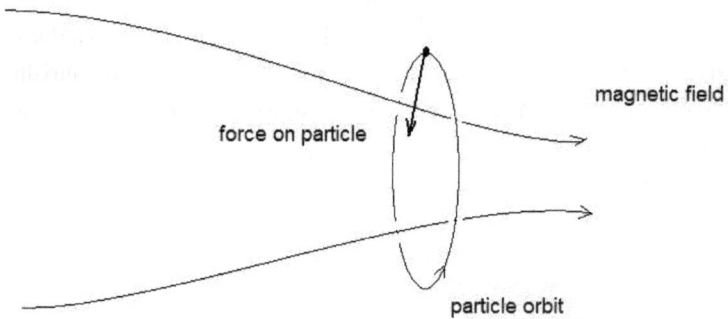

force on particle

magnetic field

particle orbit

Figure 20.7—The magnetic mirror effect

If the magnetic field is bunched together on both ends of a confinement device, particles with the proper ratio of velocity components will bounce from one end to the other and be contained within the device. Such a device is called a magnetic mirror. If B_o is the magnitude of the magnetic field in the middle of the mirror and B_m is the magnitude of the magnetic field in the mirror regions, then the condition for confinement is

$$\frac{v_\perp^2}{v^2} > \frac{B_o}{B_m},$$

where v and $v_\perp$ are the values of the total velocity (in any direction) and its component perpendicular to the axis at the middle of the mirror. For the set of all particles just meeting the confinement criterion, the velocity vectors form a cone around the axis, which is called the loss cone. Particles whose velocity vectors lie inside the loss cone are not confined. Fusion reactor concepts based on mirror confinement have not been successful, because most collisions among plasma particles do not induce fusion, but instead just cause the particles to change direction. Particles are scattered into the loss cone more often than they undergo fusion, so their average confinement time does not meet the Lawson criterion. Also, simple mirror configurations suffer instability, a phenomenon explored further below. Nevertheless, the mirror effect is important in practical fusion devices because regions of converging magnetic field lines occur in such devices, so mirror-like reflection governs some aspects of the motion of particles in them.

20.2.3 Two-fluid models and plasma waves

The idea of models was explained in Chapter 1. Models are not imagined to represent reality accurately, but only to provide convenient ways of representing phenomena. A plasma can often be modeled as a system of two or more interpenetrating fluids: the electrons, the ion species, and the neutral particles if there are any. In fusion plasmas, the ion species are often lumped together and there are no neutral particles, so the plasma is then regarded as composed of two fluids, the electrons and the ions.

Two-fluid models are especially useful in understanding phenomena called plasma waves. There are many kinds of plasma waves, including electron waves, ion waves, electromagnetic waves, hydromagnetic waves, and magnetosonic waves. Some of these waves behave differently if they propagate along or across a magnetic field. Detailed discussions of plasma waves are beyond the scope of this book. However, a sufficient understanding of plasma wave phenomena can be obtained by considering the simplest kind of wave, the electrostatic electron wave.

Imagine an infinite plasma in which the ions are uniformly distributed and fixed in space, and in which there is no magnetic field. There can be no such plasma, because the ions have thermal motion, but the ions are so much more massive than the electrons that their thermal velocity is relatively much slower, and to regard the ions as fixed in space is sometimes a useful approximation. Suppose that the electrons are disturbed in such a way that a significant portion of them at a position x along some axis (the horizontal direction in Figure 20.8 below) are displaced some small distance Δx. Suppose that this happens at all distances from the x-axis at that position. Then an electric field in the x-direction is created by the nonuniform electron distribution. This creates an unbalanced force on the displaced electrons, which attempts to pull them back into place. The electrons will overshoot their original undisplaced position, and they will oscillate back and forth until collisions dissipate the oscillation.

In magnetic confinement fusion plasmas, the particle densities are very low. As noted above, typical particle densities are on the order of 10^{14} particles per cubic centimeter, whereas the density in standard air is about 10^{19} molecules per cubic centimeter. Because of the very low densities in fusion plasmas, particles travel much farther between collisions than they do in ordinary gases, and for some purposes collisions can be neglected altogether. Plasmas in which collisions may be ignored are called collisionless plasmas, even though collisions actually do occur. In collisionless plasmas, the electron oscillations can persist through many cycles.

If the electrons had no thermal motion, the disturbed electrons would just oscillate in place around the initial disturbance. However, the thermal motion of the electrons provides a mechanism by which to transfer the disturbance along the x-axis. You may or may not find this behavior intuitive, but the phenomenon can be explained mathematically. This transfer amounts to a wave that travels along the x-axis, much like a planar sound wave in an ordinary gas. The electrostatic electron wave is shown schematically in Figure 20.8.

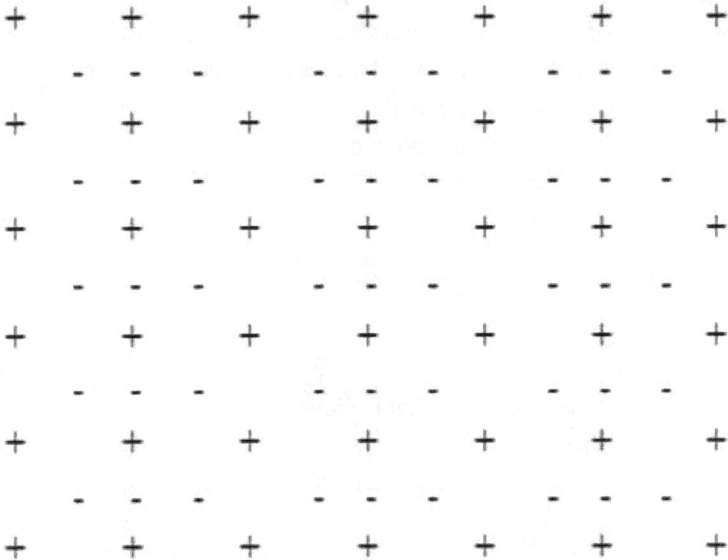

Figure 20.8—The electrostatic electron wave in a plasma. Note the uniformly spaced ions and the bunched-up electrons.

20.2.4 Magnetohydrodynamic models, plasma equilibrium, and confinement configurations

For some purposes, the two-fluid models described above are most suitable. For other purposes, a plasma can best be regarded as a single electrically conducting fluid, like liquid mercury, but much more rarefied. The study of electrically conducting fluids is called magnetohydrodynamics (MHD), or hydromagnetics.

The electrical conductivity of very hot plasmas is extremely high, greater than that of pure copper. It is often a good approximation to regard the conductivity as infinite, and some very useful approximate results follow from that viewpoint. One such result is the equilibrium equation, which applies when the plasma configuration does not vary with time. The equilibrium equation tells us that pressure variations in the plasma are balanced by magnetic forces. If the pressure varies, electric currents are induced in the plasma in such a way that the unbalanced pressure forces are balanced by magnetic forces.

For the special case of straight magnetic field lines, the equilibrium equation reduces to

$$p + \frac{B^2}{2\mu_o} = constant,$$

where μ_o is the magnetic permeability of free space.[c] The quantity $B^2/2\mu_o$ is called the magnetic pressure, so this equation tells us that the sum of fluid pressure and magnetic pressure is constant. When the field lines are curved, another term in the equation can be interpreted to account for a tensile force in the magnetic field lines, which also resists the fluid pressure.

In a machine called a theta-pinch, the magnetic field is axial, and the fluid pressure varies from zero near the walls to a maximum in the interior where the magnetic field is zero, as shown in Figure 20.9. The magnetic field in the vacuum around the plasma is induced by an azimuthal current I (that is, a current running essentially in a circular path around the axis) in a surrounding solenoid (a conducting cylinder or coil of wire). The azimuthal coordinate is customarily labeled by the lower-case Greek letter theta (θ), which is the origin of the name of the machine. In practice, the current

---

[c] This is $4\pi \times 10^{-7}$ kg-m/A-s^2, where A denotes the ampere, the unit of electric current (1A=1 coulomb/sec). The unit of magnetic permeability is usually written in henries per meter (H/m), where the henry is equal to 1 kg-m^2/A-s^2.

in the solenoid is increased very rapidly from zero, and the whole process is very dynamic so the equilibrium equations do not apply exactly. But the plasma inertia is low because of the rarefied plasma density, so the fluid and magnetic pressures still balance closely. The rising magnetic field induces azimuthal currents in the plasma that shield out the external field and permit the external field to compress the plasma.

Another linear device is the Z-pinch, in which an axial current in the plasma (i.e., in the z-direction) induces a surrounding azimuthal magnetic field that creates external magnetic pressure, as shown in Figure 20.10. Simple linear pinches are not practical fusion devices, because particles run out the ends of the confinement region too rapidly. Z-pinches also suffer from hydromagnetic instabilities as discussed in the next subsection. Nevertheless, some valuable experimental results have been obtained from pinch devices. Schemes have been proposed for plugging the ends of theta pinches with more complicated magnetic field configurations. One of them, the field-reversed theta pinch,[8] may eventually be shown to be practical.

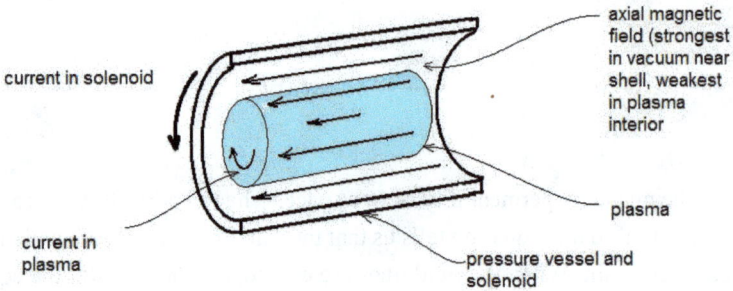

Figure 20.9—A simple theta pinch

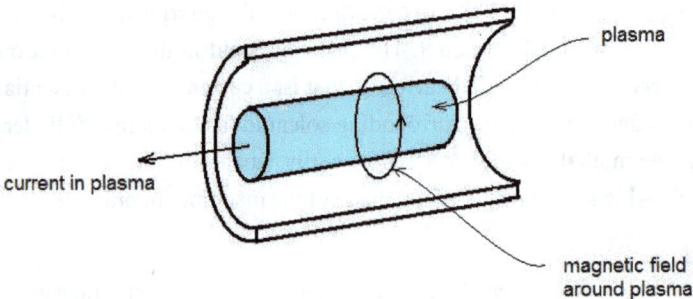

Figure 20.10—A simple Z-pinch

334

As we have seen for pinches and magnetic mirrors, most confinement schemes with straight containment vessels suffer from particle losses out the ends that make it difficult to achieve confinement times sufficient for net fusion energy production. A more successful approach has been to eliminate the ends of the device by bending it around into a torus, or a doughnut-shaped configuration.

Several approaches have been tried for magnetic confinement in toroidal devices. The most successful of these has been the tokamak (or tokomak).[9] "Tokamak" is a transliteration of a Russian acronym meaning either "toroidal chamber with magnetic coils" or "toroidal chamber with axial magnetic field." It was invented by the Soviet physicists Igor Tamm[10] and Andrei Sakharov[11] based on an idea of Oleg Lavrentiev.[12] The basic principles of the tokamak are discussed here. Specific features of the ITER tokamak are discussed in Section 20.3. To discuss toroidal devices efficiently, we need some terminology.

If you lay a doughnut flat on a table, the major axis of the doughnut is defined to be an axis perpendicular to the table, running through the center of the hole in the doughnut. You can think of the shape of the doughnut as being generated by rotating a closed curve (usually a circle, more or less) around the major axis. This closed curve is the generatrix of the toroidal surface of the doughnut. The toroidal direction follows a circle around the major axis—in other words, it goes the long way around the doughnut. The major radius of the torus is the distance from the major axis to the geometric center of the area enclosed by the generatrix. The minor axis is a circle around the major axis, passing through the geometric center of the area enclosed by the generatrix. If the generatrix is a circle, the minor radius is the radius of that circle. The poloidal direction is the direction around the minor axis—in other words, the short way around the doughnut.

In a theta-pinch, the fluid pressure in the plasma is balanced by the magnetic pressure from the axial magnetic field in the vacuum around the plasma. One might think that when a straight solenoid is bent around into a torus, the toroidal magnetic field would exert the magnetic pressure required to confine the plasma. However, that is not the case. Because of the effects of the curvature of the magnetic field, it is impossible to achieve an equilibrium condition when the magnetic field is purely toroidal.[13] Instead, the plasma pressure is balanced by a poloidal magnetic field. In various toroidal plasma confinement devices, the poloidal field is generated in different ways. In tokamaks, it is generated by toroidal electric currents in the plasma. A tokamak is more like a Z-pinch bent around into a torus with stability provided by the toroidal magnetic field.

In the original tokamak design and in many experimental devices, the toroidal plasma current is induced by exploiting Faraday's Law with a steadily increasing magnetic field in one or more magnetic cores that loop around the plasma in the poloidal direction. This changing field is induced in turn by a changing electric current in coils around the core. The whole arrangement is illustrated schematically in Figure 20.11.

As in a linear Z-pinch, a toroidal plasma confined by a purely poloidal magnetic field would be unstable. Fortunately, a plasma confined by a magnetic field combining toroidal and poloidal components can both achieve equilibrium and be stable. It turns out that even though the poloidal component of the field is providing the confinement, the toroidal component needs to be much stronger to provide adequate stability. In addition to these components, a field whose direction in the plasma is primarily parallel to the major axis, called the vertical field, is imposed for added stability. The vertical field is generated by the vertical field coils shown in Figure 20.11.

A tokamak in which the toroidal current is induced by a steadily increasing magnetic field is inherently a pulsed system, because you can't keep increasing that field forever. Alternate systems, such as beams of radiofrequency (RF) waves, are being developed to permit steady-state operation of tokamaks, including ITER.[14]

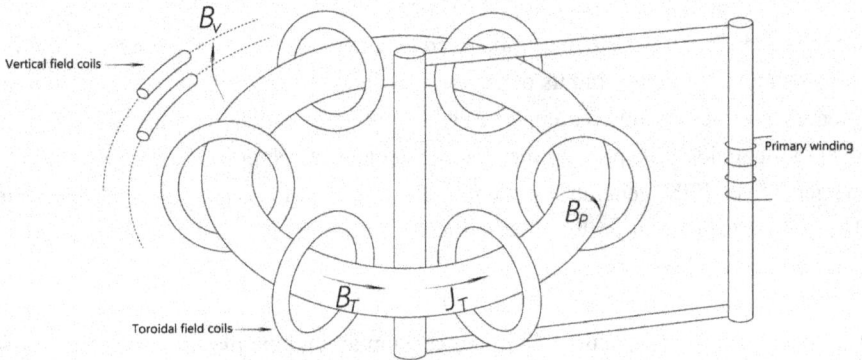

Figure 20.11—The tokamak in principle

Some interesting phenomena follow from having all these different magnetic field components in combination. It can be shown that magnetic field lines lie in surfaces of constant fluid pressure. Such surfaces are thus also called magnetic flux surfaces.[15] The direction of the resultant magnetic field at a point is unique—i.e., two magnetic field lines cannot cross, because the magnetic force on a moving charged particle cannot be in two directions at once. As shown by Maxwell's second equation, field lines are ultimately closed curves. However, it is possible for a single field line to cover an entire flux surface—in other words, to wind around infinitely many times before reconnecting with itself. Such behavior of a magnetic field line is called ergodic.

20.2.5 MHD instabilities

The basic ideas of equilibrium and stability are easy to grasp. For a system in equilibrium, all the forces are balanced, and it maintains its state of motion. In static equilibrium, the system stays in the same position in some inertial reference frame. But a system in equilibrium is not necessarily stable.

In principle, it should be possible to balance a sharpened pencil on its point on a flat table. If the center of mass of the pencil is exactly above the point, the gravitational force is perfectly aligned with the reaction force exerted by the table, and there is no side force to make the pencil fall over. You will not succeed if you try to do this. In practice, the slightest deviation from perfect vertical alignment will create a moment (i.e., a torque) that makes the pencil topple. The ideal equilibrium of the pencil balanced on its point is not a stable configuration.

On the other hand, a nail driven partway into the table is stable. Any tendency for the nail to fall over sideways is immediately resisted by the rigid table material, and the nail stays put.

If the nail, or the pencil, is lying on the table, it is said to be neutrally stable. It will lie in place unless something disturbs its position, in which case it will just stay in its new position.

The usual depictions of positively stable, neutrally stable, and negatively stable (unstable) configurations are a ball in a bowl, on a table, and on a mound, as shown in Figure 20.12.

Figure 20.12—Positive, neutral, and negative stability

Equilibria in plasmas are highly susceptible to many kinds of instabilities. Some of them involve motions of the plasma or part of it in bulk, and others are more akin to turbulence in ordinary fluids. First we will look at bulk instabilities, or magnetohy-drodynamic (MHD) instabilities.

Entire books have been written about MHD instabilities, so we won't attempt to consider all types. The easiest instability to understand is called the Rayleigh-Taylor instability, or the gravitational instability. This type of instability is the same as that affecting two immiscible fluids of different density, such as honey and air. Honey, being heavier than air, occupies the bottom of a honey jar. If you turn the jar upside down, the honey and air will switch positions; the switch will begin with a wave at the interface that grows until the exchange is complete. Watch it yourself; if it happens too quickly, put the jar in a refrigerator for a while. In a magnetically confined plasma, the surrounding vacuum magnetic field is analogous to the lighter fluid; the vacuum region and the plasma region can change places under certain conditions.

Two of the instabilities affecting Z-pinches can be explained easily, and these will give a good idea of the basic character of all MHD instabilities.

The strength of the magnetic field around a straight conductor carrying a current is inversely proportional to the distance from the centerline of the conductor. Consider a Z-pinch in which the current is carried in a thin "skin layer" on the surface of the plasma. (This is the normal condition in a Z-pinch.) Then there is no current inside the plasma, there is thus no magnetic field inside the plasma, the fluid pressure in the plasma is uniform, and the azimuthal magnetic field around the plasma, which con-fines the plasma, is strongest at the plasma surface and weaker away from the surface. Now consider what happens if some perturbation displaces the surface of the plasma inwards at some location along the axis. Such perturbations are normal occurrences, since the plasma is composed of many particles all flying around with a distribution of energies. The magnetic field strength will increase on the inwardly perturbed sur-face, because that portion of the surface is now closer to the centerline. This will cause the magnetic force on the surface of the plasma to increase in that location, but the fluid pressure will not increase. Therefore, the balance of forces required for equi-librium is lost, and the plasma surface in that location will be displaced inwards even more. Conversely, if the perturbation is outwards, the magnetic force will decrease, and the constant fluid pressure will push the surface outwards. The perturbations will increase until the current is pinched off completely at the inward perturbation or the plasma hits the wall at the outward perturbation. This type of instability is called the sausage instability; it is shown in Figure 20.13.

Another characteristic instability in a Z-pinch is called the kink instability. If a per-turbation in the equilibrium plasma moves an entire section laterally, the plasma devel-ops a bend with curvature. The magnetic field continues to be perpendicular to the local current, so that it is bunched up on the inside of the curve and spread out on the outside of the curve. Thus, the magnetic field becomes stronger on the inside of the curve and weaker on the outside. The unbalanced magnetic forces will then push the displaced plasma even farther from the equilibrium position. The perturbation will continue to grow until the plasma hits the wall. The kink instability is shown in Figure 20.14.

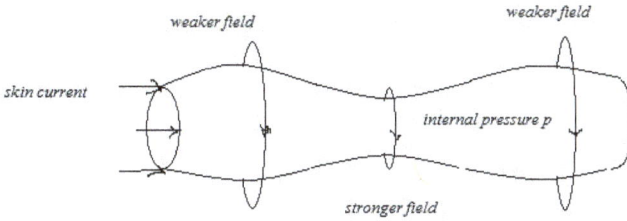

Figure 20.13—The sausage instability in a Z-pinch

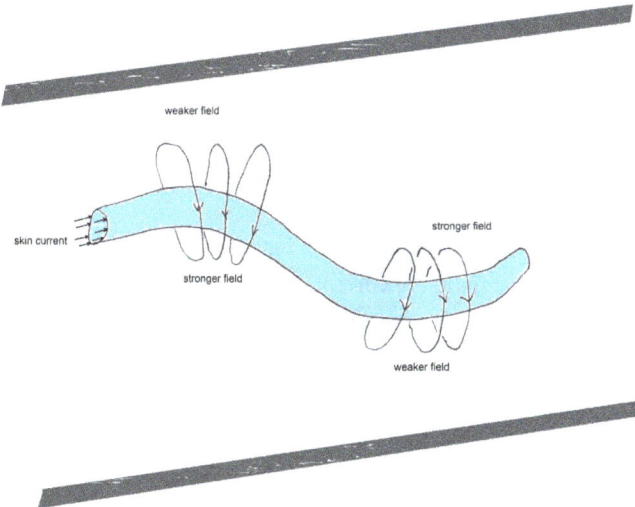

Figure 20.14—The kink instability in a Z-pinch

From these examples, you can see how MHD instabilities move a plasma around or distort it bodily. There are many other types of MHD instabilities, but they all share this same general character. Another type of instability, as noted above, is more akin to turbulence.

20.2.6 Microinstabilities

Turbulence in ordinary fluids occurs when smooth, or "laminar," flow over an object is disrupted and the flow breaks up into turbulent eddies, which often take the form of a turbulent wake behind the object. The flow in turbulent eddies is unsteady and chaotic, with the fluid velocity at any point varying wildly with time in magnitude and direction. Turbulence also occurs in pipes and channels. For a fluid of given density and viscosity, the onset of turbulence depends on the shape and size of the object and the fluid velocity.

In plasmas, phenomena called microinstabilities can cause chaotic fluid motions somewhat like turbulence. Microinstabilities, rather than moving the entire plasma in bulk, affect certain classes of plasma particles—for examples, electrons moving in particular velocity regimes. A prototypical case is the two-stream instability.

Consider a uniform unmagnetized plasma in which the electrons are moving with a certain velocity with respect to the ions. If an electrostatic electron wave arises in the plasma, it can propagate because of the relative velocity, just as thermal motion of electrons can cause an electrostatic electron wave to propagate in a quiescent plasma. If the combination of wavelength and relative velocity lies within proper bounds, the wave can grow. Eventually, the wave can become large enough to break up and give rise to chaotic motion.[16] For a given wavelength, if the velocity exceeds a certain limit, the oscillations are stable, and if it is less than a certain other limit, the oscillations are damped by a collisionless mechanism called Landau damping, after the great Soviet physicist Lev Landau.[17]

The relative velocity of the electrons and ions constitutes an electric current, and the instability serves to degrade the current. Since currents give rise to the magnetic forces that confine plasmas, one can readily see how the two-stream instability can lead to loss of confinement.

There are many kinds of microinstabilities, but they share with the two-stream instability the property that they are instabilities in various kinds of plasma waves and they similarly degrade confinement.

20.2.7 Classical and neoclassical diffusion

Consider a glass of water into which a drop of ink falls. Upon hitting the water's surface, the ink drop no longer falls with the acceleration imparted by gravity, but instead spreads out slowly until it is uniformly distributed throughout the water. The spread of the ink is an example of the process of diffusion. Diffusion of the ink is controlled by the collisions between the ink molecules and the water molecules.

The thermal motion of the ink molecules is randomly oriented. Molecules are moving in essentially every possible direction. When they undergo collisions, their directions change randomly; they take a "random walk" whose step size is the average distance between collisions (the "mean free path"). This random reorientation results in an overall flux of ink particles given mathematically by Fick's Law:

$$\Gamma_i = -D\nabla n_i \, ,$$

where D is called the diffusion coefficient and n_i is the number density of ink molecules—the number of molecules per unit volume. Recall that the gradient operator ∇ produces a vector in the direction of maximum increase of the quantity operated on, with magnitude equal to that increase, as discussed in Chapters 2 and 6. This equation says that the net flux (which has units of ink molecules per unit area per unit time—e.g., molecules per square centimeter per second) is in the direction from higher concentrations of ink to lower concentrations (i.e., because of the minus sign, in the direction opposite from the gradient vector). It can be shown that the diffusion coefficient is approximately equal to the square of the mean free path l divided by the average time τ_c between collisions:

$$D = \frac{l^2}{\tau_c}.$$

In a magnetized plasma, collisions cause migration of the ions and electrons across the magnetic field in a process that is very analogous to the diffusion of ink molecules in water. However, there are some key differences caused by the gyration of the plasma particles around the magnetic field lines and by the nature of collisions in plasmas.

Each particle orbits around its guiding center at a distance equal to its Larmor radius R_L. When it undergoes a collision, it will acquire a new guiding center. The particle normally makes many orbits around its guiding center between collisions, but the step size of its random walk is the Larmor radius, not the total "odometer" distance travelled in its many trips around its guiding center. Plasma diffusion in which the step size is the Larmor radius is called classical diffusion.

In a gas of neutral molecules, elastic collisions are much like collisions between billiard balls. They happen between two particles at a time, and they require the molecules to be very close to each other before significant forces are felt between the molecules. In plasmas, collisions are very different. The electrons and ions are not neutral, and their electrostatic forces are exerted over long distances. The scale of the distance over which an individual particle's electric field is felt is called the Debye length, $\lambda_D = (\varepsilon_o kT/q^2 n)^{1/2}$, where ε_o is the electric permittivity of free space,[d] q is the particle's charge, and n is the particle number density. The Debye length in a typical fusion plasma is about a millimeter. This distance is much greater than the average distance between particles. A particle is subjected simultaneously to the electric fields from all the other particles within its "Debye sphere," the sphere around it with radius λ_D. As a result of the simultaneous interaction of each particle with many others, each particle constantly undergoes minor changes in direction, and it only rarely undergoes an abrupt change of direction from an interaction with one other particle. In plasmas, then, a particle is considered to undergo a collision when it has a cumulative 90-degree deflection from all the simultaneous interactions it has with all the other particles within a Debye length.

Another feature of classical diffusion in plasmas is that collisions between like particles do not cause diffusion. Consider the collision shown in Figure 20.15 between two ions. For simplicity, the figure shows a direct hit resulting in a 180-degree deflection for each, but the conclusion is the same for less abrupt collisions. Each particle's guiding center shifts, but the shifts are equal and opposite and cancel each other out.

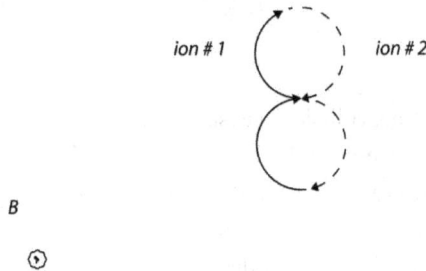

Figure 20.15—Collision between two ions. The magnetic
field lines are directed out of the page.

---

[d] This is 8.854×10^{-12} C^2/N-m^2, where C denotes the coulomb, N the newton, and m the meter. The unit C^2/N-m is also called the farad (F), after Michael Faraday, so permittivity may also be written in units of F/m.

On the other hand, in collisions between ions and electrons, the guiding centers shift in the same direction, as shown in Figure 20.16. The electron Larmor radius is smaller than the ion Larmor radius, but because electrons are much less massive, they undergo cumulative 90-degree direction changes more rapidly than ions. It turns out that the electrons and ions diffuse at the same rate, so that diffusion does not cause charge separation.

Thus, the classical cross-field diffusion coefficient for electrons is approximately equal to the square of the electron Larmor radius divided by the average cumulative 90-degree deflection time τ_{ei} for electrons in collisions with ions, and the classical cross-field diffusion coefficient for ions is approximately equal to the square of the ion Larmor radius divided by the average cumulative 90-degree deflection time τ_{ie} for ions in collisions with electrons, and the two coefficients are equal:

$$D_{\perp e} = \alpha \frac{R_{L_e}^2}{\tau_{ei}} = D_{\perp i} = \alpha \frac{R_{L_i}^2}{\tau_{ie}},$$

where α is a correction factor that accounts for some of the details of the diffusion process.

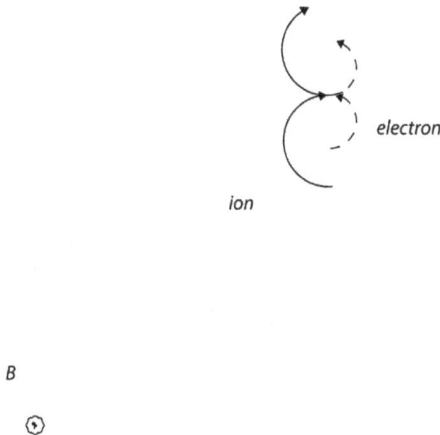

electron

ion

B

Figure 20.16—Collision between and electron and an ion. The magnetic field lines are directed out of the page.

The calculation of exact values for α and the collision times is a major portion of a postgraduate-level course, and the results are too complicated to present here. The key point is the qualitative scaling of the classical diffusion coefficients with the square of the Larmor radius and the inverse collision time.

Diffusion and electrical resistance are both caused by collisions between electrons and ions, and the diffusion coefficient is related to the electrical resistivity.

In real magnetic confinement devices, the diffusion rates are much higher than predicted by classical diffusion theory. One reason for this is microinstabilities, which as we have seen degrade the electric currents that confine plasmas. Another reason is that some processes occur that produce diffusion step sizes much larger than the Larmor radii. Diffusion driven by these processes is called neoclassical diffusion. One of the processes that drive neoclassical diffusion is the magnetic mirror effect discussed in Section 20.2.2.

In toroidal confinement devices, both the toroidal and poloidal components of the magnetic field are stronger on the inside of the torus than on the outside. The net field wraps helically around the torus, and particles gyrating around a field line will follow the field line from regions of weaker magnetic field to regions of stronger magnetic field. Thus, some of them are subject to the magnetic mirror effect. As they are reflected back and forth by the mirror effect, their guiding centers also undergo particle drifts such as the grad-**B** drift, and the locus of points through which the guiding center passes, projected onto a single cross section, looks like a banana, as shown in Figure 20.17. Collisions successively put such trapped particles into and out of their loss cones, which moves them through a sequence of banana orbits. Thus, the random walk size is not the Larmor radius, but the width of the average banana orbit, Δ_b, which is much larger. This "banana diffusion" increases the diffusion coefficient by about a factor of 10 in tokamaks.

The discussion in this section barely scratches the surface of the fascinating topic of plasma physics. But it is enough to provide an insight into the behavior of plasmas in magnetic confinement fusion devices and to show why it is so difficult to make them behave the way we want them to. In the next section, actual fusion reactor concepts are discussed.

20.3 PROPOSED FUSION REACTORS

Especially in the early years of fusion research, numerous and varied magnetic configurations were proposed for fusion reactors. Most of them have been abandoned,

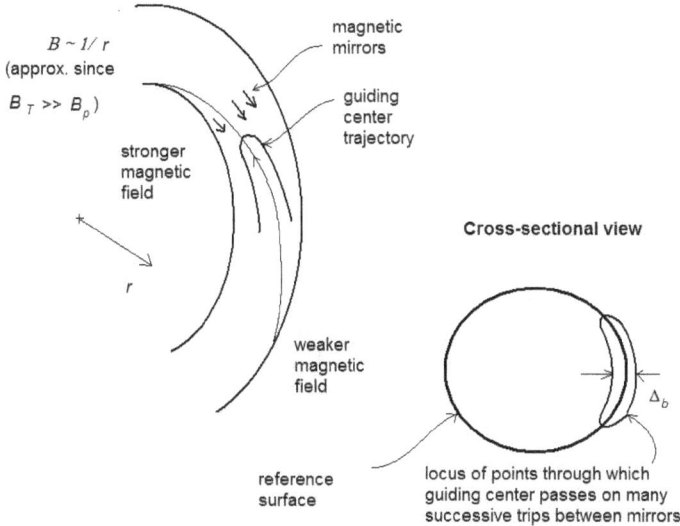

Figure 20.17—Banana diffusion in toroidal confinement devices

although interest still persists in some. In this chapter, I will provide details only for the latest inertial confinement design and for ITER. For other configurations I will only give names, references, and sometimes minimal descriptions. The interested reader may check the given references for them or do an Internet search to find more information.

Magnetic confinement schemes come in two varieties: linear and toroidal systems. Toroidal systems are geometrically complicated, with concomitant complexity in the required physical hardware. In contrast, linear systems are geometrically simple, and so is the hardware associated with them. Linear systems include the theta-pinch and Z-pinch configurations described above, along with variations such as cusp-ended sole-noids,[18] stabilized magnetic mirrors,[19] and field-reversed theta pinches or field-reversed configurations (FRCs).[20, 21] Most linear systems suffer excessive end-losses, MHD instability, or both, but FRCs are based on closed, elongated poloidal field lines contained within a straight solenoid, so they have superior confinement; they also have controllable stability properties.[22] Magnetically, they are toroidal systems, but mechanically, they are linear systems. Of all the linear systems, FRCs show the greatest promise.

Toroidal systems include the tokamak, the stellarator,[23] in which both the poloidal and toroidal components of the magnetic field are provided by currents in external coils, the reversed-field pinch,[24] and the spheromak.[25, 26] Like the FRC, the spheromak is a toroidal magnetic configuration in which no hardware passes along the major axis, so that it is mechanically simpler than other toroidal systems. However, the Sustained Spheromak Physics Experiment (SSPX)[27, 28] at Lawrence Livermore National Laboratory failed to suppress instabilities and did not achieve adequate confinement.[29] Another toroidal system, the National Spherical Torus Experiment (NSTX)[30] at the Princeton Plasma Physics Laboratory, has a "cored apple" configuration, with a conducting core along the major axis as in a tokamak, but a more spherical shape as in a spheromak. This configuration offers greater stability and higher plasma pressure than in a tokamak.[31]

20.3.1 Inertial confinement fusion and the National Ignition Facility

The physics and technology of inertial confinement fusion are summarized in Reference 32. Conceptually, as stated in Section 20.1, a small fuel pellet is irradiated from all sides by intense laser beams, and the reaction force from the ablating surface material compresses and heats the pellet to fusion conditions. During the compression, shock waves are sent inwards through the pellet and increase the density and temperature further. At a temperature of 100,000,000 K and a density 20 times that of lead, ignition occurs: Energy deposited in the pellet by fusion alpha particles sustains the temperature and leads to the fusion of a large fraction of the fuel nuclei before the pellet explodes. The particles ejected from the explosion impinge on the walls of the confinement vessel, heating the vessel and providing a means for collection of the fusion energy for conversion into electricity by an ordinary thermal cycle.

Inertial confinement schemes were first proposed in the 1960s, and serious experimental efforts began in the 1970s, mostly at the Lawrence Livermore National Laboratory. The LLNL experiments progressed with a series of increasingly powerful lasers, known as Janus, Cyclops, Argus, Shiva, and Nova. If the pellet is not almost perfectly spherical, and if the irradiation is not almost perfectly symmetrical over the pellet surface, the pellet can be blown apart by Rayleigh-Taylor instabilities as discussed in Section 20.2.5. Great progress was made in target design and irradiation technology in the course of this experimental program. The follow-up to Nova is the National Ignition Facility (NIF), which was completed in 2009, is currently being operated for test shots, and was scheduled to achieve ignition (self-sustaining fusion) in 2012.[33] As of February 2017, it had not yet achieved ignition, although NIF researchers were still hopeful. Meanwhile, the NIF is being used for research in a variety of fields involving extreme temperatures and densities.[34]

The NIF was conceived as a dual-purpose experiment. Not only will it explore the practicality of inertial confinement fusion as a source of electric power, but it will also allow simulation of thermonuclear weapons explosions. The Comprehensive Test Ban Treaty (see Chapter 23) bans all testing of nuclear weapons by signatory nations, including the United States, but some kind of experimental verification of weapons designs is needed. The NIF will permit such experimentation without the explosion of actual nuclear weapons.

The layout of the NIF facility is shown in Figure 20.18. Most of the building is occupied by the laser equipment. The laser pulse begins as a single nanojoule infrared pulse in an ytterbium-based laser called the Master Oscillator. The pulse from the Master Oscillator is divided into 48 fiber optic channels, each of which goes into a separate Preamplifier Module (PAM). In each PAM, the light is passed four times through a neodymium amplifier laser; the total output of the PAMs is about 6 J, still in the form of infrared radiation.

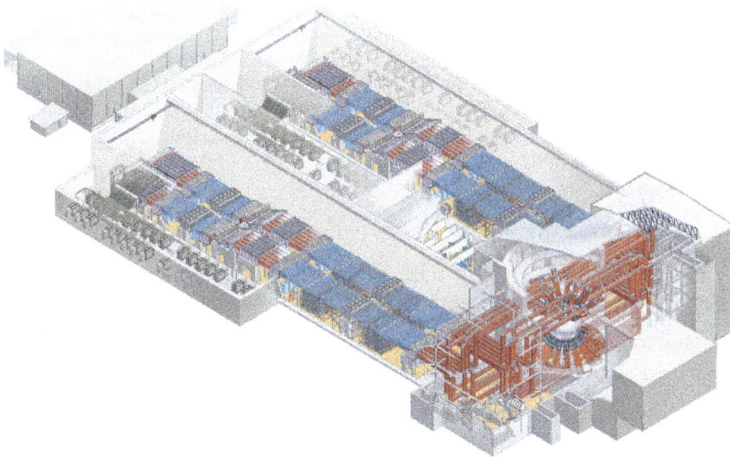

Figure 20.18—National Ignition Facility building layout. This illustration was provided to the Wikimedia Commons by LLNL, and is reproduced here under the Creative Commons Attribution-ShareAlike 3.0 license and the GNU Free Documentation License.

The amplified light from each PAM is split into four beamlets, each of which enters a separate beamline, for a total of 192 beamlines. Figure 20.19 shows one beamline.[35]

In the beamline, the light first enters the power amplifier and then goes through an optical polarization switch called a plasma electrode Pockels cell (PEPC) into the main amplifier, in which it is trapped by the PEPC until it has passed four times through the main amplifier's 11 glass amplifier slabs. It goes one more time through the power amplifier, and then into a system of transport optics called the switchyard—basically a channel of twists and turns accomplished by a series of mirrors—until it is aimed into the target chamber. Each of the 192 beams is oriented to irradiate the target chamber from a different direction, so that the beams collectively provide symmetrical irradiation of the target. Just before entering the target chamber, the beam passes through a final optics assembly, where it is converted from infrared to ultraviolet radiation. The final energy of the combined beams is 4 million J.[36]

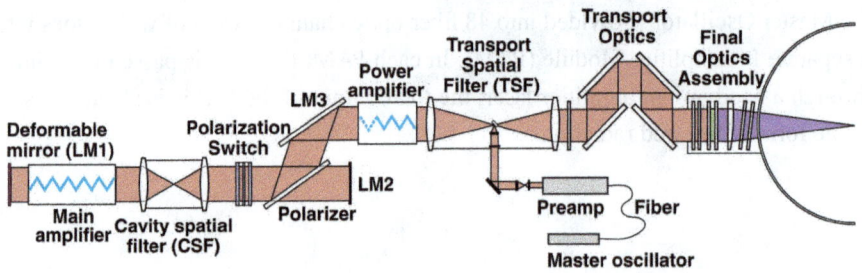

Figure 20.19—One beamline in the National Ignition Facility. This illustration was derived from works owned by Lawrence Livermore National Laboratory and may be reproduced with acknowledgement as stated in Reference 35.

The diameter of the target chamber is 10 m—over 30 feet. Figure 20.20 shows the target chamber before it was installed in the target chamber bay. The size of the target chamber ensures that the debris from the pellet explosion will become diffuse enough not to damage the chamber wall.

The pellet is contained in a metal cylinder the size of a pencil eraser called a *Hohlraum* (German for "hollow room"), and the laser beams actually heat the Hohlraum instead of the pellet inside it. The heated Hohlraum generates intense X-rays that in turn heat the pellet; the X-rays heat the pellet more uniformly than even 192 distributed laser beams could. The current pellet design is a hollow beryllium sphere about 2 mm in diameter, containing a deuterium-tritium mixture. The pellet is held in the middle of the Hohlraum by thin plastic webbing. The assembly is cooled to 18 K before placement in the target chamber, and the D-T mixture freezes onto the inner surface of the beryllium shell. A little of the D-T mixture remains in vapor form in the interior of the pellet.

Figure 20.20—NIF target chamber. From Lawrence Livermore National Laboratory; this photograph is in the public domain.

Although NIF was initially scheduled to achieve ignition in 2012, that goal has been more difficult to achieve than it was hoped to be. Nevertheless, NIF reached an important milestone by 2014, by producing more energy from fusion than was deposited in the fuel by the lasers. This was the first time a net gain of energy has been achieved in any fusion device.[37]

Various pellet designs will be tested in the NIF, not only to find how best to achieve controlled thermonuclear fusion, but also to simulate various nuclear weapons designs. NIF will be a very useful experimental facility, but it is a long way from being a prototype nuclear fusion power plant reactor. The closest thing to that is the planned tokamak experiment, ITER, which is discussed in the next section.

20.3.2 The ITER program

The first tokamaks were built in the Soviet Union by a team led by Lev Artsimovich at the Kurchatov Institute near Moscow. A long series of tokamak experiments followed their initial successes, with teams all over the world picking up the idea. Currently, over 20 tokamaks are in operation worldwide.[9] As the experiments became larger and more complicated, they also became more expensive, and international collaborations emerged to share the costs. The most successful such collaboration so far has been the Joint European Torus (JET), located in Culham, England.[38] Construction on JET began in 1978, and experiments began in 1983. JET is still in operation as of early 2019. JET has produced fusion power, but the highest fusion gain (Q) it has achieved so far is 0.7, attained in 1997.

The next step beyond JET is ITER (pronounced "eater"),[2] currently under construction and scheduled to begin operation in 2025. ITER was first proposed by Evgeny Velikhov of the Soviet Union, who became the first head of the ITER Council, the supervisory body in charge of the ITER program. ITER was originally an acronym for International Thermonuclear Experimental Reactor, but negative public perceptions of the term "thermonuclear," based on the association of the term with nuclear weapons, led to an official abandonment of the acronym. Now "ITER" alludes to the Latin word *iter*, meaning journey, direction, or way. Opposition to ITER, led by groups such as Greenpeace, is based on two arguments. First, they assert that a fusion reactor, being nuclear, is inherently dangerous. This argument is ignorant and foolish; as noted above, the problem with fusion is not a runaway reaction, but to sustain a reaction at all. Meltdown-type accidents in a fusion reactor are simply impossible. Furthermore, radioactive waste produced by a D-T fusion reactor would be orders of magnitude less in volume than that from a fission reactor, and it would all be short-lived. The second argument is that the money ITER would cost (currently projected to be € 10 billion for construction and € 5 billion for maintenance and research over a 35-year lifetime) would be better spent on renewable energy, which can be deployed immediately. I discuss renewable energy in Chapter 8, but the proposed costs of ITER are really trivial. In contrast, as of February 2010, the cost of the Iraq war was about $700 billion, and the net cost of the Troubled Asset Relief Program (TARP), as of 12 April 2010, was $89 billion.[39] Anyway, pretending that ITER doesn't mean International Thermonuclear Experimental Reactor seems to me a concession to ignorance and Luddism and a sad reflection on the level of scientific and technological awareness of so-called advanced nations. Such backwardness should be confronted head-on.

Currently, the participants in the ITER program are the European Union, India, Japan, China, Russia, South Korea, and the United States. Canada was formerly a participant, but withdrew for lack of funding. The United States was an original participant, but dropped out in 1999 and returned in 2003. Australia became a non-member participant in 2016.[40] The intermittent participation by the United States reflects the general fickle nature of support by the U.S. government of any high-budget technical program; new administrations often cancel programs that were enthusiastically supported by their predecessors.

ITER will be a tokamak reactor capable of producing 500 MW of fusion power for up to 1000 seconds per pulse. A conceptual sketch of the current ITER design is shown in Figure 20.21. Notice the human figure standing on the base of the pressure vessel at the lower left corner of the drawing; this person would not be standing there if the reactor were operating, but is only provided to show scale.

The translucent purple region in the figure is the plasma, which is contained in a vacuum chamber of about 840 m³. The magnets will be made of superconducting niobium-tin and niobium-titanium alloys cooled to 4 K; the toroidal field of 11.8 tesla (T) will be generated by 18 toroidal field coils (encircling the plasma poloidally), and the smaller poloidal field will be generated both by plasma currents and by 6 poloidal field coils (encircling the plasma toroidally). In addition, there is a central solenoid system that inductively drives the plasma current, as shown in Figure 20.11.

The tesla is a large unit of magnetic field. A field of 11.8 T is very large; by contrast, the magnetic field of the Earth at the equator is 3×10^{-5} T and the fields of magnets in typical electric motors (as in a refrigerator) are about 3×10^{-3} T. Medical MRI systems generate fields of 1.5-3 T.[41]

The ITER organization website[42] has excellent interactive graphics that explain the entire machine. Up-to-date information on the progress of ITER can also be found there.

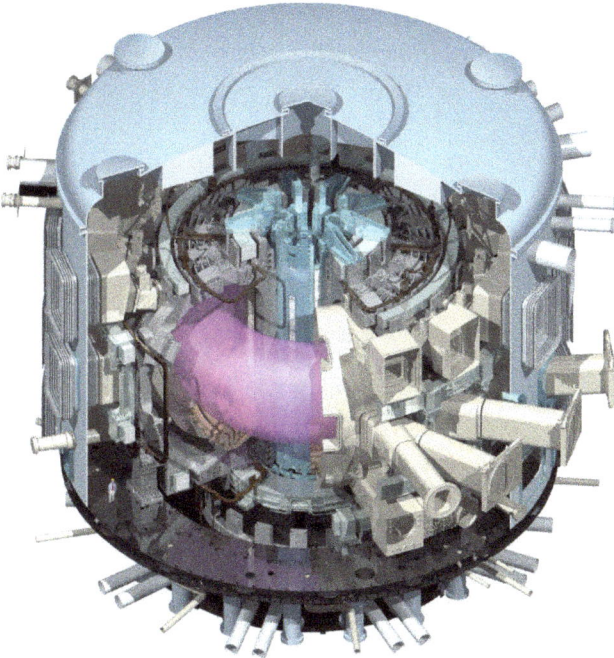

Figure 20.21—Cutaway view of ITER design. This image was put on the Wikimedia Commons by the ITER organization with permission to be used freely for any purpose.

Another important experimental tokamak is the Anhei tokamak in China.[43] This experiment, which is playing a supporting role in the ITER project, has achieved notable milestones in plasma confinement and plasma heating.

20.4 THE FUTURE OF FUSION

If ITER meets its experimental goals, and if funding agencies can be persuaded to continue fusion research and development, the next proposed project after ITER will be another international collaboration called DEMO.[44] According to the present schedule, DEMO will begin operating in 2033. Such schedules are almost always optimistic, but perhaps DEMO might actually be running by 2050. DEMO will be an actual prototype fusion power plant, supplying electricity to the grid. Successful operation of DEMO would pave the way to a fusion power economy.

To me, the potential rewards of a fusion economy are so great that I hope politicians can be talked into supporting it to completion. But proponents of such technologies as wind and solar power argue that their energy sources are available now, and fusion power is too expensive and too late. I disagree, and I address the merits and drawbacks of competing energy sources in Chapter 8.

CHAPTER 21
THE NUCLEAR FUEL CYCLE

The complete nuclear fuel cycle is the sequence of processes the uranium or thorium nuclear fuel resource undergoes from its extraction from the ground to its final disposal in a geologic repository. Although the thorium fuel cycle has some significant attractions, including resistance to nuclear weapons proliferation (see Section 23.5), most reactors are based on the uranium fuel cycle. This chapter primarily discusses the uranium fuel cycle, but many of the same principles would also apply to thorium. A major phase of the nuclear fuel cycle is nuclear waste disposal. That topic is broad enough to deserve its own chapter, which is the next one.

21.1 URANIUM AND THORIUM RESOURCES

Whether a mineral resource is economically recoverable depends on the cost of extraction and the price of the mineral. These costs are not fixed in time, so what is not economically recoverable at one time may become so if the technology becomes cheaper or the mineral price rises. As of 2007, the world's estimated recoverable uranium resources were about 5.5 million t (i.e., tonnes, or metric tons, defined as 1000 kg), of which the largest portions are in Australia (23%), Kazakhstan (15%), Russia (10%), South Africa (8%), Canada (8%), and the United States (6%).[1] This estimate was based on a uranium price of $130/kg, in 2007 U.S. dollars.

One cannot be sure of the amount of uranium residing in a geological deposit until the deposit is exhausted, but modern knowledge of geology enables mineral explorers to make accurate estimates. However, at any given time only a small fraction of the Earth's crust has been explored, and it is reasonable to expect that known deposits are only a small fraction of total recoverable deposits. In fact, known deposits have tripled since 1975 as more money has been devoted to exploration. Future discoveries are almost assured.

Nevertheless, it is informative to calculate how currently estimated resources compare with potential demand. At the current usage of 65,000 t/yr, current estimated uranium resources are only adequate to last 80 years in once-through fuel cycles using conventional reactors. (Once-through fuel cycles are those in which spent fuel is taken to a geologic repository without reprocessing after removal from a reactor.) Conventional reactors run on U-235, which comprises only 0.7% of natural uranium, and on Pu-239, which builds in the fuel from U-238 as the fuel is exposed to fission neutrons. Reprocessing, discussed in Section 21.5, extracts unused uranium and plutonium from spent fuel for use in new fuel assemblies. Reprocessing of spent fuel from conventional reactors could extend the resource significantly. However, use of breeder reactors specifically designed to convert U-238 to plutonium (see Chapter 17) could enable almost all of the uranium resource to be used—i.e., more than 100 times as much as in the once-through cycle. A fission reactor infrastructure based on breeder reactors to make plutonium fuel and thermal reactors to use it could power the whole world's economy for centuries.

Thorium is more abundant than uranium, so a thorium-based reactor infrastructure could power the world for even longer.

21.2 MINING

Mining is the part of the nuclear fuel cycle with the greatest environmental impact. Compared to coal, which occurs in seams of concentrated carbon, or crude oil, which comes out of the ground (or into the Gulf of Mexico) in concentrated form, uranium and thorium deposits are more dilute. Very-high-grade uranium ore is 20% or higher in uranium content, high-grade ore is from 20% to 2% uranium, and low-grade ore is 2% to 0.1% uranium.[1] However, as pointed out in Chapter 6, the ratio of the energy in uranium to the energy in coal is 47 million to one per atom. Thus, even low-grade ore requires the removal of only $1000/47,000,000 \approx 2 \times 10^{-5}$ as much uranium ore as coal to produce a given amount of energy.

Uranium deposits occur both in sedimentary and igneous rocks, and in hydrothermal deposits.[2, 3] Uranium and thorium are both almost ubiquitous, but mostly in very low concentrations that are not now economically recoverable. Two ores that occur in economically recoverable concentrations in sedimentary rock are pitchblende, also called uraninite,[4] and carnotite.[5] Ores of igneous origin include pegmatite,[6] nepheline syenite,[7] and disseminated deposits. Hydrothermal deposits occur in veins.

Uranium is extracted from the ground by three methods: open pit mining, box cut mining, and in-situ leaching.[2] The uranium radioactive decay chain includes radon (see Section 4.6), which is a hazardous radioactive gas. Open pit and box cut mining expose miners to radon, so precautions must be taken to minimize their exposure. For example, in open pit mining, workers do much of their work in enclosed cabins in the excavating machinery and dump trucks, and water is sprayed into the air to suppress dust. Uranium-bearing dust releases radon into miners' lungs when inhaled, so precautions must be taken to minimize dust inhalation.

In open pit uranium mining, as in surface mining for coal, a certain amount of "overburden" must be removed to get at the ore. How much must be removed depends on the details of the specific deposit, but because so much less ore needs to be removed than in coal mining, the amount of overburden to be removed must be commensurately less.

Box cut mining is similar to any other type of underground mining such as coal mining. A vertical shaft is dug from the surface to the level of the ore bodies, and the ore is reached and extracted through a system of tunnels. The details of the tunnels depend on the nature of the ore bodies. In any case, the ore is brought to the surface, where it is treated to remove the valuable minerals.

One method of extracting the uranium from its ore, used mostly when the uranium in the ore is in oxide form, is heap leaching. Heap leaching is a method also commonly used in gold mining. In heap leaching, a suitable area is leveled, sloped, and lined with a thick plastic sheet, and the extracted ore is crushed and placed in heaps on the plastic liner. A leaching agent, usually sulfuric acid, is sprayed onto the heaps, and as it trickles down through the heaps it dissolves the uranium out of the ore. The solution is collected in collecting pools and then pumped to a processing plant for removal of the uranium from the solution.

Heap leaching imposes risks of sulfuric acid leaks into groundwater, so the process must be carefully monitored for environmental protection. Current law requires mining companies to set aside reclamation money before beginning mining operations, so that the public will not be left to clean up mining sites if mining companies go bankrupt.

The third type of mining is in-situ leaching.[8] In this method, the leaching agents are pumped into the ground, dissolve the uranium-bearing minerals as they pass through the ore bodies where the minerals lie, and are then sucked out of the ground again bearing the dissolved minerals with them. The leaching agents usually include either sulfuric acid or an alkaline agent, depending on the type of mineral, and sometimes include an oxidizing agent such as hydrogen peroxide. Because in-situ leach mining is the least environmentally destructive mining method, it is to be preferred when it is practical.

Seawater contains uranium at a concentration of about 3 parts per billion. This is a very low concentration, but there is a lot of water in the ocean, so that the total uranium resource there is huge—about 4.5 billion tonnes. Currently it can be extracted at a cost of about \$250-300 per kg of uranium, which is only about twice the price of uranium cited above for 2007.[9] Perhaps in the future the cost of extraction will come down or the price of conventionally mined uranium will go up, and extraction from seawater will become competitive.

21.3 MANUFACTURING FUEL

The product of the leaching or milling operation carried out at the mining site is usually U_3O_8, called "yellowcake," in which the uranium composition is natural uranium, i.e., 0.7% U-235 and 99.3% U-238. In most thermal reactors, natural uranium is too low in U-235, although the CANDU reactor (see Chapter 15) operates on natural uranium. Fresh fuel in typical water-cooled thermal reactors contains 2-5% U-235, and HTGRs will need 10-20% U-235.[10] Fast breeder reactors normally run on plutonium, which has to be extracted from spent fuel by reprocessing,[11] which is discussed in more detail in Section 21.5. However, if uranium fuel is used in an LMFBR, it must be enriched to about 20%.[12] For all the reactors requiring enriched fuel, enrichment is the manufacturing step that follows the production of yellowcake.

Two principal means of enrichment have been applied successfully in the past: gaseous diffusion and gas centrifugal processing.[13] There are several other technically feasible processes, but only one of them, laser enrichment, has been developed enough to become economically viable. In the U.S., General Electric-Hitachi has been granted a license by the Nuclear Regulatory Commission to build and operate a laser enrichment test facility, and they have applied for a license to build and operate a commercial facility.[14] Laser enrichment is expected to be more efficient and thus more economical than gaseous diffusion or centrifugal processing.[15] These advantages have raised concerns that laser enrichment could make clandestine production of nuclear weapons by rogue states easier, but the technology is extremely complicated and the details are classified.[16]

All three processes require the uranium to be contained in a gaseous form, so the first step in enrichment is conversion of yellowcake to uranium hexafluoride (UF_6), which is a gas at standard atmospheric pressure at temperatures above 56.5 °C.[17] This conversion is accomplished in fluidized-bed chemical reactors. First, the yellowcake is ground into powder of appropriate grain size, and then in the first fluidized bed the powder is reduced by hydrogen gas at a temperature between 540 °C and 650 °C to uranium dioxide (UO_2). In two further fluidized beds, at temperatures of 480 °C to 540 °C and 540 °C to 650 °C, respectively, the UO_2 reacts with anhydrous hydrogen fluoride to form uranium tetrafluoride (HF_4), a solid salt. Finally, the uranium tetra-fluoride is treated with fluorine gas at a temperature between 340 °C and 480 °C to produce uranium hexafluoride.[18]

Laser enrichment takes advantage of differences in absorption of laser light by U-235 and U-238,[19] while the other two methods are based on the difference in the masses of U-235 and U-238. In gaseous diffusion, the mixture of $^{238}UF_6$ and $^{235}UF_6$ diffuses across a semipermeable membrane from one chamber into another, as shown in Figure 21.1. A molecule of $^{238}UF_6$ is heavier than a molecule of $^{235}UF_6$, so it passes more slowly through the membrane. The mixture on the low-pressure side of the membrane is then higher in $^{235}UF_6$ concentration than the mixture on the high-pressure side.

⊚   U-235 hexafluoride molecule

●   U-238 hexafluoride molecule

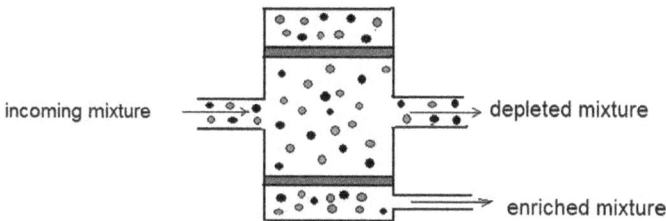

Figure 21.1—Gaseous diffusion enrichment.

In the gas centrifuge method, the heavier $^{238}UF_6$ tends to settle to the wall of a spinning cylinder, or centrifuge, more than $^{235}UF_6$, as shown in Figure 21.2. The separation provided by either of these methods is small, so both methods require many stages to raise the concentration of U-235 from 0.7% in natural uranium to a level high enough even for an LWR (for gaseous diffusion, hundreds of stages are needed). All enrichment in the U.S. is currently performed by the United States Enrichment Corporation (USEC), at a plant in Paducah, Kentucky. Currently only the gaseous diffusion method is used in the U.S., although it is an obsolete technology and centrifugal processing is more efficient. However, both USEC and Louisiana Energy Services have received licenses to build and operate gas centrifuge plants, and AREVA Enrichment Services has applied for such a license.

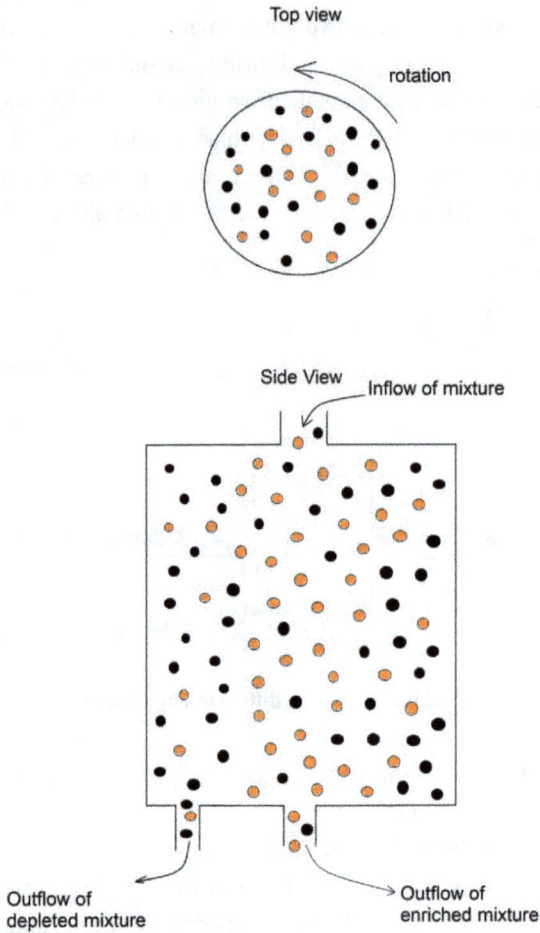

Figure 21.2—Gas centrifuge enrichment

Separation of a given mass of feed into an enriched product and depleted tails is expressed by a quantity called separative work, which is not actually work, but has units of mass. The definition of the separative work unit (SWU) is complicated.

Specifically, if a mass F of feed material containing a fraction x_f of U-235 is separated into a mass P of product containing a fraction x_p of U-235 and a mass T of tailing material containing a fraction x_t of U-235, then the separative work units required to achieve this separation are given by

$$SWU = PV(x_p) + TV(x_t) - FV(x_f),$$

where $V(x)$ is a mathematical function called the value function, given by

$$V(x) = (1 - 2x)\ln\left(\frac{1-x}{x}\right).$$

The value function is shown in Figure 21.3. The symbol ln denotes the natural logarithm.[a] One can see that an equal mixture has no "value" in this function (although a 50% enriched mixture would actually have great monetary value), and highly dilute and highly concentrated mixtures have high value.

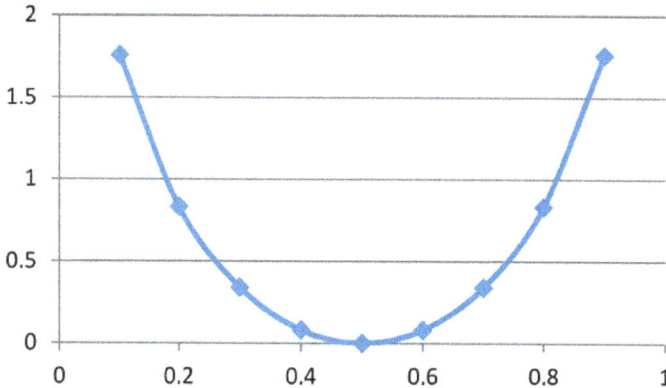

Figure 21.3—The value function

---

[a] The natural logarithm is the inverse function of the exponential function encountered now and then in previous chapters (e.g., Section 4.1). If $y=e^x$, then $x=\ln y$, where $e=2.718281828...$ The number e is called the base of natural logarithms. In plain English, the logarithm of a number is the exponent to which one must raise the base to get that number. Logarithms have many useful properties, such as converting multiplication to addition: $\ln ab = \ln a + \ln b$. Logarithms can be defined in terms of any base desired, and a very useful base is 10. So $\log_{10} ab = \log_{10} a + \log_{10} b$. This principle was the basis of the slide rule, a calculating device that was used by all engineers until the advent of the electronic calculator.

The monetary cost of a separative work unit depends on the energy consumption per SWU. Gaseous diffusion plants are much more energy-intensive than gas centrifuge plants; for a gaseous diffusion plant, an SWU costs 2400-2500 kW-h, while for a gas centrifuge plant, an SWU costs only 50-60 kW-h. Laser enrichment is expected to use even less energy.

Reference 13 gives an example for illustrative purposes: A large nuclear power plant generating 1300 MWe using fuel enriched to 3.75% U-235 uses about 25 t/yr of fuel. This fuel is produced from about 210 t of natural uranium with about 120,000 SWU.

The process for manufacturing a nuclear fuel element depends on the type of fuel being used. For example, the microsphere for HTGR fuel (see Chapter 16) consists of layers vapor-deposited onto a kernel of UO_2 or UCO. However, most nuclear power plant fuel is based on UO_2 pellets stacked in tubes of cladding material; in LWRs, the cladding is an alloy of zirconium called Zircaloy.

In the process of manufacturing the pellets, the enriched uranium hexafluoride gas is first converted back into uranium dioxide, in the form of a ceramic powder.[20] This conversion process takes several steps.[15, 21, 22] In one approach, the uranium hexafluoride is hydrolyzed by water to produce uranyl fluoride (UO_2F_2), and then the addition of aqueous ammonia produces ammonium diuranate, $(NH_4)_2U_2O_7$. This is reduced with hydrogen to UO_2. The UO_2 is pulverized to powder, then mixed with an organic binder and pressed into pellets. The pellets are sintered[23] at high temperature (1900 °C) to bind the powder grains together.[24]

The fuel pellets are loaded into fuel rods, which are placed in fuel assemblies as described in Chapter 11.

21.4 IN-CORE FUEL MANAGEMENT IN REACTOR OPERATIONS

Consumption of fuel during reactor operation is characterized in several ways.[25] The term "burnup" is used to refer to such consumption, although of course literal burning has nothing to do with the consumption process. One unit for burnup just quantifies energy produced per mass of fuel: GWd/MTHM, or gigawatt-days per metric ton of heavy metal (heavy metal being the uranium or plutonium fuel). MWd/MTHM, or megawatt-days per metric ton of heavy metal, is also used. The term "heavy metal" refers to all the fissionable nuclides in the fuel, not just the fissile content. Typical values for spent fuel from LWRs are in the range of 40-60 GWd/MTHM, while advanced LWR designs are expected to achieve 90 GWd/MTHM. Past

HTGRs, such as the Fort St. Vrain reactor, were designed to achieve about 100 GWd/MTHM,[26] but some Generation IV HTGRs are hoped to achieve ultra-high burnup in the range of 700 GWd/MTHM.[27]

Another way to characterize burnup is in the percentage of the initial heavy metal atoms that gets fissioned, or % FIMA. A burnup of 100% FIMA in a uranium-fueled reactor would require consumption of all the U-238 as well as the fissile U-235. Consumption of the U-238 takes place either by fast fission or by conversion to plutonium and subsequent thermal fission. Neither process is likely to enable burnup anywhere near 100% FIMA in uranium-fueled thermal reactors. The ultra-high burnup hoped for in HTGRs would occur in plutonium-fueled reactors, in which all of the fuel is fissile.

The connection between burnup in GWd/MTHM and % FIMA comes through the usable energy per fission of a heavy metal atom, which is about 200 MeV. Then 100% FIMA turns out to be about 938 GWd/MTHM, or roughly 1000 GWd/MTHM. So typical LWRs achieve about 5% FIMA, and the advanced LWR designs mentioned above should achieve about 9% FIMA. For HTGR fuel with low-enriched uranium, the burnup of 19% FIMA achieved by researchers at the Idaho National Laboratory was a record.[28]

In a reactor core of uniform composition, the neutron flux distribution is not uniform. In the idealized case of a bare (unreflected) slab reactor, which is infinite in two dimensions and of a constant thickness in the other dimension, the one-speed neutron diffusion equation (see Section 6.2.1) predicts a cosine-shaped neutron flux distribution,[b] as shown in Figure 21.4(a). Adding reflectors to the sides makes some of the neutrons that leak out of the core bounce back in, increasing the flux near the edges, as shown in Figure 21.4 (b). But the flux distribution is still not uniform, and therefore the fission heating is not released uniformly. Therefore, the temperature increase in coolant channels in different parts of the core is also uneven. This uneven heating would make a reactor thermodynamically inefficient. Also, the burnup in different fuel assemblies is uneven. To increase efficiency and achieve maximum burnup in all fuel assemblies, reactor operators shut down their reactors periodically (typically once a year) and rearrange the fuel assemblies, removing fuel that has achieved maximum

---

[b] The cosine is a trigonometric function; trigonometry can be defined as the study of triangles and functions based on them. If α is an acute angle in a right triangle, the cosine of α (written as cos α) is defined as the ratio of the leg adjacent to α to the hypotenuse. The sine of α (written as sin α) is the complement of the cosine; it is the ratio of the leg opposite from α to the hypotenuse. For angles greater than 90°, appropriate complementary right triangles are constructed to define the cosine. But the quantity whose cosine (or sine) is taken doesn't have to be an angle. In the expressions cos x or sin x, x can be any variable. In Figure 20.4(a), the flux distribution is a cosine about the midline of the slab.

burnup, placing fresh fuel in regions of lower neutron flux, and placing more highly burned fuel in regions of higher neutron flux. Fuel assemblies typically remain in a reactor for three or four loading cycles, moving from place to place as needed for a uniform power distribution. These reshuffling operations are known as in-core fuel management,[29] and there are sophisticated reactor physics codes that enable operators to reshuffle fuel to the best economical advantage. Control rod placement is also used to promote uniform power distribution by making the flux distribution more uniform.

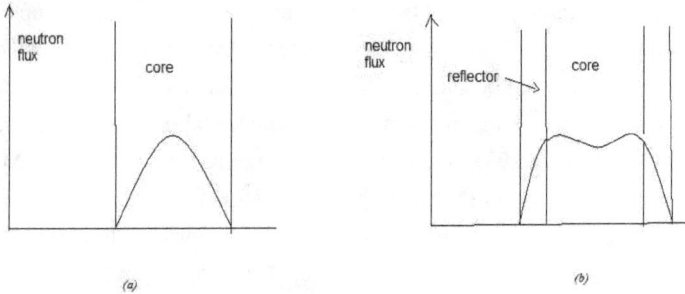

Figure 21.4 – Neutron flux distributions in bare (a) and reflected (b) infinite slab reactors

21.5 NUCLEAR FUEL REPROCESSING

Nuclear fuel reprocessing[30] entails the removal of some or all of the actinides from spent fuel for reuse. Actinides are elements with atomic number from 89—i.e., actinium—to 103 (lawrencium).[31] In spent reactor fuel, the actinides are mostly uranium and plutonium. Some higher actinides, such as americium,[32] also build up in reactor fuel, and some proposed fuel cycle schemes are designed to consume these contaminants.[33]

As noted above, typical LWR fuel is enriched to about 5%, and LWRs typically achieve about 5% FIMA. If only U-235 were consumed, this would imply that all of the burnup comes from U-235 fission. However, some of the U-238 undergoes fast fission, and some of the Pu-239 that builds in undergoes fission itself. So some of the burnup comes from other nuclides than U-235, and not all of the U-235 is consumed; in fact, the U-235 concentration in spent fuel is usually higher than that in natural uranium.[34] Excluding oxygen, the composition of spent LWR fuel is generally about 3% fission products, 1% plutonium, and 96% uranium, with traces of higher actinides.[35]

The uranium is mostly U-238; about 1% of the uranium is U-235 and about ½% of it is U-236, which builds in by (n,γ) reactions in about 18% of neutron absorptions in U-235.[36] The plutonium is about 80% Pu-239, which is an excellent fuel for both nuclear reactors and nuclear weapons; most of the rest is Pu-240, which, because of its high rate of spontaneous fission, makes reactor-grade plutonium (i.e., plutonium extracted from spent power reactor fuel) very difficult to use in nuclear weapons. Pu-240 builds in by (n,γ) reactions in Pu-239, so the longer the fuel is left in a reactor the more Pu-240 it acquires. Reactors specifically designed for the production of weapons plutonium operate with short fuel cycles to prevent the buildup of Pu-240. Some Pu-242, which also has a high rate of spontaneous fission, builds in by successive (n,γ) reactions from Pu-240.

It is important to understand the role of Pu-240 and Pu-242 when considering the pros and cons of reprocessing, because one of the arguments against reprocessing is that it promotes nuclear weapons proliferation. For this reason, President Gerald Ford suspended commercial reprocessing in the U.S. in 1976, and President Jimmy Carter banned it altogether in 1977. The fission products in spent fuel emit such intense radiation that handling spent fuel requires special equipment and radiation shielding. It would not be practical for either clandestinely acting agents or overtly acting commandos to divert spent fuel and extract plutonium from it for use in weapons. However, the fission products are removed in reprocessing and (mostly) discarded as nuclear waste. The remaining uranium and plutonium are much less radioactive, and they can be handled safely if proper precautions are taken. If they come out of reprocessing in a combined form (mixed-oxide, or MOX, fuel) for use in LWRs, the fissile content will be of low enrichment, but the plutonium can be extracted chemically and used to make a nuclear weapon. It is much more practical for terrorist groups to perform such chemical separation processes than it is for them to enrich uranium or to remove plutonium from raw spent fuel. If the plutonium comes out of reprocessing already separated, as appropriate for the ultra-high-burnup HTGRs mentioned above, the terrorists wouldn't even need to perform that step. But the high rates of spontaneous fission sustained in Pu-240 and Pu-242 make reactor-grade plutonium very difficult to use in a nuclear weapon. In a simple fission bomb, pieces of the nuclear fuel have to be separated until the time of detonation, and at that time they are thrust together forcefully by chemical explosives. In a bomb made of plutonium extracted from power reactors (i.e., reactor-grade plutonium), excessive spontaneous fission causes premature detonation, and the bomb is said to "fizzle." A bomb made from reactor-grade plutonium was successfully tested at the Nevada test site in 1962,[37] but such bombs are difficult to make, and the significantly shorter half-life of Pu-240

(6564 years, versus 24,000 years for Pu-239) makes the material more radioactive than weapons-grade plutonium, so that remote handling facilities are required in the manufacture of a nuclear weapon from reactor-grade plutonium. Terrorists are unlikely to be able to obtain and operate such facilities, but since they use suicide bombers they could conceivably use "suicide reprocessors." Clearly, reprocessing and the use of reprocessed fuel require secure facilities.

Other objections to reprocessing are that it is expensive and it is environmentally very dirty. The economic issue depends on the price of uranium. As high-grade ores are depleted, the price of uranium will increase and make reprocessing more economically competitive. The environmental issue must be taken in context. Because so much energy is contained in such small volumes of nuclear fuel, reprocessing is inherently on a small scale compared with facilities for other types of energy production, and pollution resulting from reprocessing activities can be controlled. Compare this with the pollution and other environmental devastation caused by coal and petroleum (see Chapter 8).

Against these alleged drawbacks, reprocessing offers some compelling advantages. First, by enabling some or all of the unused fissile and fertile nuclides to be recovered and used, reprocessing extends the energy resource. If the remaining uranium is converted to plutonium in breeder reactors, essentially all of the uranium can eventually be used, rather than only the U-235. Since U-235 constitutes only 0.7% of natural uranium, use of the U-238 multiplies the effective energy resource by as much as 99.3/0.7=141.8 times.

Even if only the plutonium is used, the energy resource is extended by roughly 20%, since about 1% of the heavy metal in spent fuel is plutonium, compared to the 5% U-235 in the original fuel. However, removal of the plutonium makes waste disposal much easier. The half-life of Pu-239 is 24,000 years. This is short enough to make its radioactivity hazardous (remember, the shorter the half-life the more intense the radioactivity), so that it must be isolated from drinking water, but it is long enough to need a long time to decay. Pu-239 alpha-decays to U-235, which has a half-life of 703.8 million years, and that is hardly radioactive at all. In ten half-lives, only 0.1% of a radioactive substance is undecayed, and in 20 half-lives, only one-millionth of it is undecayed. This is the sort of timescale over which one must plan to keep Pu-239 out of the biosphere. That works out to about a half-million years, which is a long time to promise the integrity of a repository. If the actinides, including Pu-239, are removed by reprocessing, and only the fission products (which are highly radioactive and therefore decay rapidly) are buried in a waste repository, then the activity of the waste form is reduced to the level of the original uranium ore within a few centuries.[38]

President Ronald Reagan rescinded Carter's ban on reprocessing in 1981, but the economics of reprocessing have not been favorable for private enterprise to undertake unsubsidized reprocessing programs. Government support has not been provided for general commercial reprocessing programs, but in 1999 the National Nuclear Security Administration (NNSA) initiated a program with a consortium now called Shaw AREVA MOX Services, LLC, to build a plant at the Department of Energy's site at Savannah River, South Carolina, to manufacture MOX fuel.[39] The purpose of this program was to take weapons-grade plutonium from excess nuclear warheads discarded as part of the arms reduction agreements concluded by the U.S. and Russia, and consume it as nuclear reactor fuel. Construction of this facility began in 2005. However, cost overruns and a desire by some in the NNSA to convert the facility to plutonium pit production for nuclear weapons led to the cancellation of the project in October 2018; closure of the project was expected to take a year.[40] South Carolina opposed the closure because of its economic impact on the state, and as of April 2019, as this is being written, the state was considering taking its case to the U.S. Supreme Court.[41]

Several chemical processes have been developed to accomplish reprocessing. The first such process, developed in World War II for the extraction of plutonium for nuclear weapons, was the bismuth phosphate process, which quickly became obsolete. The current standard method is the PUREX process (for Plutonium-Uranium Extraction), developed by Herbert H. Anderson and Larned B. Asprey in the Manhattan Project.[42] PUREX is an aqueous process—i.e., it takes place in a water-based solution. First, the spent fuel pellets are dissolved in nitric acid. Some constituents of the spent fuel are not soluble, but remain as fine suspended solids; these are removed to prevent them from interfering with the subsequent reactions. Next, an organic solvent, consisting of 30% tributyl phosphate in a hydrocarbon such as kerosene, is introduced to separate the uranium and plutonium from the fission products. The organic solvent, which takes up the uranium and plutonium, forms an emulsion in the solution, and the fission products remain in the nitric acid component of the mixture. Irradiation of the mixture by fission products degrades the organic solvent, allowing some of the fission products to be taken up in it. Thus, it is necessary to use more than one stage in the process. Finally, the plutonium and the uranium are separated from each other. Some of the fission products, such as technetium, are useful, and they can be separated from the other fission products in further processing.

Some modifications of the PUREX process are discussed in Reference 30. These include the UREX process, which is intended to prevent the separation of the uranium and plutonium in order to reduce the potential for use of the product in weapons.

Aqueous processes contain fissile nuclides and water, which carries the potential for criticality accidents, as happened at the Tokai nuclear fuel plant in Japan in 1999.[43] A newer process, which is not aqueous, is pyroprocessing, or processing at high temperatures, in which the solvents are molten salts. Pyroprocessing is not yet in large-scale use, but it has been demonstrated at Argonne National Laboratory as part of its Integral Fast Reactor (IFR) project.[44] The "Integral" in IFR referred to the inclusion of reprocessing facilities on the same site as the reactor, to eliminate the transportation of plutonium fuel. Pyroprocessing facilities are more compact than aqueous processing facilities, which makes them more practical for collocation with the reactor.

21.6 NUCLEAR WASTE DISPOSAL

This section is just a brief overview. Nuclear waste disposal is treated at length in Chapter 22.

After the decision by President Jimmy Carter in 1977 to halt all nuclear fuel reprocessing for commercial reactors in the U.S., the nuclear fuel cycle for commercial reactors devolved into the once-through cycle, in which spent fuel is stored at the reactor site for a few years to let its radiation level subside and then taken to an underground repository. The Yucca Mountain site in Nevada was chosen as the site for a repository, but opposition from Nevada eventually led to the apparent cancellation of the project. Currently, no repository site has been designated in the U.S. Several candidate sites exist, two of which were studied in some detail before the selection of Yucca Mountain. These sites are discussed in Chapter 22.

As noted above, the time scale required for isolation of spent fuel is hundreds of thousands of years, but reprocessing could greatly reduce the amount of waste material to be discarded and also greatly reduce the time during which the waste material must be isolated from the environment.

21.7 THE NUCLEAR FUSION FUEL CYCLE

The first nuclear fusion reactors will use the deuterium-tritium (D-T) reaction, which produces a 14.1-MeV neutron. This neutron creates activation products in the structure surrounding the fusion plasma chamber, which are radioactive waste that needs to be disposed of, but it also provides the mechanism for producing tritium for fuel. Deuterium is present in nature in large quantities; one out of every 6500 hydrogen atoms in water on Earth is deuterium. Tritium is radioactive, with a half-life of only 12 years, so it is not found in nature in sufficient quantities for use in power generation. However, the element lithium is abundant in the Earth's crust, and the reactions

$$^6\text{Li}(n,\alpha)\text{T} + 4.78 \text{ MeV}$$

and

$$^7\text{Li}(n,n'\alpha)\text{T} - 2.47 \text{ MeV}$$

can be applied to breed tritium in a region around the plasma called the blanket. Natural lithium comprises 7.4% ^{6}Li and 92.6% ^{7}Li. The neutron absorption cross section for ^{6}Li is highest in the energy range below 1 MeV, with a value of about 950 barns for thermal neutrons, whereas the cross section for ^{7}Li has a threshold at 2.47 MeV and reaches its peak of about 430 millibarns at 8 MeV. Whether the tritium breeding is mainly accomplished in ^{6}Li or ^{7}Li depends on the neutron energy spectrum in the blanket. The reaction in ^{7}Li is endothermic, so it reduces the energy available for power production, but it uses the more abundant isotope.[45]

In any case, I regard D-T fusion reactors as only a stepping-stone to reactors that apply cleaner fusion reactions. The D-D reaction has two branches,

$$\text{D} + \text{D} \rightarrow {}^3\text{He} + \text{n} + 3.2 \text{ MeV}$$

and

$$\text{D} + \text{D} \rightarrow \text{T} + \text{p} + 4.0 \text{ MeV}.$$

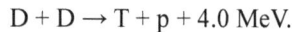

The first reaction produces a neutron, which activates surrounding structure and produces some radioactive waste, but the waste volume is much less than that arising from the D-T reaction. The second reaction produces tritium, which gives the D-T reaction and its 14.1 MeV neutron, but this is a secondary effect, and, again, the resulting waste is much less than that from a D-T reactor. The D-^{3}He reaction,

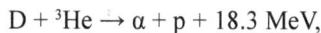

$$\text{D} + {}^3\text{He} \rightarrow \alpha + \text{p} + 18.3 \text{ MeV},$$

itself produces no neutrons, and the energy gain is high, but there is a D-D side reaction and ^{3}He is a scarce resource on the Earth. It has actually been proposed to mine

^{3}He on the Moon, where it is more abundant![46]

Ultimately, a fusion reactor using the p-boron-11 reaction (boron is widely available, and boron-11 constitutes 80% of natural boron) would produce no neutrons at all:

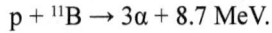

$$p + {}^{11}B \rightarrow 3\alpha + 8.7 \text{ MeV}.$$

There are other possible "aneutronic" fusion reactions, but the p-Boron-11 reaction seems the most feasible. Even for this reaction, the requirements are formidable: the temperature needs to be about ten times as high as for D-T fusion, and the confinement has to be better because of the lower energy released.[47] But progress is made in small steps, and eventually a p-Boron-11 reactor may be achieved. Such a reactor would produce no radioactive waste at all and satisfy the world's need for abundant clean energy.

CHAPTER 22
NUCLEAR WASTE DISPOSAL

In 1982, the U.S. Congress passed the Nuclear Waste Policy Act,[1] which directed the U.S. Department of Energy (USDOE) to identify a site for the permanent isolation from the biosphere of approximately 70,000 t of spent fuel and high-level nuclear waste from nuclear power reactors, and to build and operate a waste repository at that site. Funding for the project has been provided by a tax of about $300 million to $500 million per year collected from nuclear electric utilities. The Act established January 31, 1998 as the target date for opening the repository, but a series of missteps prevented that from happening, and no repository for nuclear power reactor waste yet exists. A repository for high-level wastes generated by the U.S. government does exist, however, as discussed in Section 22.5.

Since the termination of commercial reprocessing by President Jimmy Carter in 1977 (see Section 21.5), all reactor waste disposal programs in the U.S. have been directed towards the disposal of spent fuel assemblies in the once-through fuel cycle. If general reprocessing is resumed, a repository will have to receive the high-level waste (mostly fission products) remaining after the removal of the actinides. Some details of a repository for high-level waste will differ from those of a repository for spent fuel. However, the overarching requirement that must be met by a repository site will be the same: The repository must isolate radionuclides from the biosphere until they decay to levels that do not pose threats to public or environmental health. Because all of the experience acquired by the U.S. in nuclear power reactor waste disposal has been aimed at the disposal of spent fuel, this chapter focuses primarily on that technology, and in particular on three sites that have been studied in the U.S. for spent-fuel disposal. First, we review the specific requirements that must be satisfied by a repository.

22.1 REQUIREMENTS FOR WASTE REPOSITORIES

After radioactive waste is buried in the ground, there are several possible ways by which it can emerge into the biosphere. One is geological activity—earthquakes and volcanism. Therefore, a geological site selected for a repository must show convincing evidence that such activity has not happened there for a very long time and is very unlikely to happen for a very long time in the future. Recall from Section 21.5 that the 24,000-year half-life of Pu-239 imposes a requirement of about 500,000 years on the containment time of spent fuel in a repository. Therefore, a repository site for spent fuel must be shown with high confidence to be geologically stable for at least a half-million years into the future. A statutory requirement of one million years was established for the proposed repository at Yucca Mountain, Nevada.[2] Specifically, the site must limit exposure to any member of the public to 15 mrem/year for the first 10,000 years and to 100 mrem/year from 10,000 to one million years.

Another path for emergence of buried radioactive waste into the biosphere is dissolution in groundwater[a] and subsequent advective transport into the biosphere, such as by flow into an aquifer tapped by wells. This path is blocked by selection of a site that has not had water in it for a very long time and can be shown to be very unlikely to contain water for a very long time into the future, or a site in which the presence of water is so infrequent that the waste form will not dissolve during the required containment time, or a site in which the water is so anoxic that it cannot dissolve the waste form during the required containment time. Sites with each of those three qualities have been studied as potential repositories in the U.S.

A final path for emergence of buried waste into the biosphere is intrusion into the repository by human activity. For example, drilling into the site during exploration for minerals may inadvertently penetrate a repository. Efforts to prevent such intrusion may take the form of marking the site with symbols that could be understood by technically literate people who have no knowledge of current languages, but to expect such symbols to last a half-million years is optimistic. Realistically, one can only hope that the amount of radioactive waste brought into the biosphere by human activity will be limited before its dangerous nature is recognized.

The Environmental Protection Agency (EPA) has prescribed a methodology for determining the permissible concentrations of individual radionuclides in soil and water.[3] This methodology has been applied to a list of radionuclides that have

---

[a] Groundwater is water located beneath the surface of the ground. Water above the surface, as in rivers or lakes, is called surface water.

a potential for damage to human or environmental health if released into the biosphere.[4] Table 22.1 contains concentrations of these radionuclides that would result in a one-in-a-million chance per year of harm to individuals exposed to them via the indicated pathways. The acronym CERCLA in the heading of Table 22.1 refers to the Comprehensive Environmental Response, Compensation, and Liability Act of 1980, also known as Superfund, a law intended to identify and clean up sites contaminated with hazardous substances, including radioactive materials.[5] Radionuclide releases from a repository must be shown with a high level of confidence to remain below the concentrations shown in Table 22.1 forever.

Table 22.1—CERCLA Risk-Based Radionuclide Detection Limits

Risk level – 1.0E-6

Detection Limits

Nuclide	Water (pCi/mL)	Soil (ingestion) (pCi/g)	Soil (external) (pCi/g)
H-3	8.8E-01	1.5E+04	
C-14	5.3E-02	8.8E+02	
K-40	4.3E-03	7.2E+01	1.5E+00
Co-60	3.2E-03	5.3E+01	9.7E-02
Sr-89	1.6E-02	2.6E+02	1.8E+03
Sr-90	1.4E-03	2.4E+01	
Y-90	1.5E-02	2.5E+02	
Tc-99	3.7E-02	6.1E+02	1.4E+06
Ru-106	5.0E-03	8.4E+01	
Sb-125	5.7E-02	9.4E+02	7.0E-01
I-129	2.5E-04	4.2E+00	2.0E+02
Cs-134	1.2E-03	1.9E+01	1.6E-01
Cs-137*	1.7E-03	2.8E+01	4.2E-01
Eu-152	2.3E-02	3.8E+02	2.3E-01
Eu-154	1.6E-02	2.6E+02	2.0E-01
Eu-155	1.1E-01	1.8E+03	1.4E+01
Pb-210†	9.3E-05	1.6E+00	6.4E+03
Ra-226	4.0E-04	6.6E+00	7.0E+01

Ra-228	4.8E-04	7.9E+00	
Ac-228†	9.5E-02	1.6E+03	2.9E-01
Th-228†	4.3E-03	7.2E+01	1.5E+03
Th230†	3.7E-03	6.1E+01	1.5E+04
Th-232†	4.0E-03	6.6E+01	3.2E+04
U-233	3.0E-03	5.0E+01	2.0E+04
U-234	3.0E-03	5.0E+01	2.8E+04
U-235	3.0E-03	5.0E+01	3.5E+00
U-238†	3.0E-03	5.0E+01	4.0E+04
Np-237	2.2E-04	3.6E+00	1.1E+02
Pu-238	2.2E-04	3.6E+00	3.0E+04
Pu-239	2.1E-04	3.5E+00	4.9E+04
Pu-240	2.1E-04	3.5E+00	3.1E+04
Pu-242	2.2E-04	3.6E+00	3.6E+04
Am-241	2.0E-04	3.3E+00	1.7E+02
Am-243	2.0E-04	3.3E+00	3.5E+01
Cm-243	2.5E-04	4.2E+00	5.2E+00
Cm-244	3.0E-04	5.0E+00	2.8E+04
Cm-245	2.0E-04	3.3E+00	1.6E+01
Cm-246	2.0E-04	3.3E+00	3.1E+04
Cm-247	2.2E-04	3.6E+00	9.1E-01
Cm-248	5.2E-05	8.7E-01	3.8E+04

*includes gamma emissions from Ba-137m.
†in natural uranium (U-238) or thorium (Th-232) decay series.

22.2 FEATURES OF POTENTIAL GEOLOGICAL REPOSITORY SITES

Three types of geological formations are conducive to meeting the requirements for repositories. Salt formations are inherently dry, because salt dissolves in water and frequent or continuous exposure to water would have removed the salt. Formations in arid country sufficiently separated from the water table (the level of groundwater in the Earth's crust) may have only occasional exposure to water from precipitation. Finally, formations deep below the Earth's surface saturated with groundwater may have so little dissolved oxygen in the water that corrosion is extremely slow.

In the 1980s, three different sites were considered as potential geological repositories for spent fuel: the Yucca Mountain site in Nevada, a salt dome in the Texas Panhandle, and the DOE's Hanford Site in southeastern Washington. In addition, a waste repository for transuranic waste from the U.S. nuclear weapons program, the Waste Isolation Pilot Plant (WIPP), has been operating in a salt bed a half-mile below the ground near Carlsbad, New Mexico, since 1999.[6]

The Yucca Mountain site is located in a formation of consolidated volcanic ash called tuff.[7] The formation consists of alternating layers of welded tuff, non-welded tuff, and semi-welded tuff laid down by a series of eruptions from a now-extinct caldera volcano.[8] Yucca Mountain is actually a ridge line formed by a fault that tilted the remnant of the volcano. The layers are highly fractured, but the fractures are mostly contained within the individual layers of the tuff. The repository site is perched above an aquifer, and transport of water through the repository site to the aquifer below takes place primarily through the fractures. Currently, the site is arid, and entry of water into the site is infrequent. Predictions of radionuclide releases from the site by the Yucca Mountain research staff, based on reasonable assumptions about groundwater intrusion and about the corrosion and radionuclide transport properties of the waste package and the site, indicate that exposure to individuals in the public will be less than 1 mrem/year for the first million years.[2] Because of the periodic intrusion of groundwater into the site, retention of radionuclides by the repository would be highly dependent on the engineered barriers, which are primarily a waste package shell of nickel alloy and a titanium drip shield perched above the waste packages.[9]

The Texas Panhandle site, located in Deaf Smith County near the New Mexico border, is inherently free of intrusion by groundwater because it is in a salt bed. However, it is perched above the Ogallala Aquifer, one of the most important sources of water in the U.S. for agriculture and a source of drinking water for millions of people. This vast aquifer starts at a depth between 100 and 400 feet and has a thickness between a few feet and more than a thousand feet.[10] The salt beds under consideration lie between 200 and 400 feet below the ground surface.[11] So nuclear waste buried in this repository site would be in close proximity to the aquifer, which makes the Deaf Smith County site less favorable than the other two.

The Hanford Site, near Richland, Washington, was studied in the Basalt Waste Isolation Project (BWIP).[12] The repository would be located in the Cohasset flow of the Grande Ronde Basalt formation, about 3,000 feet underground. This repository location is saturated with groundwater, but because of the depth of the water beneath the Earth's surface and the slow rate of movement of the groundwater through the basalt rock, this water is almost devoid of oxygen. Without oxygen, corrosion would be extremely slow; transport of radionuclides would also be very slow because of the slow movement of the water and because of features of the waste package design that would retain radionuclides dissolved in the water.

These three sites were being studied in parallel during the 1980s, and so-called site characterization plans were being developed for all the sites. The U.S. Congress found the cost of performing three studies to be objectionable, so in 1987 it decided to continue funding only the study that then appeared to have the best chance for success, which was the Yucca Mountain study. The State of Nevada considered this decision to be purely political, with the nuclear waste being forced down their throats because they had little political clout, so they began a policy of resistance to the repository that by 2009 had apparently succeeded. In that year, the Obama administration announced its plans to terminate the Yucca Mountain project and to devote $197 million to the exploration of alternative measures for the disposal of waste from nuclear power reactors.[13] However, on 29 June 2010, the Nuclear Regulatory Commission (NRC) ruled that the termination of the Yucca Mountain project was in violation of the Nuclear Waste Policy Act as amended in 1997, and that only Congress, and not the administration, has the legal authority to terminate the project.[14] Furthermore, legislation was introduced in the U.S. Senate to require the USDOE to refund the approximately $30 billion that utilities had by then paid into the fund to develop the repository.[15] The status of the project remains in turmoil in early 2019.

22.3 THE BASALT WASTE ISOLATION PROJECT (BWIP)

I worked on the BWIP from 1985 to 1988, so I have more personal knowledge of the BWIP layout and design than I have of the other projects. Since the future of the Yucca Mountain project is so uncertain, the BWIP site may be reconsidered, so it is appropriate to discuss it here.

As noted above, the BWIP repository[16] was to have been located about 3,000 feet below the ground, in layers of fractured basalt saturated with groundwater. Access to the repository would have been provided by vertical shafts to the repository. In the repository design, the repository itself consisted of four large sections, or compartments, to which access was provided by horizontal tunnels, called the main access drifts, bored into the basalt. Side drifts, tunnels running off to the sides of the main access drifts, each led to a so-called emplacement panel, and each panel comprised four emplacement rooms. The emplacement rooms were tunnels about a half-mile long from which boreholes for individual waste packages were drilled on each side at intervals of about 20 feet. The horizontal extent of the repository design was about 1 mile by 2½ miles.

The large size of the repository design was not due to an enormous volume of waste that needed to be buried there. The total mass of the waste to be buried is about 70,000 t, but if that were all consolidated into a cube with the density of steel, it would only occupy a cube about 70 feet on a side. The waste needed to be spread out to limit the heat load on the rock.

The waste packages were to be inserted into the boreholes, and a packing material consisting of bentonite clay and crushed basalt was to be tamped around the waste package within a surrounding shell in the borehole. Bentonite clay has a very useful adsorptive property—a broad variety of materials dissolved in water soaking into the clay, including many of the radionuclides in the waste package, are taken out of solution and adsorbed onto the surfaces of the clay particles. This property greatly retards the transport of any radionuclides that would be dissolved if groundwater corroded through the waste package and immersed the waste form.

The waste form could have been intact spent fuel assemblies, the fuel rods from spent fuel assemblies, vitrified high-level waste, or any other nuclear waste composition to be buried. The waste package was assumed to be eventually invaded by groundwater, which is a neutron moderator, so calculations had to be performed to ensure that the fissile nuclides were not so densely packed that the arrangement could become critical. Whatever the waste form would have been, it was to be held in a metal vessel called the canister. The canister was to be placed in a sturdier, corrosion-resistant barrier called the container. The packing was to be put into the region between the container and a surrounding shell, which would have rested against the basalt host rock.

Analyses of corrosion, radionuclide dissolution, heat transport, groundwater transport, and criticality were under way when the BWIP was terminated in 1987 as noted above. If the Yucca Mountain project is indeed terminated, then a search will resume for an alternate site, and the Hanford site and the Deaf Smith County site will probably be considered once again.

22.4 THE YUCCA MOUNTAIN DESIGN

The repository design proposed for the Yucca Mountain site[2] is the most highly developed repository design for spent fuel. The design layout consists of a main tunnel, five miles long and 25 feet wide, with 40 miles of smaller tunnels, called emplacement drifts, branching off from the main tunnel. The waste packages are placed along the sides of the emplacement drifts, as shown in Figure 22.1.[17] In the figure, flag no. 1 shows the delivery of canisters of waste contained in shipping casks to the repository by rail, flag no. 2 shows the facility where the canisters are removed from the casks and placed in permanent storage containers, flag no. 3 shows an automated system that delivers the storage containers to the emplacement drifts, and flag no. 4 shows the placement of a storage container in an emplacement drift.

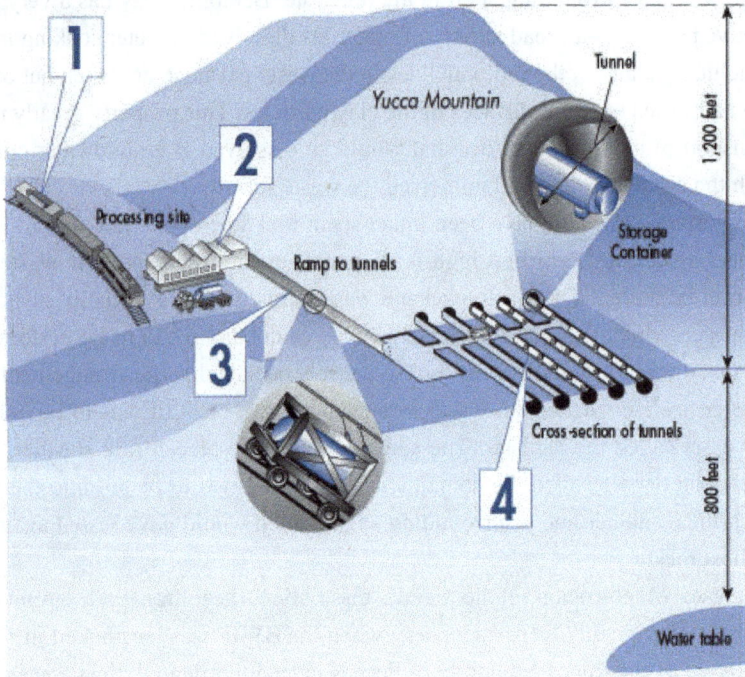

Figure 22.1—Layout of Yucca Mountain nuclear waste repository. This figure was taken from Reference 17, which links to a publication of the U.S. Nuclear Regulatory Commission; it is in the public domain under the terms of Title 17, Chapter 1, Section 105 of the U.S. Code.

As noted above, the 1987 congressional legislation to terminate the BWIP and Deaf Smith projects and fund only the Yucca Mountain project was met with vehement opposition. In Nevada, this legislation was termed the "Screw Nevada Bill." Nevadans have had to live with the Nevada Test Site, which is adjacent to the Yucca Mountain site, where over 1000 nuclear weapons have been tested and where two low-level radioactive waste disposal facilities already exist, and they felt that they were being used as the nation's nuclear waste dumping ground because they lacked the power to stop it. But in 2006, Nevada Senator Harry Reid became the Senate Majority Leader, and Nevada suddenly acquired a great deal of political clout. His sworn opposition to the Yucca Mountain project was probably a large element in the eventual suspension of the project.

Technical objections to the Yucca Mountain site have been based on questions about the quality of the research findings on infiltration of the site by water. However, a review in 2006 by the DOE's Office of Civilian Radioactive Waste Management confirmed that the modeling work done by Yucca Mountain research staff was technically sound and that the Yucca Mountain site would be suitable for long-term high-level nuclear waste storage.

Also, the Yucca Mountain site is seismically quite active. In 2007, it was discovered that a fault line runs right underneath a storage pad where spent fuel canisters were to be cooled before being moved into the emplacement drifts. This discovery required the relocation of several planned structures in the repository design. However, earthquakes at the repository location are not expected to compromise the integrity of the containment of radionuclides.[18]

In hindsight, it is clear that the top-down imposition of the repository project on Nevadans was ill-advised. In Sweden, siting of nuclear waste disposal facilities has involved local communities from the beginning, granting them veto power at every step in the siting process. This approach has been much more successful. As stated by Rodney Ewing and Frank von Hippel, communities near proposed repository sites "should have early and continued involvement in the process, including funding that would allow them to retain technical experts."[19]

22.5 THE WASTE ISOLATION PILOT PLANT (WIPP)

The Waste Isolation Pilot Plant (WIPP) is located about 26 miles east of Carlsbad, New Mexico, in salt beds about 3,000 feet thick lying about 2,000 feet under the Earth's surface.[20] Figure 22.2 shows the layout of the WIPP facility.

WIPP is devoted to the disposal of high-level radioactive waste from the U.S. nuclear weapons program. The waste accepted for disposal in WIPP must have activity from alpha-emitting transuranic radionuclides with half-lives greater than 20 years that exceeds 100 nCi/g in intensity. The Salado and Castille salt formations in which the WIPP repository is located have been geologically stable for more than 250 million years. Salt flows plastically under pressure, and the cavities excavated for the waste packages will close around the waste over time, producing a solid mass of salt in which diffusion of water and radionuclides will be extremely slow if the waste packages themselves are ever breached. A warning system will be emplaced on the ground above the repository to notify future explorers of the underlying hazard; this system will include an array of granite pillars and a roofless granite room with warnings in English, Spanish, Russian, French, Chinese, Arabic, and the local Navajo language. The repository is licensed to secure the waste for the next 10,000 years.

WIPP Facility and Stratigraphic Sequence

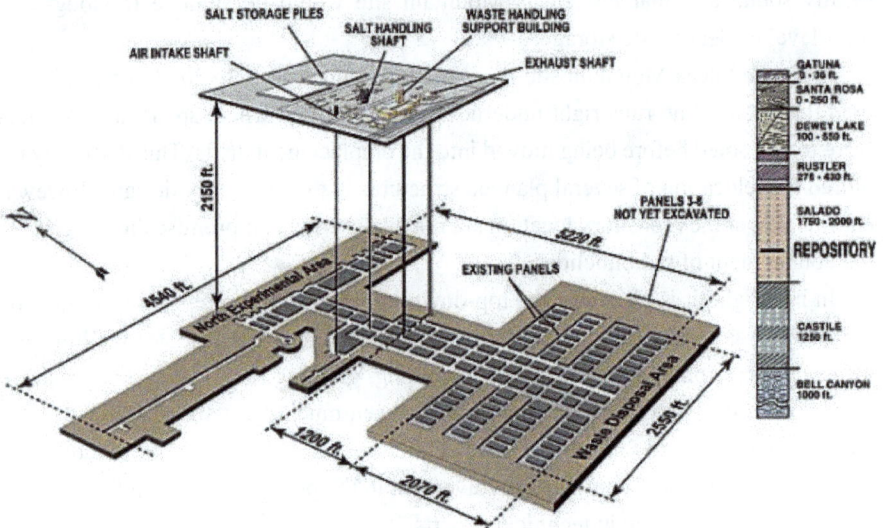

Figure 22.2—Layout of WIPP repository. This figure was taken from the Wikimedia Commons and is in the public domain.

22.6 NUCLEAR WASTE DISPOSAL OUTSIDE OF THE U.S.

Countries outside of the U.S. that use nuclear power do not yet have operating repositories for nuclear power plant waste, but Finland and Sweden have selected sites and France, Switzerland and the United Kingdom are conducting site selection protocols.[21] In 2003, a site at Krasnokamensk was proposed in Russia. As pointed out to the United Nations General Assembly in 2003 by Dr. Mohamed ElBaradei (then Director General of the U.N.'s International Atomic Energy Agency), not all countries have either the geological features required for a high-level nuclear waste disposal site or the financial resources required to develop one, so an international site would be desirable for use by such countries. In the 1990s a group called Pangea Resources identified sites in Australia, southern Africa, Argentina and western China as suitable for such repositories, with Australia preferred for economic and political reasons.[22]

22.7 SUBSEABED DISPOSAL

In the 1970s and 1980s, Rip Anderson of Sandia Laboratories and Charles Hollister of Woods Hole Oceanographic Institution led a study of a location in the Pacific Ocean north of Hawaii that has very attractive properties for the disposal of spent nuclear fuel assemblies or high-level nuclear waste. This location covers 39,000 square miles around the geographic coordinates 32N164W. It is a layer of viscous clay sediment 325 feet thick lying under four miles of water. This clay is a mixture of detritus from oceanic organisms, volcanic ash, dust, and micrometeorites that has accumulated over millions of years and lain undisturbed through all that time by any geologic activity. Waste packages dropped into this muck would sink through it and be entombed safely. Like the groundwater in the Hanford site, the water in the muck is anoxic, and corrosion would be almost nonexistent. The waste would be irretrievable because of the depth of the water. This method of disposal is probably the most secure method of all for disposal of nuclear waste, but the project was terminated by political pressure from advocates of the Yucca Mountain project.[23] But the seabed location is still there, and if the Yucca Mountain project is indeed eventually terminated, the subseabed proposal may be revived.

CHAPTER 23
NUCLEAR WEAPONS

I wonder how many Americans lie awake at night worrying about nuclear weapons. Not many, I would guess. But I do. I think the gravest danger our nation faces in the near future is the detonation of a nuclear weapon by terrorists in an American city, probably New York or Washington. Such an attack would drastically change America, and in fact the whole world, for the worse. But fanatic Islamic fundamentalists would like nothing better.

The basic principles of nuclear weaponry have been public knowledge for a long time. Nevertheless, it is not a simple matter to build a nuclear weapon. In this chapter, I want to tell enough about how nuclear weapons work, and how they are made, to assure the reader that nuclear energy can be deployed widely, even with fuel reprocessing, without increasing the risk that nuclear weapons will fall into the hands of terrorists. For more detail on some aspects of nuclear weapons, see References 1 and 2.

In particular, it should become clear that building even a simple nuclear weapon is a project requiring industry on a national scale, not something that can be done by fugitive terrorists hiding in caves. The most likely way for terrorists to obtain a nuclear weapon is simply to buy one, either from a hostile regime or from a sympathetic individual in a country where nuclear weapons security is lax. The United States government should be focusing intense diplomatic efforts to prevent such an event, as it is doing in many ways, some of which are discussed below.

There is very little quantitative information in this chapter. For one thing, I do not know such specific information. I have never worked in weapons technology, and I spent a year looking for work because I passed up an opportunity to do so. For another thing, most such information is classified (kept secret for the sake of national security), and even if I did know it I would be put in jail if I revealed it.

Ironically, the legacy of nuclear weapons up to the present time is mostly positive. The intense arms race between the two superpowers created some seriously contaminated production sites, but at least in the United States these were localized, and some of them, such as the Rocky Flats site near Boulder, Colorado, have been cleaned up.[a, 3] The nuclear standoff between the United States and the Soviet Union from the late 1940s until the collapse of the Soviet Union in 1991 probably prevented a third world war. Two nuclear bombs were dropped on Japan to end World War II, with great devastation to the cities of Nagasaki and Hiroshima, but the ensuing Japanese surrender made an Allied invasion of Japan unnecessary, which surely saved both Allied and net Japanese lives.

However, the story is not over, and I am not optimistic about how it will end. The Cold War remained cold because a nuclear attack on either superpower by the other would have been suicidal to the attacker. But against whom would one retaliate if a terrorist trucked a bomb into a city and set it off? In 2006, Jacques Chirac, then president of France, said that France was prepared to launch a nuclear strike against any country that sponsored a nuclear terror attack against French interests.[4] But it might not be possible to determine who the sponsor was. Then what?

So I worry, but not about terrorists building bombs from scratch. To see why not, read on.

23.1 FISSION BOMBS

Fission bombs, or "atomic bombs," run on a neutron chain reaction, just like fission reactors. The effective detonation of a fission bomb requires the consumption of a substantial fraction of the fissile material in the bomb in a very short time. As the temperature and pressure in the exploding weapon increase, the bomb blows itself apart. Unless the chain reaction consumes enough of the fissile material before this explosive disassembly occurs, the bomb is said to "fizzle." Some radioactive fission products are dispersed and some blast energy is released, but the huge release of energy in an extensive detonation does not occur.

In order to achieve detonation in a sufficiently short time, most of the fission neutrons must go on to induce subsequent fission reactions. In other words, the bomb

---

[a] Some plutonium contamination remains in the soils around the Rocky Flats site, which has led to considerable resistance in the surrounding communities to the development of a highway through the Rocky Flats National Wildlife Refuge, as the Rocky Flats site is now designated, and to opening the Refuge to the public (https://www.cpr.org/2019/09/03/plutonium-test-results-stall-plan-for-toll-road-near-rocky-flats/, https://www.coloradohometownweekly.com/2019/09/03/plaintiffs-in-rocky-flats-national-wildlife-refuge-lawsuit-file-brief-in-support-of-claim/).

must be highly prompt supercritical. This requirement is achieved by two design properties. First, the bomb material must be highly concentrated in fissile nuclides. Second, the bomb must be large enough and configured properly to prevent excessive neutron leakage. The necessary size is called the "critical mass," which depends on the bomb material and configuration.

Fission bombs have been made from two materials: U-235 and Pu-239. If the bomb uses U-235, it must be enriched to about 90% or more in U-235. If it uses Pu-239, the Pu-239 must be sufficiently pure.

The bomb cannot be stored with the critical mass assembled, because it would explode spontaneously. The critical mass must be brought together during the detonation process. There are three means by which this has been successfully accomplished.[2] In the first of these means, called the gun-type assembly method, the bomb is divided into two hemispheres held on opposite ends of a tube analogous to a gun barrel. A chemical explosive on one end of the tube drives the hemisphere on that end down the tube to the other end, where it collides with the other hemisphere to assemble the critical mass. When the two hemispheres are separated at the ends of the tube, neutron leakage dominates and the device is subcritical. In the second means, called the implosion method, the bomb material is formed as a sphere or cylinder, called a pit, surrounded by a chemical explosive. The pit may be hollow. The dimensions and normal density of the pit ensure enough leakage to produce subcriticality. When the chemical explosive detonates, the pit is driven radially into a much smaller volume—its density increases above normal solid density—and the configuration becomes prompt supercritical. This method is more difficult to achieve, because it is difficult to ensure that the implosion will be radially symmetrical. If the implosion is too asymmetric, part of the imploding shell will be squirted out in a hydrodynamic instability like those that make nuclear fusion challenging, as discussed in Chapter 20. In the third means, called the two-point linear implosion method, the bomb material is initially formed as an elongated solid ellipsoid, which once again is subcritical because of leakage. The ellipsoid is stored in a tube with chemical explosive detonators at each end. The shock waves sent out by the exploding detonators are shaped by wave shapers that cause the ellipsoid to be squeezed into a sphere, which suffers from less neutron leakage and becomes prompt supercritical; the density of the material is increased during the implosion to increase the multiplication factor further. The gun-type method is suitable only for U-235 weapons; the other two may also be used for Pu-239 bombs.

23.1.1 Uranium bombs

Bombs using U-235 require enrichment facilities, as described in Chapter 21. Either gaseous diffusion enrichment or centrifugal enrichment facilities are large, expensive, and technically sophisticated. The Iranian enrichment program, which the Iranian government dubiously claimed to be intended for the manufacture of reactor fuel, was taking one of the most industrially advanced nations in the Islamic world years to build (as of 2015, Iran's enrichment program was suspended in accordance with an international agreement,[5] although Donald Trump's withdrawal of the U.S. from the agreement has led Iran to violate its commitments and puts the future of the agreement in jeopardy).[6] Terrorists do not have access to the facilities required for enrichment of uranium. However, as the Iranian example shows, any nation determined to build nuclear weapons can develop the required facilities. Impoverished North Korea evidently used both enrichment of uranium and reactor production of plutonium in its nuclear weapons program.[7]

Research reactors using highly enriched uranium fuel are considered to be a vulnerable source of weapons material, especially those reactors located in places like university campuses. Unirradiated fuel could be stolen by terrorists and used to build at least a fizzle-bomb, in which enough nuclear fission occurs to do localized damage and scatter radioactive contamination over a significant area. To reduce the possibility of such events, the U.S. Department of Energy initiated the Reduced Enrichment for Research and Test Reactors (RERTR) Program in 1978.[8] This program has now become the Global Threat Reduction Initiative (GTRI), managed by the National Nuclear Security Agency. The goal of the program was to complete the conversion of all domestic civilian nuclear research and test reactors that use highly enriched fuel (>20% enriched) to configurations that can use low-enriched fuel (<20% enriched) by 2014. The GTRI also assists other countries in converting their research and test reactors to low-enriched fuel. Such conversion requires redesign of the fuel elements, so it is not a straightforward matter of simply reducing the U-235 content of the fuel. As of 2019, 103 research reactors and medical radioisotope production facilities have either converted to low-enrichment fuel or been shut down,[9] but the work is not complete.

23.1.2 Plutonium bombs

As discussed in Chapter 17, plutonium is naturally produced in any nuclear reactor containing U-238. The initial reaction is

$$^{238}\text{U}(n,\gamma)^{239}\text{U} \rightarrow \beta + ^{239}\text{Np}; \; ^{239}\text{Np} \rightarrow \beta + ^{239}\text{Pu}.$$

However, subsequent reactions produce other plutonium isotopes:

$$^{239}\text{Pu}(n,\gamma)^{240}\text{Pu},$$

$$^{240}\text{Pu}(n,\gamma)^{241}\text{Pu},$$

$$^{241}\text{Pu}(n,\gamma)^{242}\text{Pu},$$

All of these plutonium isotopes are radioactive, but their half-lives are so long that their decay does not affect the amounts of the isotopes accumulated during reactor operation. In particular, since they do not decay significantly, the longer a sample of U-238 remains in the reactor, the greater becomes the amount of these higher plutonium isotopes that accumulate in the sample. That's a very bad thing if you want to build a bomb, but it's a very good thing to people who worry about nuclear power reactors being used for the production of nuclear weapons. The property of the higher plutonium isotopes that makes things that way is that ^{240}Pu and ^{242}Pu exhibit relatively high rates of spontaneous fission. This property prevents the gun-type plutonium bomb from working at all, and it tends to cause bombs to fizzle if they are made from reactor-grade plutonium—i.e., plutonium extracted from spent fuel.[b] The neutron chain reaction begins before the critical mass is fully assembled. In order to produce weapons-grade plutonium, one must irradiate U-238 "targets" for relatively short periods of time, so that the higher plutonium isotopes do not build in. Most power reactor pressure vessels are sealed during their fuel cycles, and the reactor must be shut down and the pressure vessel must be opened to extract fuel assemblies. It would be very inefficient and expensive to keep doing this frequently for production of weapons-grade plutonium. It is much more practical to build reactors for the specific purpose of producing weapons-grade plutonium, such as the Hanford N-Reactor that produced much of the plutonium for the U.S. nuclear weapons arsenal.[10] The N-Reactor also produced electricity for the local power grid, but its primary mission was plutonium production for weapons.

---

[b] It is possible to make a successful fission bomb from reactor-grade plutonium, but it is much more difficult. Terrorists trying to build a bomb from scratch, even using ideal materials, would have a very high probability of failure. Using reactor-grade plutonium would increase the failure probability much more.

Special consideration is needed for reactors like the CANDU reactor and the pebble-bed reactor, which are fueled continuously, and from which fuel elements can be diverted after brief irradiation in the reactor. My colleagues and I at the INL performed an analysis of the potential for diversion of lightly irradiated pebbles from a PBR to produce weapons.[11] We found that the production rate of clean plutonium (that might seem an oxymoron to some people, but I mean sufficiently pure ^{239}Pu) would be very slow, and it would be easy to detect by shortfalls in the production of power. If a country could produce its own pebble-bed fuel, such detection would not deter it from using its reactors for weapons production. However, pebble-bed fuel is difficult to produce, and most nations would opt to buy it from qualified suppliers. Then the supply could be cut off if evidence of clandestine use for weapons production were detected. Furthermore, any nation sophisticated enough to make pebble-bed fuel could easily build a reactor better suited for its purpose.

23.1.3 Fuel reprocessing and nuclear security

Fuel reprocessing, discussed in detail in Section 21.5, is the process of removing usable fuel materials from spent fuel. Some of the points made there are reiterated here for convenience. These materials include U-235 that has not been consumed by fission, U-238 that has not been transmuted into plutonium, and Pu-239 that has been created by the breeding reaction presented above. Because of the highly radioactive fission products present in spent fuel, handling and transportation of spent fuel require specialized equipment, and extraction of fissile materials from spent fuel requires major industrial facilities. Such facilities are not available to terrorists.

However, after reprocessing, these usable materials have been separated from the fission products. Their half-lives are very long, so they are not very radioactive. The half-life of Pu-239 is 24,000 years, that of U-235 is 700 million years, and that of U-238 is 4.47 billion years. Furthermore, they are all alpha-emitters, which are safe to handle in solid form. People who are concerned about the security of nuclear weapons materials worry that terrorists might capture recycled reactor fuel in transit and use it to make nuclear weapons.

Such concerns need to be addressed, but they can be. First, reactor fuel made of uranium is a mixture of U-235 and U-238 enriched to only a few percent in U-235. Weapons-grade uranium is enriched to 90% or more. To enrich reactor uranium fuel to weapons grade would require just the same kinds of enrichment facilities that were taking the Iranians years to develop. Terrorists don't have such facilities.

Second, the plutonium in spent reactor fuel is difficult to use in weapons because it is too rich in Pu-240 and Pu-242. Furthermore, it is not necessary to transport it at all. Reprocessing could be accomplished at reactor sites and reprocessed fuel could be used there without shipping. The Integral Fast Reactor program was a proposed project to demonstrate exactly that concept.[12] Finally, if reprocessed fuel were really an easy way for terrorists to obtain nuclear weapons, they would have taken that way already. Great Britain, France, Russia and Japan have been reprocessing fuel for decades, and they have never had any attacks on their reprocessed fuel shipments.[13]

23.2 FUSION BOMBS

Fusion bombs, or "hydrogen bombs," are the ultimate inertial confinement fusion reactors. They usually use the D-T fusion reaction, which releases both energy and a high-energy neutron. In a single-stage fusion bomb, the deuterium-tritium mixture is contained inside a hollow fission pit, and the detonating pit compresses and heats the D-T mixture to fusion temperatures at high density. In addition to the additional energy released by the fusion reactions, an energy gain is achieved as each fusion neutron initiates an additional fission chain reaction in the surrounding fission pit. Usually, most of the energy in such bombs is actually obtained from fission reactions, and the bombs are properly called fusion-boosted fission weapons.

More powerful fusion bombs are constructed in two stages. The first stage is as described above. The second stage, contained within the same cavity as the first, comprises a cylindrical annulus of plutonium, called the sparkplug, a surrounding layer of polystyrene foam, and an inner mixture of deuterium and tritium. The second stage of the explosion is initiated by X-rays emitted by the first stage. These X-rays are guided by the inner surface of the cavity to impinge on the foam, heating and compressing it to extreme temperature and pressure. The pressure implodes the plutonium sparkplug to much higher temperature and pressure than the chemical explosion achieved in the first stage. This sparkplug implodes in turn and creates higher temperature and pressure in the enclosed D-T mixture than is achieved in the first stage. Thus, the secondary explosion releases much more energy than the first stage. Still, most of the energy released comes from fission reactions, even in the second stage.

Design and construction of such weapons are challenging undertakings, and I do not foresee terrorists being able to do them.

23.3 ARMS REDUCTION AND NON-PROLIFERATION TREATIES

The arms race and the Cold War between the United States and the Soviet Union lasted essentially from the end of World War II until the dissolution of the Soviet Union in 1991. Both nations deployed more than ten thousand nuclear weapons on land-based intercontinental ballistic missiles (ICBMs), strategic bombers, and missiles carried on nuclear-powered submarines.

The insanity of such numbers was apparent to both nations, and beginning in 1963 numerous treaties were established to limit or reduce nuclear arsenals and prohibit testing of nuclear weapons.[14] The Strategic Arms Reductions Treaties I and II (START I and II, 1991 and 1993, respectively) required both nations to reduce their warhead numbers from over 10,000 to 6,000 (START I) and then from 6,000 to 3,500-3,000 (START II). The Comprehensive Test Ban Treaty (CTBT, 1996) bans all testing of nuclear weapons; it was signed by 180 nations (including the United States) and ratified by 148 (not including the United States).

The most important treaty concerning the acquisition of nuclear weapons by additional nations and by terrorists is the Nuclear Non-Proliferation Treaty (NNPT), which first became effective in 1970.[15, 16] Concern over the widespread acquisition of nuclear weapons, or nuclear weapons proliferation, arose almost as soon as the post-World-War-II arms race began, and the formulation of a treaty to address that concern was launched in 1958 by the Irish Minister for External Affairs, Frank Aiken. But it took another ten years before a treaty was ready for signing, and another two years before it took effect. Finland was the first nation to sign; since then 190 nations have signed the treaty, although one, North Korea, signed it, then violated it, and finally, in 2003, withdrew from it.

Five parties to the treaty already possessed nuclear weapons before the treaty was written: the United States, the Soviet Union, France, Great Britain, and China. These nations are called Nuclear Weapons States (NWS) in the treaty; now, Russia has replaced the Soviet Union as an NWS. All other signatories are called Non-Nuclear Weapons States (NNWS). Four other nations who are not parties to the treaty also have, or are believed to have, nuclear weapons: India, Pakistan, Israel, and North Korea. All but Israel have openly declared their weapons status; Israel has an official policy of ambiguity, but is believed to possess between 75 and 400 nuclear warheads and the capability to deliver them by intercontinental ballistic missiles, aircraft, or submarine-launched missiles.[17] South Africa had a nuclear weapons program and is believed to have tested a nuclear bomb, but it signed the treaty in 1991, admitted its weapons program in 1993, and dismantled its weapons and opened itself to inspection by 1994. Libya signed the treaty but had a secret weapons program in violation of the treaty; however, Libya terminated its program in 2003 without ever building a nuclear weapon.

The treaty is reviewed every five years in Review Conferences of the Parties to the NNPT; sessions of the Preparatory Committee for the Review Conferences take place annually. The treaty was originally agreed to expire after 25 years, but at the Review Conference of May 1995 the signatories extended it indefinitely and unconditionally.

The NNPT is considered to comprise three "pillars," or overarching principles: non-proliferation, disarmament, and peaceful use. The treaty relies on the International Atomic Energy Agency (IAEA)[18] to perform inspections stipulated by the treaty.

The non-proliferation pillar comprises Articles I-III of the treaty. Article I obligates NWS signatories not to transfer weapons or other nuclear explosive devices to any recipient whatsoever, nor to assist, encourage, or induce any NNWS to acquire, manufacture, or obtain control over such weapons or devices. Article II obligates NNWS signatories not to receive such weapons or devices, nor to manufacture them, nor to receive assistance in manufacturing them. Article III obligates NNWS signatories to accept "safeguards" in accordance with the IAEA's safeguards system—essentially, to submit to inspections of its nuclear facilities in order to ensure that they are being used solely for peaceful purposes.

The peaceful uses pillar comprises Articles IV and V of the treaty. Article IV guarantees every party to the treaty an inalienable right to use nuclear energy for peaceful purposes; this right includes the right to manufacture nuclear fuel, build nuclear reactors, and perform research in nuclear technology. Article V guarantees that any benefits from peaceful applications of nuclear explosions will be shared with NNWS signatories. Although such activities as excavating harbors and digging canals by nuclear explosions are feasible, after some tests in the 1960s and 1970s by the United States and the Soviet Union, such programs were terminated.[19] Article V is intended to address such possible uses.

The disarmament pillar comprises Articles VI and VII of the treaty. These articles are very succinct, and they are quoted below, verbatim from the treaty:

ARTICLE VI

Each of the Parties to the Treaty undertakes to pursue negotiations in good faith on effective measures relating to cessation of the nuclear arms race at an early date and to nuclear disarmament, and on a treaty on general and complete disarmament under strict and effective international control.

ARTICLE VII

Nothing in this Treaty affects the right of any group of States to conclude regional treaties in order to assure the total absence of nuclear
weapons in their respective territories.

I present these articles verbatim so that you can see for yourself how far-reaching
they are. Not only does Article VI obligate signatories to stop the nuclear arms race,
it also obligates NWS signatories (by implication) to divest themselves of nuclear
weapons. But it doesn't stop there. It also obligates all signatories to the treaty to
negotiate in good faith towards "complete disarmament under strict and effective
international control."

You can form your own opinion of such goals. A good discussion of the pros
and cons is presented in Reference 16. So far, the United States has dismantled over
13,000 nuclear warheads, and Russia has dismantled similar numbers. Much of the
uranium taken from Russian warheads is now being used as fuel in nuclear power
plants in the United States. Thus, both nations have gone a long way towards nuclear
disarmament. However, complete elimination of nuclear weapons by both nations
might actually be destabilizing. Also, complete disarmament, with all military polic-
ing powers handed to a group like the United Nations, would be anathema to many
Americans and seems very improbable to me.

Before the signing of the "Iran nuclear deal,"[5] the peaceful use pillar was apparently providing Iran, a signatory to the treaty, with a cover for the development of nuclear
weapons. Because the same equipment can be used to enrich uranium for reactor fuels
and weapons, Iran claimed that its enrichment program was peaceful in its intentions.
However, Iran has kept much of its nuclear technology secret, and an investigation by
the IAEA in 2003 concluded that its activities were systematically in violation of its
safeguards obligations under the NNPT. On 19 February 2010, a news story on cnn.
com reported that the new director general of the IAEA, Yukia Amano, was preparing
a report stating that Iran was defying United Nations orders about its nuclear program
and may have been working secretly to develop a nuclear warhead for a missile.[20]
The same story quotes Iran's supreme leader, Ayatollah Ali Khamenei, as saying that
nuclear weapons are forbidden for religious reasons. Such considerations have not
prevented Pakistan, another Muslim nation, from building nuclear weapons.

Iran claims that it has had to keep its activities secret because pressure from the
United States caused contracts with several foreign suppliers to be cancelled. There
is some basis for this assertion. Before the Islamic Revolution in 1979, companies in
the United States, Germany, France, Belgium, Spain, and Sweden had contracted with

Iran to help build nuclear power stations there, and after the revolution these countries terminated their agreements without paying back any of the investments Iran had made in those companies. The German company, Kraftwerk Union, based its withdrawal on non-payment of contractual obligations by Iran.[21] Under U.S. pressure, Argentina cancelled some nuclear-technology-related contracts with Iran. Perhaps the hostile behavior of the Iranian regime in holding 53 hostages from the American Embassy in Tehran from 4 November 1979 to 20 January 1981, in violation of international law granting diplomatic immunity,[22] made western countries unwilling to supply Iran with nuclear technology. Also, the Iran-Iraq war of 1980-1988 made work in Iran too dangerous; the partially completed reactors at Bushehr were bombed by Iraq.

However, from 1987 to 1993 Argentina helped Iran convert a research reactor in Tehran from highly enriched uranium to low-enriched uranium (19.75% U-235). Russia formed a joint venture with Iran called Persepolis in the early 1990s to help Iran develop peaceful nuclear technology. (Russia has also helped Iran develop missile technology, which doesn't seem so peaceful.) Work on the Bushehr I power plant resumed with Russian help. The plant began producing electricity on 3 September 2011, and by February 2012 it was producing 75% of its rated power.[23] So Iran is able to obtain nuclear technology for peaceful purposes, although perhaps not from friends of the United States.

Reference 21 chronicles the progress of the Iranian nuclear program, Iran's obfuscation and noncooperation with the IAEA, and the reaction of western powers, particularly the United States. The Iranian actions certainly aroused suspicion of clandestine weapons development, but the reactions of the United States were also inflammatory. Under President George W. Bush, the United States threatened a first nuclear strike against Iran, in clear violation of its obligations under the NNPT. But the progress of Iranian nuclear technology certainly suggests weapons development. In February 2010, Iran announced that it had begun to enrich uranium to 20% U-235, ostensibly for a medical reactor. However, it is much easier to go from 20% to the 90% needed for weapons than it is to go from the natural 0.7% to 20%. Although Iranian intent to develop weapons cannot be proven until Iran builds a weapon, all signs pointed towards that ultimate goal. As of 2017, Iran's effort to build nuclear weapons has apparently been suspended,[24] although Iran claims that increased sanctions imposed by the U.S. over Iran's ballistic missile program and other unrelated activities constituted a breach of the Iran nuclear deal.[25] As mentioned in Section 23.1.1, the U.S. has withdrawn from the agreement, Iran is violating some of its commitments under the agreement, and the future of the agreement is uncertain as of July 2019.

Whether or not Iran builds nuclear weapons, Iran does not pose a threat to the

United States through a direct nuclear attack by a missile or an aircraft. For any nation to attack the United States that way would be suicidal; despite recent arms reductions, the United States still possesses several thousand nuclear warheads and would retaliate massively against such an attack. However, as nuclear weapons proliferate, the chance that terrorists can obtain them obviously increases, particularly when states that acquire nuclear weapons are hostile to the United States, as Iran certainly is.

As noted above, some have called for pre-emptive military strikes against Iran's nuclear facilities to prevent them from completing their development of nuclear weapons. I believe that such actions would be very counterproductive. The young people of Iran have recently massed together in dissent against their repressive, totalitarian government. Attacks against Iran would force the dissenters to rally behind the existing government and abandon their demands for freedom. Instead, western powers should covertly support the dissenters, hoping that a regime change from within can return Iran to responsible citizenship in the community of nations.

While military actions against al-Qaeda and the Taliban, which harbored al-Qaeda before the infamous attacks on the United States on 11 September 2001,[26] seem justified and necessary, terrorism as a threat against the United States in particular and western civilization in general is unlikely to be defeated by force. In my opinion, the war in Iraq has probably been the best recruiting tool for terrorists that al-Qaeda could have asked for. In *How to Win a Cosmic War*, Reza Aslan, an American of Iranian descent, explains the motivations of terrorists in terms of service to God.[27] His view on how to win the war: Don't fight it! Change the context of the conflict. A school for girls in Afghanistan will do a lot more good than an army division, although the army may be needed to protect the school.

23.4 MEASURES TO PROTECT AGAINST TERRORISM

There are two pathways along which a nuclear weapon could be smuggled into the United States: by standard shipping routes and by clandestine movement across a border or onto an uninhabited coastal region.

Nuclear weapons emit very little radiation, but U-235, U-238, and Pu-239 all sustain low rates of spontaneous fission. Some of the neutrons emitted in spontaneous fission go on to induce subsequent fissions (but not enough for criticality), and others are absorbed in (n,γ) reactions, which emit gamma rays. Gamma rays penetrate matter deeply, and they can be detected by instrumentation outside whatever container a bomb is hidden within. Some neutrons also wander out through container walls and can be detected. Also, X-ray machines, or even gamma-ray and muon emitters, can direct high-energy radiation into containers, and such radiation scattered from heavy nuclei such as uranium or plutonium has different properties from those of radiation scattered from ordinary materials. Considerable effort is being directed at the development of such detection systems and their deployment both at ports of entry to the United States and at foreign ports of origin for ships bound for the United States.[28]

Some of these techniques are even applicable at distances of the order of a kilometer. The ability to detect nuclear weapons at a distance is useful for surveillance of smuggling routes or ships at sea. Still, the Navy and the Coast Guard can't be everywhere at once, and I fear that our coasts are vulnerable. I hope that our defense agencies know more about how to defend against this threat than I do.

23.5 THE THORIUM FUEL CYCLE AS A PATH TO NON-PROLIFERATION

Thorium has been considered as a source of nuclear energy for a long time.[29] The resistance of the thorium fuel cycle to nuclear weapons proliferation is one factor in its favor.

The naturally occurring isotope of thorium is Th-232, which is not a fissile nuclide. However, it is fertile, with a neutron absorption and a series of beta-decay reactions leading to the fissile nuclide U-233:

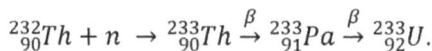

$$^{232}_{90}Th + n \ \rightarrow \ ^{233}_{90}Th \overset{\beta}{\rightarrow} \ ^{233}_{91}Pa \overset{\beta}{\rightarrow} \ ^{233}_{92}U.$$

In addition to undergoing fission upon absorbing thermal neutrons, U-233 produces U-232 by an $(n,2n)$ reaction. U-232 is also produced by two different paths following the absorption of neutrons by Th-232 nuclei. U-232 is relatively short-lived (its half-life is 73.6 years), so it has high specific activity (radioactive decay rate per unit mass), and its decay chain produces some fairly high-energy gamma rays. These qualities make spent thorium fuel dangerous to handle by anyone not properly equipped. This makes theft of spent thorium fuel impractical except by people who are willing to die in the process; unfortunately, there seems to be no shortage of such people. Also, U-232, which cannot be separated chemically from U-233, is a contaminant inherently present in the U-233, rendering the mixture difficult to use in nuclear weapons. Furthermore, thorium fuel can be mixed with small amounts of depleted uranium (essentially all U-238) or natural uranium to keep the U-233 concentration too low for use in weapons. Again, the U-233 cannot be separated from the mixture by chemical processing.

Some thorium-fueled reactors have been operated, such as the THTR[30] mentioned in Chapter 16, but various technical challenges make the thorium fuel cycle more difficult to apply than the uranium fuel cycle. Still, the more proliferation-resistant technology confers enough appeal to the thorium fuel cycle to warrant further development of thorium-fueled reactors.

We have now reached the end of our survey of nuclear technology. Although the treatment has necessarily been superficial, it has been comprehensive, and you should now be aware of the foundational science behind all areas of nuclear technology and you should now understand how nuclear systems work.

MATHEMATICAL DESCRIPTION OF NUCLEAR REACTOR FEEDBACK

In Section 6.3, the equations governing nuclear reactor feedback were stated in words rather than in mathematical symbols. In order to understand a complicated mathematical expression, you really have to be able to translate the mathematical symbols into such verbal interpretations, but for practical application the verbal interpretations are unwieldy. This section restates the equations of Section 6.3 in mathematical shorthand for the interested reader. We begin this reformulation by writing the power coefficient of reactivity α_p (alpha-sub-p) as the partial derivative

$$\alpha_P = \frac{\partial \rho}{\partial P}.$$

Remember that this just means the change in reactivity ρ per unit change in power P, for very small changes in power. For stability, α_p must be negative: When power increases, reactivity must decrease, and vice versa. The power coefficient of reactivity may be written in terms of the relevant physical properties:

$$\alpha_P = \frac{\partial \rho}{\partial T_f}\frac{\partial T_f}{\partial P} + \frac{\partial \rho}{\partial T_m}\frac{\partial T_m}{\partial P} + \frac{\partial \rho}{\partial d_m}\frac{\partial d_m}{\partial P} + \frac{\partial \rho}{\partial d_c}\frac{\partial d_c}{\partial P} \ ,$$

where
T_f = fuel temperature,
T_m = moderator temperature,

d_m = moderator density, and

d_c = coolant density.

This equation is an example of the chain rule of differentiation. Each term in the equation is the contribution to the power coefficient from a change in a specific property of the reactor, such as the fuel temperature, and when you add up all the contributions you get the total change.

We can define the partial derivatives of the reactivity with respect to the physical properties as coefficients of reactivity themselves:

$$\alpha_{T_f} = \frac{\partial \rho}{\partial T_f} = \text{fuel temperature coefficient of reactivity,}$$

$$\alpha_{T_m} = \frac{\partial \rho}{\partial T_m} = \text{moderator temperature coefficient of reactivity,}$$

$$\alpha_{d_m} = \frac{\partial \rho}{\partial d_m} = \text{moderator density coefficient of reactivity, and}$$

$$\alpha_{d_c} = \frac{\partial \rho}{\partial d_c} = \text{coolant density coefficient of reactivity.}$$

Then the equation for α_p becomes

$$\alpha_p = \alpha_{T_f} \frac{\partial T_f}{\partial P} + \alpha_{T_m} \frac{\partial T_m}{\partial P} + \alpha_{d_m} \frac{\partial d_m}{\partial P} + \alpha_{d_c} \frac{\partial d_c}{\partial P}.$$

An increase of power heats things up and makes them expand, so inherently the rates of change of the temperatures with respect to power are positive and the rates of change of the densities with respect to power are negative:

$$\frac{\partial T_f}{\partial P} > 0, \frac{\partial T_m}{\partial P} > 0, \frac{\partial d_m}{\partial P} < 0, \text{and} \frac{\partial d_c}{\partial P} < 0.$$

Therefore, in order for α_p to be negative as required, it is sufficient that

$$\alpha_{T_f} < 0, \alpha_{T_m} < 0, \alpha_{d_m} > 0, \text{and} \alpha_{d_c} > 0.$$

For BWRs, in which the coolant void fraction f_v is usually used in place of the coolant density, α_{d_c} is replaced by the void coefficient of reactivity,

$$\alpha_{f_v} = \frac{\partial \rho}{\partial f_v},$$

and the corresponding term in the equation for α_p is replaced by

$$\alpha_{f_v} \frac{\partial f_v}{\partial P}.$$

An increase in power causes an increase in void fraction, so for stability α_{f_v} must be negative.

MAXWELL'S EQUATIONS

In Section 20.2.1, the essential meanings of Maxwell's equations are discussed, but the equations themselves are not actually presented. Yet Maxwell's equations are one of the most elegant mathematical descriptions of physical phenomena in all of physics, and I would like to present them here for the interested reader. Some of the discussion in Section 20.2.1 is repeated here to make the presentation self-contained.

There are several systems of units for electromagnetic quantities, and the choice of units affects the coefficients in the equations. The equations presented below are expressed in SI units (cf. Chapter 2).

Maxwell's equations tell us about the behavior of electric and magnetic fields. An electric field, denoted by E, is a measure of the electrostatic force F_e (both its magnitude and its direction) exerted on a stationary charge q at any point in space by other charged particles: $F_e = qE$. As you go from point to point in space, you can follow the direction of the force and construct curves, called electric field lines, connecting the points. A magnetic field B is a measure of the magnetic force F_m exerted at any point of space on a moving particle of charge q in addition to the electrostatic force; $F_m = qv \times B$, where v is the velocity vector of the particle. Strictly speaking, we should call B the magnetic flux density vector, but it is common to call it the magnetic field. The symbol x denotes a special way to multiply two vectors; it is called the cross product, and it is a vector perpendicular to both of the vectors being multiplied. Its magnitude is proportional to the sine[a] of the angle between the two vectors; if the vectors are parallel, the cross product is zero, and if they are perpendicular, the cross

---

[a] The sine is a mathematical function defined in trigonometry. If α is an acute angle in a right triangle, the sine of α (written as sin α) is defined as the ratio of the side opposite α to the hypotenuse. For angles greater than 90°, appropriate complementary right triangles are constructed to define the sine.

product has its maximum value, which is the product of the magnitudes of the two vectors. The cross product $v \times B$ is read as "v cross B." In the same way as for electric field lines, you can construct magnetic field lines in space. The total electromagnetic force on a charged particle is given by the sum of F_m and F_e:

$$F_{em} = q(E + v \times B).$$

This equation is called the Lorentz force equation.

The SI unit of charge is the coulomb (C). The magnitude of the charge on an electron or a proton, called the elementary charge, is 1.602×10^{-19} C, so there are about 6.242×10^{18} electrons or protons in a coulomb of charge. The SI unit for electric field is 1 newton/coulomb, which is also obviously equal to 1 N-m/C-m. Defining the volt (V) as 1 J/C, or 1 N-m/C, we can also express the SI unit for electric field as 1 V/m, which is the usual way of doing it. The volt is a unit of something called electric potential, which is electric potential energy per unit charge. This is analogous to gravitational potential energy per unit mass (see Section 4.2). If you raise the electric potential energy of a coulomb of charge in an electric field by one joule, you raise the electric potential of the charge, or any part of it, by one volt.

The SI unit for B is called the tesla (T), after the fascinating character Nikola Tesla.[1] An older unit is the gauss (G), named for the German mathematician and physicist Carl Friedrich Gauss;[2] 1T = 10,000 G. Magnetic flux is measured in webers (Wb), named for the German physicist Wilhelm Eduard Weber;[3] together with Gauss, Weber invented the first practical telegraph. The units of magnetic flux and magnetic flux density are related as 1 T = 1 Wb/m^2. From the Lorentz force equation, you can determine that 1 Wb = 1 V-s. The weber is defined from Faraday's Law, as discussed below.

Maxwell's first equation is $\nabla \cdot D = \rho_e$, where ρ_e is the net electric charge density (in units of coulombs per cubic meter) and D is called the electric flux density vector (coulombs per square meter). This quantity is related to the electric field by the constitutive equation $D = \varepsilon E$, where ε is a material property called the electric permittivity of the material (in units of C^2/N-m^2). The "dot product" is another way of multiplying two vectors, which gives a scalar result (just an ordinary number with only magnitude and not direction). Actually, because the gradient operator ∇ (defined in Section 6.2.1) is a differential operator—that is, it takes appropriate derivatives of the quantity following it—the quantity $\nabla \cdot D$ (read as "del dot D") is not merely a multiplicative product, but it is constructed just like the dot product of two ordinary vectors. What this equation says physically is that electric fields emanate from charged particles, as shown in Figure AII.1 for a single charged particle, repeated below from Chapter 20.

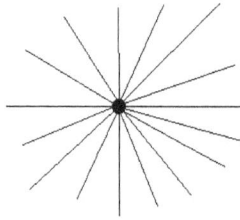

Figure AII.1—Electric field lines emanating from a single point charge

Electric charge has sign, so if a positive charge and an equal negative charge are placed next to each other, except very near them their combined electric field will be zero.

Maxwell's second equation is $\nabla \cdot \mathbf{B}=0$. This equation says that there are no magnetic monopoles, or magnetic charges from which magnetic field lines emanate. That means that ultimately all magnetic field lines are closed curves.

The great 20th-Century theoretical physicist Paul Dirac[4] showed that magnetic monopoles should exist, which would change Maxwell's second equation to be analogous to the first, and it would add a term in the fourth equation analogous to the current term in the third equation, as discussed next. However, no magnetic monopoles have ever been found, despite some experimental efforts.[5] Perhaps the absence of magnetic monopoles is due to the extremely high mass a magnetic monopole would have to have. Current theory predicts that a magnetic monopole would have a mass as high as 10^{14} TeV, where a TeV, for tera electron volt, is 10^{12} eV, or 10^6 MeV. The mass of a proton in energy units is 938 MeV, so a magnetic monopole would have the mass of about 10^{17} protons.[6]

Maxwell's third equation is $\nabla \times \mathbf{H}=\mathbf{j}+\partial \mathbf{D}/\partial t$. It is actually $\mathbf{H}$ that is properly called the magnetic field intensity, and it is related to the magnetic flux density by another constitutive equation, $\mathbf{B}=\mu \mathbf{H}$, where μ is called the magnetic permeability of the material. Electric current is the net charge passing through a surface per unit time, in units of coulombs per second. This unit is also called the ampere (A), after the French mathematician and physicist André-Marie Ampère.[7] The quantity $\mathbf{j}$ is called the particle current density, in units of amperes per square meter. Since the gradient operator has units of inverse length, one can see that $\mathbf{H}$ has units of amperes per meter. Then μ has units of webers per ampere-meter. Still another unit is defined in the SI, the henry (H), for the American scientist Joseph Henry.[8] The henry is equal to a weber per ampere. Then μ is usually given in H/m, which is the same thing. As usual, t denotes time in units of seconds.

Maxwell's third equation says that magnetic fields are created by the motion of charged particles. There are two terms on the right-hand side of the equation; in analogy to the particle current density j, the quantity $\partial D/\partial t$ is called the displacement current density. The form of this equation makes it appear that there are two different sources of magnetic fields, particle currents and changing electric fields, but they are actually two complementary ways of accounting for the motion of charged particles.[9] Maxwell's third equation also tells us that magnetic field lines induced by a single moving charged particle are perpendicular to the velocity of the moving charged particle. The magnetic field created by an electric current in a straight wire is illustrated in Figure AII.2, also repeated from Chapter 20. The sign of the electric current is defined so that a flow of positive charges is a positive current. Since a current in a wire arises from the flow of electrons, which are negatively charged, the current is in the direction opposite from the direction of the electron flow. The direction of the magnetic field follows a right-hand rule: If you align your right thumb with the current, the direction of the magnetic field is the direction in which your fingers curl around the wire.

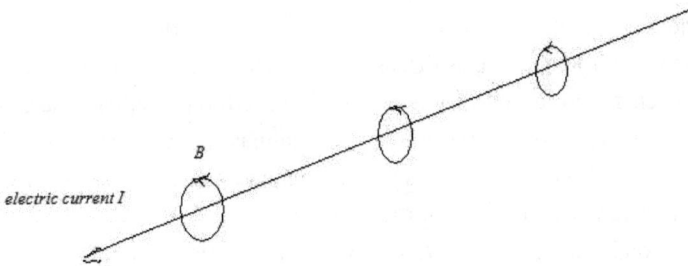

Figure AII.2—The magnetic field induced by a current in a straight wire.

Maxwell's fourth equation is $\nabla \times E = -\partial B/\partial t$. This equation is a form of Faraday's Law, named for the English scientist Michael Faraday (1791-1867), whom some historians of science consider the best experimentalist in the history of science.[10] This equation says that a magnetic field changing in time induces an electric field. This principle is applied ubiquitously in electric generators and transformers. Figure AII.3, repeated from Chapter 20, illustrates the basic principle. As noted above, Faraday's Law can be used to define the weber, the SI unit of magnetic flux: If a current loop enclosing one Wb of magnetic flux decays to zero in one second, the electric field induced in the loop creates an electric potential increase of one volt around the loop.

oscillating magnetic field

oscillating electric field

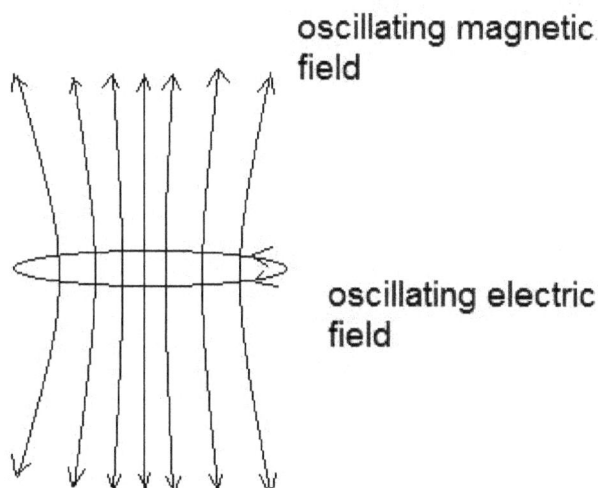

Figure AII.3—Electric field induced by a changing magnetic field

In summary, then, Maxwell's equations are

$$\mathbf{\nabla} \cdot \mathbf{D} = \rho_e$$

$$\mathbf{\nabla} \cdot \mathbf{B} = 0$$

$$\mathbf{\nabla} \times \mathbf{H} = \mathbf{j} + \partial \mathbf{D}/\partial t$$

and

$$\mathbf{\nabla} \times \mathbf{E} = -\partial \mathbf{B}/\partial t \,.$$

These four simple equations, together with the constitutive equations

$$\mathbf{D} = \varepsilon \mathbf{E}$$

and

$$\mathbf{B} = \mu \mathbf{H},$$

govern all electromagnetic phenomena, from static electricity and electrical circuits to electromagnetic waves, including radio waves, light, X-rays, and gamma rays. To the mathematical physicist, this is utterly beautiful!

REFERENCES

Chapter 1

1   Richard P. Feynman, Robert B. Leighton, and Matthew Sands, *The Feynman Lectures on Physics*, Addison Wesley, 1963, Vol. 1, p.1-1; the Feynman Lectures can be download-ed at http://student.fizika.org/~jsisko/Knjige/Opca%20Fizika/Feynman%20Lectures%20 on%20Physics/

2   Sir Karl R. Popper, "Science as Falsification," http://www.stephenjaygould.org/ctrl/pop-per_falsification.html, originally in *Conjectures and Refutations*, London: Routledge and Keagan Paul, 1963, pp. 33-39; from Theodore Schick, ed., *Readings in the Philosophy of Science*, Mountain View, CA: Mayfield Publishing Company, 2000, pp. 9-13. The first URL should be required reading in all science curricula.

3   http://www.math10.com/en/algebra/introduction-to-algebra/algebra-axioms.html

4   Richard T. Weidner and Robert L. Sells, *Elementary Modern Physics*, Alternate Second Edition, Allyn and Bacon, Boston, 1973, p.10

5   Bernard G. Harvey, *Introduction to Nuclear Physics and Chemistry*, 2nd Edition, Prentice-Hall, 1969, pp, 301-2

Chapter 2

1   http://en.wikipedia.org/wiki/Newton%27s_laws_of_motion

2   http://en.wikipedia.org/wiki/Convection

3   http://en.wikipedia.org/wiki/Advection

4   John H. Lienhard, *A Heat Transfer Textbook*, Prentice-Hall, Inc., Englewood Cliffs, N.J., 1981, pp. 27-28.

5   http://en.wikipedia.org/wiki/Ludwig_Boltzmann

6   http://en.wikipedia.org/wiki/Exponential_function

7   Gordon J. Van Wylen and Richard E. Sonntag, *Fundamentals of Classical Thermodynamics*, John Wiley and Sons, New York, 1965.

8   http://en.wikipedia.org/wiki/Bernoulli%27s_principle

9   http://en.wikipedia.org/wiki/Venturi_effect

10 http://en.wikipedia.org/wiki/Proton

11 http://en.wikipedia.org/wiki/Observable_universe

12 http://en.wikipedia.org/wiki/SI_prefix

Chapter 3

1. https://en.wikipedia.org/wiki/File:Atomic_resolution_Au100.JPG
2. http://www.nndc.bnl.gov/chart/reCenter.jsp?z=43&n=56, http://atom.kaeri.re.kr/
3. http://www.vectraondemand.com/lockheed/
4. http://en.wikipedia.org/wiki/Periodic_table
5. http://en.wikipedia.org/wiki/Quantum_mechanics
6. http://en.wikipedia.org/wiki/Photoelectric_effect
7. http://en.wikipedia.org/wiki/Schr%C3%B6dinger_equation
8. http://en.wikipedia.org/wiki/File:HAtomOrbitals.png
9. http://commons.wikimedia.org/wiki/Category:Electron_shell_diagrams
10. http://creativecommons.org/licenses/by-sa/3.0/
11. http://en.wikipedia.org/wiki/Strong_interaction
12. http://en.wikipedia.org/wiki/Quark
13. Sean Carroll, *The Particle at the End of the Universe: How the Hunt for the Higgs Boson Leads Us to the Edge of a New World*, Penguin, 2012, pp. 101-102.
14. http://en.wikipedia.org/wiki/Gluon

Chapter 4

1. http://en.wikipedia.org/wiki/Curie
2. http://en.wikipedia.org/wiki/Electromagnetic_spectrum
3. http://images.google.com/imgres?imgurl=http://mynasadata.larc.nasa.gov/images/EM_Spectrum3-new.jpg&imgrefurl=http://mynasadata.larc.nasa.gov/ElectroMag.html&usg=__9ammTA9VuxqxMyaHTsYB4Y3qfV4=&h=688&w=962&sz=108&hl=en&start=7&um=1&tbnid=e65qHUkp2-GCeM:&tbnh=106&tbnw=148&prev=/images%3Fq%3Delectromagnetic%2Bspectrum%26um%3D1%26hl%3Den%26rlz%3D1T4DKUS_enUS305%26sa%3DX
4. http://en.wikipedia.org/wiki/Thermal_radiation
5. http://commons.wikimedia.org/wiki/Commons:GNU_Free_Documentation_License_1.2
6. https://en.wikipedia.org/wiki/Positron
7. http://en.wikipedia.org/wiki/Neutrino
8. Frank Close, *Neutrino*, Oxford University Press, 2010.
9. http://en.wikipedia.org/wiki/Cosmic_ray
10. Samuel Glasstone and Alexander Sesonske, *Nuclear Reactor Engineering*, Third Edition, Van Nostrand Reihnold, New York, 1981, p. 574-575.
11. Samuel Glasstone and Alexander Sesonske, *Nuclear Reactor Engineering*, Third Edition, Van Nostrand Reihnold, New York, 1981, p. 590.
12. Alan E. Waltar, *Radiation and Modern Life*, Prometheus Books, 2004, p. 43.
13. Samuel Glasstone and Alexander Sesonske, *Nuclear Reactor Engineering*, Third Edition, Van Nostrand Reihnold, New York, 1981, p. 587.
14. http://en.wikipedia.org/wiki/Radium
15. http://www.cancer.gov/cancertopics/factsheet/Risk/heterocyclic-amines
16. Gwyneth Cravens, *Power to Save the World: The Truth About Nuclear Energy*, Alfred A. Knopf, 2007, pp. 115-116.
17. http://en.wikipedia.org/wiki/Radiation_hormesis
18. *Nuclear News*, September 2012, pp. 46-47.

[19] http://www.nrc.gov/reading-rm/doc-collections/cfr/part020/
[20] http://www.ieer.org/ensec/no-4/limits.html
[21] http://en.wikipedia.org/wiki/Foodborne_illness#Global_impact
[22] http://russp.org/BLC-3.html
[23] http://www.epa.gov/radon/zonemap.html
[24] http://www.newworldencyclopedia.org/entry/Plutonium#Toxicity
[25] http://en.wikipedia.org/wiki/Plutonium
[26] http://hypertextbook.com/facts/MichaelPhillip.shtml

Chapter 5

[1] Samuel Glasstone and Alexander Sesonske, *Nuclear Reactor Engineering*, Third Edition, Van Nostrand Reihnold, New York, 1981, Chapter 2

[2] http://www.nndc.bnl.gov/exfor/endf00.jsp; this URL is the access portal to on-line data from the Evaluated Nuclear Data File, version ENDF/B-VII, December 15, 2006. An extensive description of the ENDF/B-VII data library can be found in M.B. Chadwick, P. Oblozinsky, M. Herman at al., "ENDF/B-VII.0: Next Generation Evaluated Nuclear Data Library for Nuclear Science and Technology", Nuclear Data Sheets, vol. 107, pp. 2931-3060, 2006. The ENDF/B-VII data library is a publication of the United States government for use in worldwide research and is in the public domain.

[3] http://www.orau.org/ptp/PTP%20Library/library/DOE/Misc/h1019v1.pdf. This is the URL for the on-line version of *DOE Fundamentals Handbook: Nuclear Physics and Reactor Theory*, vol. 1 of 2, DOE-HDBK-1019/1-93, U.S. Department of Energy, January 1993, approved for unlimited public release. The figure was taken from Module 2, p. 9.

[4] http://www.orau.org/ptp/PTP%20Library/library/DOE/Misc/h1019v1.pdf. This is the URL for the on-line version of *DOE Fundamentals Handbook: Nuclear Physics and Reactor Theory*, vol. 1 of 2, DOE-HDBK-1019/1-93, U.S. Department of Energy, January 1993, approved for unlimited public release. The figure was taken from Module 1, p. 53.

[5] http://www.orau.org/ptp/PTP%20Library/library/DOE/Misc/h1019v1.pdf. This is the URL for the on-line version of *DOE Fundamentals Handbook: Nuclear Physics and Reactor Theory*, vol. 1 of 2, DOE-HDBK-1019/1-93, U.S. Department of Energy, January 1993, approved for unlimited public release. The figure was taken from Module 1, p. 57.

[6] Samuel Glasstone and Alexander Sesonske, *Nuclear Reactor Engineering*, Third Edition, Van Nostrand Reinhold, New York, 1981, p. 16.

[7] Karl O. Ott and Winfred A. Bezella, *Nuclear Reactor Statics*, American Nuclear Society, LaGrange Park, IL, p.p. 261-267

[8] Francis F. Chen, *Introduction to Plasma Physics*, Plenum Press, New York, 1974, Chapter 9

[9] http://en.wikipedia.org/wiki/Deuterium

[10] http://en.wikipedia.org/wiki/Tritium

Chapter 6

[1] http://en.wikipedia.org/wiki/SL-1

[2] John R. Lamarsh, *Introduction to Nuclear Reactor Theory*, Addison-Wesley Publishing Co., Reading, Massachusetts, 1966, Chapters 9 and 10.

[3] James. J. Duderstadt and Louis J. Hamilton, *Nuclear Reactor Analysis*, John Wiley and

Sons, New York, 1976, pp. 288-295.

[4] "Modular Pebble-bed Reactor Project: Laboratory-directed Research and Development Program, FY 2001 Annual Report," INEELEXT-2001-1623, Idaho National Engineering and Environmental Laboratory and the Massachusetts Institute of Technology, William K. Terry, editor, December 2001.

[5] http://www.hse.gov.uk/foi/internalops/nsd/tech_asst_guides/tast042.pdf

[6] William L. Oberkampf and Timothy G. Trucano, "Benchmarking of CFD Codes for Application to Nuclear Reactor Safety," *Nucl. Eng. Design 238*, No. 3, March 2008, pp. 716-743; also http://www.sciencedirect.com/science?_ob=ArticleURL&_udi=B-6V4D-4NYBMNP-C&_user=10&_rdoc=1&_fmt=&_orig=search&_sort=d&view=c&_acct=C000050221&_version=1&_urlVersion=0&_userid=10&md5=df040503cfb0af-6279c9050abf313c7e

[7] William K. Terry and David W. Nigg, "One-Dimensional Diffusion Theory Kinetics in RELAP5," *Nucl. Sci. and Eng.* 120, pp. 110-123 (1995).

[8] IRPhEP (International Reactor Physics Experiment Evaluation Project), http://irphep.inl.gov/

[9] ICSBEP (International Criticality Safety Benchmark Evaluation Project), http://icsbep.inl.gov/

[10] http://en.wikipedia.org/wiki/Multipurpose_Applied_Physics_Lattice_Experiment

[11] Ronald Allen Knief, *Nuclear Energy Technology: Theory and Practice of Commercial Nuclear Power*, Hemisphere Publishing Co., Washington, 1981, pp. 160-165.

[12] http://www.nrc.gov/reading-rm/basic-ref/glossary/design-basis-accident.html

[13] Abderrafi M. Ougouag, Richard R. Schultz, and William K. Terry, "Automatically Scramming Nuclear Reactor System," U. S. Patent No. 6,804,320 B2, October 12, 2004.

[14] Abderrafi M. Ougouag, Richard R. Schultz, and William K. Terry, "Method for Automatically Scramming a Nuclear Reactor, U. S. Patent No. 6,980,619 B2, December 27, 2005.

Chapter 7

[1] E. E. Khalil, *Power Plant Design*, CRC Press, 1990.

[2] http://en.wikipedia.org/wiki/British_thermal_unit

[3] Joseph H. Keenan and Frederick G. Keyes, *Thermodynamic Properties of Steam*, John Wiley & Sons, New York, 1937.

[4] Samuel Glasstone and Alexander Sesonske, *Nuclear Reactor Engineering*, Van Nostrand Rheinhold, New York, 3rd Edition, 1981, pp. 623-625.

Chapter 8

[1] http://en.wikipedia.org/wiki/Coal

[2] http://en.wikipedia.org/wiki/World_energy_resources_and_consumption

[3] http://en.wikipedia.org/wiki/Surface_mining

[4] http://www.epa.gov/reg3wapd/nps/mining/

[5] http://en.wikipedia.org/wiki/Mountaintop_removal_mining

[6] Katherine Bagley, "King Coal," *Audubon* magazine, November-December 2009, pp. 12-13.

[7]   http://en.wikipedia.org/wiki/File:MTR1.jpg

[8]   http://en.wikipedia.org/wiki/Coal_mining

[9]   http://wiki.answers.com/Q/How_much_coal_does_a_500_mw_power_plant_use_daily

[10] http://www.american-rails.com/hopper-cars.html

[11] http://www.ornl.gov/info/ornlreview/rev26-34/text/colmain.html

[12] http://en.wikipedia.org/wiki/Air_pollution

[13] http://www.who.int/news-room/fact-sheets/detail/household-air-pollution-and-health

[14] https://www.cdc.gov/nceh/radiation/brochure/profile_radon.htm

[15] https://www.cancerresearchuk.org/health-professional/cancer-statistics/worldwide-cancer#heading-One

[16] Si-Heon Kim, Won Ju Hwong, Jeong-Sook Cho, and Dae Ryong Kang, "Attributable risk of lung cancer deaths due to indoor radon exposure," *Annals of Occupational and Environmental Medicine 28:8*, 2016, https://aoemj.biomedcentral.com/articles/10.1186/s40557-016-0093-4

[17] http://www.newsweek.com/2009/07/17/toxic-tsunami.html

[18] http://en.wikipedia.org/wiki/Peak_coal

[19] http://en.wikipedia.org/wiki/Carbon_capture_and_storage

[20] http://en.wikipedia.org/wiki/Petroleum

[21] http://en.wikipedia.org/wiki/Light_crude_oil

[22] http://en.wikipedia.org/wiki/Heavy_crude_oil

[23] http://en.wikipedia.org/wiki/Sweet_crude_oil

[24] http://en.wikipedia.org/wiki/Oil_shale

[25] http://en.wikipedia.org/wiki/Kerogen

[26] http://en.wikipedia.org/wiki/Extraction_of_petroleum

[27] http://en.wikipedia.org/wiki/Asphalt

[28] http://en.wikipedia.org/wiki/Gasoline

[29] https://en.wikipedia.org/wiki/List_of_countries_by_proven_oil_reserves

[30] https://en.wikipedia.org/wiki/Oil_shale

[31] https://www.bp.com/content/dam/bp/en/corporate/pdf/energy-economics/statistical-review-2017/bp-statistical-review-of-world-energy-2017-full-report.pdf

[32] http://en.wikipedia.org/wiki/Fischer%E2%80%93Tropsch_process

[33] https://en.wikipedia.org/wiki/Hydraulic_fracturing

[34] http://en.wikipedia.org/wiki/Deepwater_Horizon_oil_spill

[35] http://www.sciencedaily.com/releases/2010/04/100414111018.htm

[36] http://www.epa.gov/oem/docs/oil/fss/fss02/ricepresent.pdf

[37] http://en.wikipedia.org/wiki/Smog

[38] http://commons.wikimedia.org/wiki/Commons:GNU_Free_Documentation_License,_version_1.2

[39] Barry Yeoman, "Crude Awakening," *Audubon* 112, No. 2, March-April 2010, pp. 76-83.

[40] http://en.wikipedia.org/wiki/Oil_reserves

[41] http://en.wikipedia.org/wiki/Natural_gas

[42] http://en.wikipedia.org/wiki/Hydrocarbon

[43] http://en.wikipedia.org/wiki/Standard_conditions_for_temperature_and_pressure

[44] http://en.wikipedia.org/wiki/Alkene

[45] http://en.wikipedia.org/wiki/Methanogen

[46] http://en.wikipedia.org/wiki/Archaea

[47] http://en.wikipedia.org/wiki/Three-domain_system

[48] http://www.energy.ca.gov/biomass/

[49] http://naturalgas.org/overview/resources/

[50] https://www.pennlive.com/news/2017/07/fatal_gas_explosions_pennsylva.html

[51] https://www.cbsnews.com/news/fugitive-gas-leak-pipeline-caused-house-explosion-fires-tone-colorado/

[52] https://www.foxnews.com/us/boston-area-gas-explosions-kill-at-least-1-injure-20-trigger-evacuations-officials-say

[53] https://www.usatoday.com/story/news/nation/2018/09/18/massachusetts-natural-gas-explosions-pressure-ed-markey-elizabeth-warren-columbia/1345591002/

[54] http://en.wikipedia.org/wiki/Hydroelectricity

[55] https://en.wikipedia.org/wiki/List_of_countries_by_electricity_production_from_renewable_sources

[56] http://en.wikipedia.org/wiki/Hoover_Dam

[57] http://en.wikipedia.org/wiki/Three_Gorges_Dam

[58] http://en.wikipedia.org/wiki/Itaipu_Dam

[59] http://www.fozdoiguacudestinodomundo.com.br/en/itaipu-hydroelectric-plant

[60] http://en.wikipedia.org/wiki/Guri_Dam

[61] http://en.wikipedia.org/wiki/Hydropower

[62] http://www.arkansas.com/lakes-rivers/lake.aspx?id=2

[63] http://www.bigskyfishing.com/River-Fishing/South-MT-Rivers/bighorn-river/bighorn_overview.php

[64] http://www.usbr.gov/projects/Facility.jsp?fac_Name=Yellowtail%20Dam

[65] Robert W. Adler, *Restoring Colorado River Ecosystems: A Troubled Sense of Immensity*, Island Press, Washington, D.C., 2007.

[66] http://www.usbr.gov/projects/Facility.jsp?fac_Name=Yellowtail+Afterbay+Dam

[67] http://www.usbr.gov/projects/Powerplant.jsp?fac_Name=Yellowtail Powerplant

[68] Rocky Barker, *Saving All the Parts*, Island Press, Washington, D.C., 1993, pp. 11-12, 56-108.

[69] http://fishandgame.idaho.gov/fish/sockeye/

[70] http://en.wikipedia.org/wiki/Redfish_Lake

[71] http://www.seattlepi.com/local/38065_lonesome07.shtml

[72] http://www.cbbulletin.com/397749.aspx

[73] http://www.psmfc.org/habitat/salmondam.html

[74] http://www.americanrivers.org/assets/pdfs/dam-removal-docs/lower_snake_econ_fact-sheet_06-05.pdf

[75] http://en.wikipedia.org/wiki/Teton_Dam

[76] http://www.usbr.gov/pmts/sediment/projects/TetonRiver/Reports/report.pdf

[77] http://matdl.org/failurecases/Dam%20Cases/Teton%20Dam.htm

[78] Randy Scholfield, "Staying Power," *Trout*, Summer 2009, pp. 37-43.

[79] http://matdl.org/failurecases/Dam%20Cases/malpasset_dam_failure.htm

[80] http://en.wikipedia.org/wiki/Solar_vehicle

[81] http://en.wikipedia.org/wiki/Solar_constant

[82] http://en.wikipedia.org/wiki/Insolation

[83] http://www.solarserver.com/knowledge/lexicon/s/solar-constant.html

[84] http://en.wikipedia.org/wiki/List_of_countries_by_electricity_consumption

[85] https://en.wikipedia.org/wiki/Solar_panel

[86] http://en.wikipedia.org/wiki/Photovoltaics

[87] http://www.airport-technology.com/projects/denver/

[88] http://en.wikipedia.org/wiki/Solar_water_heating

[89] http://en.wikipedia.org/wiki/Concentrated_solar_power

[90] http://www.foxnews.com/us/2014/02/15/world-largest-solar-plant-burning-up-birds-in-nevada-desert/

[91] http://svtc.org/wp-content/uploads/Silicon_Valley_Toxics_Coalition_-_Toward_a_Just_and_Sust.pdf

[92] http://en.wikipedia.org/wiki/Wind_power

[93] http://www.stanford.edu/group/efmh/winds/2004jd005462.pdf

[94] http://www.windpowerengineering.com/construction/installation/modular-towers-and-the-quest-for-stronger-wind/

[95] http://www.jvas.org/news_wpd.html

[96] http://en.wikipedia.org/wiki/Bat

[97] http://www.batcon.org/index.php/our-work/regions/usa-canada/protect-mega-populations/cab-intro

[98] http://wildlife.state.co.us/WildlifeSpecies/LivingWithWildlife/Mammals/

[99] https://www.audubon.org/magazine/spring-2018/how-new-technology-making-wind-farms-safer-birds

[100] http://www.usatoday.com/news/nation/2005-01-04-windmills-usat_x.htm

[101] http://www.foxnews.com/science/2013/09/12/study-says-wind-farms-have-killed-alarming-number-eagles/

[102] http://www.washingtontimes.com/news/2013/dec/6/obama-issues-permits-wind-farms-kill-more-eagles/?page=all

[103] http://en.wikipedia.org/wiki/Cape_Wind

[104] https://en.wikipedia.org/wiki/Cost_of_electricity_by_source

[105] https://www.newsweek.com/whats-true-cost-wind-power-321480

[106] George Crabtree and Jim Misewich, *The Grid: Ready for Renewables?*, APS News, 19, No. 11, December 2010, p.8

[107] http://www.aps.org/policy/reports/popa-reports/upload/integratingelec.pdf

[108] https://en.wikipedia.org/wiki/List_of_U.S._states_by_electricity_production_from_renewable_sources

[109] http://en.wikipedia.org/wiki/Biomass

[110] http://en.wikipedia.org/wiki/Biodiesel

[111] http://www.wired.com/autopia/2008/05/making-renewabl/

[112] https://www.usda.gov/oce/reports/energy/USbiopower_04_2014.pdf

[113] http://www.eia.doe.gov/cneaf/alternate/page/renew_energy_consump/table4.html

[114] http://zfacts.com/p/436.html

[115] http://en.wikipedia.org/wiki/Ethanol_fuel

[116] http://www.meti.go.jp/report/downloadfiles/g30819b40j.pdf

[117] http://en.wikipedia.org/wiki/Ethylene

[118] http://en.wikipedia.org/wiki/Corn_ethanol

[119] http://www.reuters.com/article/idUSTRE6AS61L20101129?utm_source=feedburner&utm_medium=feed&utm_campaign=Feed:+reuters%2FUSgreenbusinessNews+(News+%2F+US+%2F+Green+Business)&utm_content=Google+Reader

[120] http://www.nass.usda.gov/Statistics_by_Subject/result.php?0FB-5C4D8-22CD-3971-9C07-0A7CE540A42D&sector=CROPS&group=FIELD%20CROPS&comm=CORN

[121] http://www.fuel-testers.com/ethanol_fuel_disadvantages.html

[122] http://www.fuel-testers.com/ethanol_problems_damage.html

[123] http://www.fuel-testers.com/list_e10_engine_damage.html

[124] http://en.wikipedia.org/wiki/Common_ethanol_fuel_mixtures

[125] http://members.opei.org/news/detail.dot?id=11846

[126] http://detnews.com/article/20101220/AUTO01/12200364/
Detroit-automakers-join-lawsuit-over-E15-fuel-approval

[127] http://www.pacificethanol.net/site/index.php/media/straight_story_article/345/

[128] http://en.wikipedia.org/wiki/Cellulosic_ethanol

[129] http://en.wikipedia.org/wiki/Food_vs._fuel

[130] http://www.guardian.co.uk/environment/2008/jul/03/biofuels.renewableenergy

[131] Dina Cappiello and Matt Apuzzo, "Ethanol policy has high environmental cost," Associated Press, November 13, 2013, printed in the Boulder, CO, *Daily Camera*.

[132] http://www.transportation.anl.gov/pdfs/AF/265.pdf

[133] http://en.wikipedia.org/wiki/Ethanol_fuel_energy_balance

[134] http://www.energyjustice.net/files/ethanol/pimentel2003.pdf

[135] http://www.springerlink.com/content/r1552355771656v0/

[136] http://ianrnews.unl.edu/static/0901220.shtml

[137] Ted Williams, "Incite: The Edge of Insanity," *Audubon*, January-February 2014, p.58.

[138] http://en.wikipedia.org/wiki/E85_in_standard_engines

[139] http://en.wikipedia.org/wiki/E85

[140] http://www.usatoday.com/money/industries/energy/2007-11-29-e85-ethanol_N.htm

[141] http://en.wikipedia.org/wiki/Ethanol_fuel_in_the_United_States

[142] http://en.wikipedia.org/wiki/Inner_core

[143] http://en.wikipedia.org/wiki/Outer_core

[144] http://en.wikipedia.org/wiki/Mantle_(geology)

[145] http://en.wikipedia.org/wiki/Crust_(geology)

[146] http://en.wikipedia.org/wiki/Magma

[147] http://en.wikipedia.org/wiki/Volcano

[148] http://en.wikipedia.org/wiki/Geyser

[149] http://en.wikipedia.org/wiki/Hot_spring

[150] http://en.wikipedia.org/wiki/Plate_tectonics

[151] http://en.wikipedia.org/wiki/Mantle_plume

[152] http://www.columbia.edu/itc/ldeo/v1011x-1/jcm/Topic3/Topic3.html

[153] http://www.agu.org/pubs/crossref/1993/93RG01249.shtml

[154] http://geoheat.oit.edu/bulletin/bull28-3/art2.pdf

[155] http://en.wikipedia.org/wiki/Primary_energy

[156] http://en.wikipedia.org/wiki/Geothermal_energy

[157] https://www.eia.gov/todayinenergy/detail.php?id=17871&src=email

[158] http://www.geo-energy.org/reports/2016/2016%20Annual%20US%20Global%20
Geothermal%20Power%20Production.pdf

[159] https://en.wikipedia.org/wiki/Geothermal_heating

[160] http://en.wikipedia.org/wiki/Heat_pumps

[161] http://www.energysavers.gov/your_home/space_heating_cooling/index.cfm/
mytopic=12640

[162] http://en.wikipedia.org/wiki/1,1,1,2-Tetrafluoroethane

[163] http://en.wikipedia.org/wiki/File:3-ton_Slinky_Loop.jpg

[164] http://en.wikipedia.org/wiki/Coefficient_of_performance

[165] http://en.wikipedia.org/wiki/Tidal_power

[166] http://en.wikipedia.org/wiki/Tide_mill

[167] http://en.wikipedia.org/wiki/List_of_tidal_power_stations

[168] http://en.wikipedia.org/wiki/Annapolis_Royal_Generating_Station
[169] http://en.wikipedia.org/wiki/Jiangxia_Tidal_Power_Station
[170] http://en.wikipedia.org/wiki/Kislaya_Guba_Tidal_Power_Station
[171] http://en.wikipedia.org/wiki/Rance_Tidal_Power_Station
[172] http://en.wikipedia.org/wiki/Strangford_Lough
[173] https://www.theguardian.com/environment/2016/aug/29/world-first-for-shetlands-in-tidal-power-breakthrough
[174] http://ilienergy.com/2016/09/scotland-home-of-the-worlds-first-tidal-farm/
[175] http://en.wikipedia.org/wiki/Uldolmok_Tidal_Power_Station
[176] https://en.wikipedia.org/wiki/Sihwa_Lake_Tidal_Power_Station
[177] https://www.tocardo.com/eastern-scheldt-tidal-project-starts-producing-power/
[178] http://en.wikipedia.org/wiki/Penzhin_Tidal_Power_Plant
[179] http://en.wikipedia.org/wiki/Perpetual_motion

Chapter 9

[1] http://www.sciner.com/Opticsland/FS.htm
[2] http://www.gi.alaska.edu/ScienceForum/ASF8/817.html
[3] http://en.wikipedia.org/wiki/Greenhouse_gas
[4] http://pubs.giss.nasa.gov/docs/2010/2010_Schmidt_etal_1.pdf
[5] http://barrettbellamyclimate.com/page15.htm
[6] http://en.wikipedia.org/wiki/Carbon_dioxide
[7] http://www.skepticalscience.com/co2-higher-in-past.htm
[8] http://en.wikipedia.org/wiki/Jurassic
[9] https://www.e-education.psu.edu/earth103/node/1018
[10] http://www.sciencedaily.com/releases/2009/10/091008152242.htm
[11] https://en.wikipedia.org/wiki/Consilience
[12] https://en.wikipedia.org/wiki/Carbon_dioxide_in_Earth%27s_atmosphere
[13] http://commons.wikimedia.org/wiki/Commons:GNU_Free_Documentation_License,_version_1.2
[14] https://en.wikipedia.org/wiki/Atmosphere_of_Earth
[15] http://www.sciencedaily.com/releases/2007/11/071114163448.htm
[16] http://math.ucr.edu/home/baez/temperature/
[17] http://en.wikipedia.org/wiki/Global_warming
[18] https://www.climate.gov/news-features/understanding-climate/climate-change-global-temperature
[19] Petit, J.R., J. Jouzel, D. Raynaud, N.I. Barkov, J.-M. Barnola, I. Basile, M. Benders, J. Chappellaz, M. Davis, G. Delayque, M. Delmotte, V.M. Kotlyakov, M. Legrand, V.Y. Lipenkov, C. Lorius, L. Pépin, C. Ritz, E. Saltzman, and M. Stievenard. 1999. Climate and atmospheric history of the past 420,000 years from the Vostok ice core, Antarctica. *Nature* 399: 429-436, 3 June 1999.
[20] http://blogs.telegraph.co.uk/news/jamesdelingpole/100095506/there-has-been-no-global-warming-since-1998/
[21] http://www.nasa.gov/topics/earth/features/upsDownsGlobalWarming.html
[22] http://www.forbes.com/sites/petergleick/2012/02/05/global-warming-has-stopped-how-to-fool-people-using-cherry-picked-climate-data/
[23] http://www.noaanews.noaa.gov/stories2010/20100728_stateoftheclimate.html

[24] http://world.time.com/2014/01/17/australia-is-melting-under-a-horrifying-heatwave/
[25] http://culter.colorado.edu:1030/~saelias/glacier.html
[26] Nicola Scafetta and Bruce J. West, "Is climate sensitive to solar variability?" *Physics Today*, March 2008, pp. 50-51.
[27] http://www.ipcc.ch/organization/organization.htm
[28] http://en.wikipedia.org/wiki/Climate_model
[29] http://en.wikipedia.org/wiki/General_circulation_model
[30] http://en.wikipedia.org/wiki/Carbon_cycle
[31] https://en.wikipedia.org/wiki/History_of_climate_change_science
[32] http://en.wikipedia.org/wiki/Scientific_opinion_on_climate_change
[33] https://www.nytimes.com/2018/11/23/climate/us-climate-report.html
[34] https://www.nytimes.com/2018/10/07/climate/ipcc-climate-report-2040.html?module=inline
[35] http://www.telegraph.co.uk/earth/environment/climatechange/4808122/Scientists-find-bigger-than-expected-polar-ice-melt.html
[36] http://en.wikipedia.org/wiki/Kyoto_Protocol
[37] Bjorn Lomborg, Cool It: *The Skeptical Environmentalist's Guide to Global Warming, Vintage Books*, a division of Random House, New York, 2nd Edition, 2008.
[38] Ian Plimer, *Heaven and Earth: Global Warming – the Missing Science*, Taylor Trade Publishing, Lanham, Maryland, 2009.
[39] Jared Diamond, *Collapse: How Societies Choose to Fail or Succeed*, Penguin Books, New York, 2nd Edition, 2011
[40] Raymond T. Pierrehumbert, "Infrared radiation and planetary temperature," *Physics Today*, January 2011, pp. 33-38
[41] https://en.wikipedia.org/wiki/Patrick_Moore_(environmentalist).
[42] http://www.thegwpf.org/patrick-moore-should-we-celebrate-carbon-dioxide/.
[43] Joseph Romm, *Climate Change: What Everyone Needs to Know*, Oxford University Press, 2018.
[44] https://en.wikipedia.org/wiki/Permafrost
[45] https://en.wikipedia.org/wiki/Carbon_sequestration
[46] https://en.wikipedia.org/wiki/The_Population_Bomb
[47] http://www.2think.org/tpe.shtml
[48] Lester R. Brown, *Full Planet, Empty Plates: The New Geopolitics of Food Scarcity*, W. W. Norton & Company, New York, 2012
[49] https://www.populationconnection.org/

Chapter 10

[1] http://en.wikipedia.org/wiki/Nuclear_Medicine
[2] http://www.govtrack.us/congress/record.xpd?id=108-s20041004-25
[3] http://en.wikipedia.org/wiki/Positron_emission_tomography
[4] http://en.wikipedia.org/wiki/SPECT
[5] http://en.wikipedia.org/wiki/Computed_tomography
[6] http://en.wikipedia.org/wiki/Magnetic_resonance_imaging
[7] http://en.wikipedia.org/wiki/File:PET-image.jpg#file
[8] http://en.wikipedia.org/wiki/Gamma_camera
[9] http://en.wikipedia.org/wiki/File:Lung_SPECT-CT_keosys_format_dicom.JPG

10 C. Perrier and E. Segré, "Some Chemical Properties of Element 43," *J. Chem. Phys.* 5, p. 712, (1937)
11 http://en.wikipedia.org/wiki/Multipurpose_Applied_Physics_Lattice_Experiment
12 Isotopes and Radiation Briefs, *Nuclear News*, December 2016, p. 70.
13 "Draft EIS Issued for Northwest Medical, *Nuclear News*, December 2016, pp. 69-70.
14 Ralph G. Bennett, Jerry D. Christian, David A. Petti, William K. Terry, and S. Blaine Grover, "A system of 99mTc production based on distributed electron accelerators and thermal separation," *Nucl. Tech.* 126, pp. 102-121 (1999)
15 https://www.energy.gov/nnsa/nnsa-s-molybdenum-99-program-establishing-reliable-supply-mo-99-produced-without-highly
16 http://en.wikipedia.org/wiki/Magnetic_resonance_imaging
17 http://en.wikipedia.org/wiki/File:MR_Knee.jpg
18 http://commons.wikimedia.org/wiki/Commons:GNU_Free_Documentation_License_1.2
19 http://creativecommons.org/licenses/by-sa/3.0/deed.en
20 http://www.world-nuclear.org/info/inf55.html
21 http://en.wikipedia.org/wiki/Boron_Neutron_Capture_Therapy
22 http://mit.edu/nrl/www/bnct/research/research.html
23 http://www.nupecc.org/report97/report97_astrobib/node15.html
24 Dr. Charles A. Wemple, personal communication.

Chapter 11

1 http://en.wikipedia.org/wiki/Tritium
2 http://atom.kaeri.re.kr/ton/nuc1.html
3 http://en.wikipedia.org/wiki/Nuclear_reactor#Classification_by_generation
4 http://en.wikipedia.org/wiki/Passively_safe
5 Ronald Allen Knief, *Nuclear Energy Technology: Theory and Practice of Commercial Nuclear Power*, Hemisphere Publishing, Washington, 1981, p. 231.
6 Shannon Bragg-Stratton, "Development of advanced accident-tolerant fuels for commercial LWRs," *Nuclear News*, March 2014, pp. 83-91.
7 http://www.euronuclear.org/info/encyclopedia/f/fuel-rod.htm.
8 Ronald Allen Knief, *Nuclear Energy Technology: Theory and Practice of Commercial Nuclear Power*, Hemisphere Publishing, Washington, 1981, p. 234.
9 http://www.nrc.gov/reading-rm/basic-ref/glossary/design-basis-accident.html
10 http://www.directives.doe.gov/pdfs/doe/doetext/neword/421/g4211-1gpg.pdf
11 http://www.cnsc.gc.ca/pubs_catalogue/uploads/RD-310_e_PDF.pdf
12 http://en.wikipedia.org/wiki/Boiling_water_reactor_safety_systems
13 http://en.wikipedia.org/wiki/File:Bwr-rpv.svg.
14 http://www.world-nuclear.org/info/inf06.html
15 http://en.wikipedia.org/wiki/Boiling_water_reactor_safety_systems#Reactor_core_isolation_cooling_system_.28RCIC.29
16 Samuel Glasstone and Alexander Sesonske, *Nuclear Reactor Engineering*, Third Edition, Van Nostrand Reihnold, New York, 1981, Chapter 11.
17 http://en.wikipedia.org/wiki/ABWR
18 https://en.wikipedia.org/wiki/Advanced_boiling_water_reactor#Deployments.
19 http://www.winus.org/portals/0/meetings/win200607/presentations/12.pdf
20 http://www.ne.doe.gov/np2010/pdfs/ABWROverview.pdf

21 http://en.wikipedia.org/wiki/ESBWR

22 http://www.ne.doe.gov/np2010/pdfs/esbwrOverview.pdf

23 http://en.wikipedia.org/wiki/Boiling_water_reactor_safety_systems#Isolation_
Condenser_.28IC.29

24 www.foronuclear.org/.../Descripcion_general_ESBWR.pdf - *Spain*

25 http://www.gepower.com/prod_serv/products/nuclear_energy/en/downloads/abwr_plant.pdf

26 Ronald Allen Knief, *Nuclear Energy Technology: Theory and Practice of Commercial Nuclear Power*, Hemisphere Publishing, Washington, 1981, pp. 566-570.

27 http://en.wikipedia.org/wiki/Pressurized_water_reactor

28 http://www.nrc.gov/reactors/pwrs.html

29 http://en.wikipedia.org/wiki/File:Reactorvessel.gif,

30 http://www.eia.doe.gov/cneaf/nuclear/page/nuc_reactors/pwr.html

31 http://mitnse.files.wordpress.com/2011/03/nuclear_power_011.pdf

32 http://en.wikipedia.org/wiki/AP1000

33 http://www.ap1000.westinghousenuclear.com/

34 http://nuclearinfo.net/twiki/pub/Nuclearpower/WebHomeCostOfNuclearPower/AP1000Reactor.pdf

35 http://en.wikipedia.org/wiki/European_Pressurized_Reactor

36 http://en.wikipedia.org/wiki/Advanced_Pressurized_Water_Reactor

37 http://www.areva.com/servlet/lobProvider?blobcol=urluploadedfile&blobheader=application%2Fpdf&blobkey=id&blobtable=Downloads&blobwhere=1210091797174&filename=atmea%2C0.pdf

38 Han-Gong Kim, "The Design Characteristics of Advanced Power Reactor 1400," IAEA-CN-164-3S09, International Atomic Energy Agency, 2009.

39 http://nuclearstreet.com/nuclear_power_industry_news/b/nuclear_power_news/archive/2011/02/11/nrc-seeking-comments-on-ap1000-reactor-design-amendment-021101.aspx

40 https://www.nytimes.com/2017/03/29/business/westinghouse-toshiba-nuclear-bankruptcy.html

41 http://www.ap1000.westinghousenuclear.com/ap1000_psrs_pccs.html

42 http://www.sse.tulane.edu/FORUM_2003/Matzie%20Presentation.pdf

43 http://nuclearinfo.net/twiki/pub/Nuclearpower/WebHomeCostOfNuclearPower/AP1000Reactor.pdf

44 "World List of Nuclear Power Plants," *Nuclear News*, March 2014, p. 58.

45 http://www.nrc.gov/reactors/new-reactors/design-cert/ap1000.html

46 http://www.areva-np.com/common/liblocal/docs/Brochure/EPR_US_%20May%202005.pdf

47 http://www.mnes-us.com/htm/usapwrdesign.htm

48 http://www.nrc.gov/reactors/new-reactors/design-cert/apwr/dcd.html#dcd

Chapter 12

1 http://en.wikipedia.org/wiki/Shoreham_Nuclear_Power_Plant

2 http://www.historylink.org/index.cfm?DisplayPage=output.cfm&File_Id=5482

3 "Renaissance Watch, *Nuclear News*, February 2014, pp. 22-23.

4 Ronald Allen Knief, *Nuclear Energy Technology: Theory and Practice of Commercial Nuclear Power*, Hemisphere Publishing, Washington, 1981, Chapter 18.

5 http://en.wikipedia.org/wiki/File:Tmi-2_schematic.svg

6 Gianni Petrangeli, *Nuclear Safety*, Butterworth-Heinemann, 2006, Appendix 17.

7   http://www.insc.anl.gov/matprop/uo2/melt.php
8   Grigori Medvedev, *The Truth about Chernobyl*, Basic Books, 1991, pp. 10-11.
9   J. Kemeny, et al., "The Need for Change: The Legacy of TMI," Report of the President's Commission on the Accident at Three Mile Island, U. S. Government Printing Office, Washington, D.C., October 1979.

Chapter 13

1   http://en.wikipedia.org/wiki/RBMK
2   http://en.wikipedia.org/wiki/Chernobyl_accident
3   Grigori Medvedev, *The Truth About Chernobyl*, Basic Books, Inc., 1991.
4   The Chernobyl Forum: 2003-2005. Chernobyl's Legacy: Health, Environmental, and Socio-Economic Impacts and Recommendations to the Governments of Belarus, the Russian Federation and UkrainePDF. IAEA. 2nd revised version.
5   *Nuclear News*, "World List of Nuclear Power Plants," March, 2014, pp. 47-67.

Chapter 14

1   http://en.wikipedia.org/wiki/2011_T%C5%8Dhoku_earthquake_and_tsunami
2   http://en.wikipedia.org/wiki/Fukushima_Daiichi_nuclear_disaster
3   http://fukushima.ans.org: American Nuclear Society Special Committee on Fukushima Michael Corradini et al., *Fukushima Daiichi: ANS Committee Report*, American Nuclear Society, March 2012, revised June 2012.
4   http://en.wikipedia.org/wiki/Megathrust_earthquake
5   http://en.wikipedia.org/wiki/Richter_magnitude_scale
6   http://en.wikipedia.org/wiki/Moment_magnitude
7   Thorne Lay and Hiroo Kanamori, "Insights from the great 2011 Japan earthquake," *Physics Today*, December 2011, pp. 33-39.
8   http://en.wikipedia.org/wiki/Tsunami
9   http://nctr.pmel.noaa.gov/database_devel.html
10 http://en.wikipedia.org/wiki/Daiichi
11 http://www.nrc.gov/reading-rm/basic-ref/teachers/03.pdf: Reactor Concepts Manual, Chapter 3, Boiling Water Reactor (BWR) Systems, U.S. Nuclear Regulatory Commission.
12 http://en.wikipedia.org/wiki/Fukushima_Daiichi_Nuclear_Power_Plant
13 http://en.wikipedia.org/wiki/Kajima
14 http://www.reuters.com/article/2011/10/20/us-japan-nuclear-tsunami-idUSTRE-79J0B420111020
15 http://en.wikipedia.org/wiki/Fukushima_Daiichi_Nuclear_Power_Plant
16 "Additional Report of the Japanese Government to the IAEA—The Accident at TEPCO's Fukushima Nuclear Power Stations (Second Report)," Government of Japan, Nuclear Emergency Response Headquarters (September 2011).
17 http://en.wikipedia.org/wiki/PH
18 http://en.wikipedia.org/wiki/Iodine
19 http://atom.kaeri.re.kr/ton/nuc7.html
20 http://en.wikipedia.org/wiki/Caesium
21 http://en.wikipedia.org/wiki/Radiation_effects_from_Fukushima_Daiichi_nuclear_disaster

22   https://www.japan.go.jp/tomodachi/2017/autumn2017/fukushima_food.html

23   http://chernobylplace.com/current-status-of-fukushima/

24   http://www.world-nuclear-news.org/RS_Deaths_confirmed_at_Fukushima_
      Daiichi_0304111.html

25   http://www.world-nuclear.org/information-library/safety-and-security/safety-of-plants/
      fukushima-accident.aspx

26   Amory Lovins, "Soft Energy Paths for the 21st Century," https://rmi.org/insight/soft-ener-
      gy-paths-for-the-21st-century/ 2011

27   http://www.who.int/social_determinants/thecommission/kurokawa/en/index.html

28   http://spectrum.ieee.org/static/the-postfukushima-world

29   http://en.wikipedia.org/wiki/Nuclear_power_in_Germany

30   "World List of Nuclear Power Plants," *Nuclear News*, March 2014, pp. 52-53.

31   http://en.wikipedia.org/wiki/Nuclear_power_in_Switzerland

32   http://en.wikipedia.org/wiki/Nuclear_power_in_Italy

33   http://en.wikipedia.org/wiki/Nuclear_power_in_Japan

34   https://en.wikipedia.org/wiki/Paris_Agreement

35   Silverstein, Ken. "Japan Circling Back To Nuclear Power After Fukushima Disaster". forbes.
      com. Retrieved 12 January 2018.

36   http://www.nrc.gov/reactors/operating/ops-experience/japan-info.html

37   "US nuclear plants getting Fukushima-inspired safety upgrades," *Physics Today*, January
      2013, pp. 20-21.

Chapter 15

1   http://en.wikipedia.org/wiki/CANDU_reactor

2   Ronald Allen Knief, *Nuclear Energy Technology: Theory and Practice of Commercial
      Nuclear Power*, Hemisphere Publishing, Washington, 1981, Chapter 13.

3   http://en.wikipedia.org/wiki/File:CANDU_Reactor_Schematic.svg. This image was taken from
      the Wikipedia Commons, and it may be reproduced freely provided that attribution is given and
      the same conditions apply to the reproduction as to the original. See the link for details.

4   "World List of Nuclear Power Plants," *Nuclear News*, March 2019, pp. 41-42.

5   http://en.wikipedia.org/wiki/NRX

Chapter 16

1   http://en.wikipedia.org/wiki/Magnox

2   http://en.wikipedia.org/wiki/Advanced_gas-cooled_reactor

3   Ronald Allen Knief, *Nuclear Energy Technology: Theory and Practice of Commercial
      Nuclear Power*, Hemisphere Publishing, Washington, 1981, p. 333.

4   http://en.wikipedia.org/wiki/Peach_Bottom_Nuclear_Generating_Station

5   http://en.wikipedia.org/wiki/Fort_St._Vrain_Generating_Station

6   http://en.wikipedia.org/wiki/AVR_reactor

7   http://en.wikipedia.org/wiki/Thorium_High_Temperature_Reactor

8   http://en.wikipedia.org/wiki/HTR-10

9   https://www.iaea.org/NuclearPower/Downloadable/Meetings/2015/2015-10-19-10-23/
      DOC/D1_HTTR_Test_Reactor_20151015.pdf

10 http://www.world-nuclear-news.org/RS-Safety-review-sought-for-HTTR-2711144.html
11 http://www.world-nuclear-news.org/NN-Xe-100-HTGR-moves-to-conceptual-design-1703177.html
12 https://en.wikipedia.org/wiki/HTR-10
13 https://www.neimagazine.com/features/featurehtr-pm-making-dreams-come-true-7009889/
14 http://www.businessday.co.za/articles/Content.aspx?id=121307
15 http://www.ngnpalliance.org/
16 http://www.neimagazine.com/story.asp?storyCode=2055645
17 http://www.world-nuclear.org/info/inf41_US_nuclear_power_policy.html
18 http://analysis.nuclearenergyinsider.com/qa/barbara-hogan-reasons-i-pulled-plug-pbmr
19 http://www.world-nuclear-news.org/NN-Areva_modular_reactor_selected_for_NGNP_development-1502124.html
20 https://smr.inl.gov/Document.ashx?path...Int%2FANTARES.pdf
21 http://www.nrc.gov/reactors/advanced/ngnp.html
22 Press release of 24 January 2013; download PDF file via entry in https://www.ngnpalliance.org/index.php/resources
23 http://gif.inel.gov/roadmap/pdfs/technology_goals_nerac_subcommittee.pdf
24 http://en.wikipedia.org/wiki/Uranium_dioxide
25 http://en.wikipedia.org/wiki/List_of_elements_by_melting_point
26 http://en.wikipedia.org/wiki/Silicon_carbide
27 https://www.eia.gov/energyexplained/?page=us_energy_home
28 http://www.teslamotors.com/
29 https://en.wikipedia.org/wiki/Charging_station#Charging_time
30 https://en.wikipedia.org/wiki/Chevrolet_Volt
31 https://www.tesla.com/semi
32 http://en.wikipedia.org/wiki/Fuel_cell
33 http://www-formal.stanford.edu/jmc/progress/hydrogen.html
34 http://avt.inel.gov/hydrogen.shtml
35 http://www.cmt.anl.gov/oldweb/Science_and_Technology/Fuel_Cells/Publications/High_Temperature_(Steam)_Electrolysis.pdf
36 http://en.wikipedia.org/wiki/Electrolysis_of_water
37 http://74.125.95.132/search?q=cache:H2EKZtTZSjQJ:www.jlab.org/hydrogen/talks/Walters.pdf+nuclear+hydrogen+production&cd=2&hl=en&ct=clnk&gl=us
38 https://fusion.gat.com/pubs-ext/MISCONF02/A23944.pdf
39 http://docs.google.com/gview?a=v&q=cache:jg8-GB3cqe4J:www.inspi.ufl.edu/icapp08/program/abstracts/8125.pdf+nuclear+hydrogen+production+herring&hl=en&gl=us
40 http://www.enchantedlearning.com/usa/states/area.shtml
41 http://www.hydrogen.energy.gov/pdfs/review04/st_1_miliken.pdf
42 Richard A. Muller, *Physics for Future Presidents: The Science behind the Headlines*, W. W. Norton & Co., Inc., New York, 2008, pp. 69-70.
43 https://inlportal.inl.gov/portal/server.pt?open=514&objID=2251&parentname=CommunityPage&parentid=13&mode=2
44 http://en.wikipedia.org/wiki/Nuclear_fuel
45 Philip E. MacDonald, et al., "NGNP Point Design – Results of the Initial Neutronics and Thermal-Hydraulic Assessments During FY-03," INEEL/EXT-03-00870, Rev. 1, Idaho National Engineering and Environmental Laboratory, September 2003. [Note: The Idaho National Laboratory was formerly known as the Idaho National Engineering and

Environmental Laboratory, among other names in its evolution.]

[46] D. A. Petti, et al.., "Key Differences in the Fabrication, Irradiation and Safety Testing of U.S. and German TRISO-coated Particle Fuel and Their Implications on Fuel Performance," INEEL/EXT-02-00300, INEEL, June 2002.

[47] "NGNP and Hydrogen Production Preconceptual Design Report – Executive Summary Report," NGNP-ESR-RPT-001, Rev. 1, Westinghouse Electric Company LLC, June 2007, p.40.

[48] "NGNP with Hydrogen Production Preconceptual Design Studies Report – Executive Summary," Document No. 12-9052076-000, Areva NP, Inc., June 2007, p. 31.

[49] https://art.inl.gov/NGNP/INL%20Documents/Year%202007/inl-ext-07-12967.pdf

[50] Y. Muto and Y. Kato, "A New Pebble Bed Core Concept with Low Pressure Drop," GLOBAL 2003 International Conference, Atoms for Prosperity: Updating Eisenhower's Global Vision for Nuclear Energy, Vol. 1, American Nuclear Society, New Orleans, LA, November 2003.

[51] https://commons.wikimedia.org/wiki/File:Pebble_bed_reactor_scheme_(English).svg

[52] Richard L. Moore, Chang H. Oh, Brad J. Merrill, and David A. Petti, "Studies on Air Ingress for Pebble Bed Reactors," Appendix D in "Modular Pebble-Bed Reactor Project, Laboratory-Directed Research and Development Program, FY-02 Annual Report," David A. Petti, Editor, INEEL/EXT-02-01545, Idaho National Engineering and Environmental Laboratory (the former title of the Idaho National Laboratory), November 2002.

[53] J. Schaar, W. Frohling, and H. Hohn, "Status of the Experiment NACOR for Investigations on the Ingress of Air into the Core of an HTR Module" Nuclear Energy Agency Workshop on High Temperature Engineering Research Facilities and Experiments, Petten, The Netherlands, November 12-14, 1997.

[54] M. Hishida and T. Takeda, "Study on air ingress during an early stage of a primary-pipe rupture accident of a high-temperature gas-cooled reactor," *Nucl. Eng. & Design*, 126 (1991), pp. 175-187.

Chapter 17

[1] http://en.wikipedia.org/wiki/Fast_breeder_reactor

[2] http://www.inl.gov/factsheets/ebr-1.pdf

[3] Ronald Allen Knief, *Nuclear Energy Technology: Theory and Practice of Commercial Nuclear Power*, Hemisphere Publishing, Washington, 1981, p. 356.

[4] http://en.wikipedia.org/wiki/Experimental_Breeder_Reactor_II

[5] http://en.wikipedia.org/wiki/USS_Seawolf_(SSN-575)

[6] http://en.wikipedia.org/wiki/Enrico_Fermi_Nuclear_Generating_Station

[7] http://www.insc.anl.gov/cgi-bin/sql_interface?view=rx_model&qvar=id&qval=12

[8] http://www.igorr.com/home/liblocal/docs/Proceeding/Meeting%2011/Guidez.pdf

[9] http://www.jaea.go.jp/jnc/jncweb/02r-d/fast.html

[10] http://www.fzk.de/fzk/idcplg?IdcService=FZK&node=0698&lang=en

[11] http://www.tesionline.com/__PDF/23539/23539p.pdf

[12] http://www.world-nuclear.org/info/inf40.html

[13] http://en.wikipedia.org/wiki/Superph%C3%A9nix

[14] http://en.wikipedia.org/wiki/Shippingport_Atomic_Power_Station

[15] http://www.chemicalelements.com/elements/pb.html

[16] http://www.chemicalelements.com/elements/na.html

[17] http://en.wikipedia.org/wiki/Eutectic_point

[18] http://en.wikipedia.org/wiki/Lead_cooled_fast_reactor
[19] http://en.wikipedia.org/wiki/Molten_salt_reactor
[20] http://en.wikipedia.org/wiki/Aircraft_Reactor_Experiment
[21] http://en.wikipedia.org/wiki/Molten-Salt_Reactor_Experiment
[22] http://en.wikipedia.org/wiki/File:LMFBR_schematics2.svg
[23] http://en.wikipedia.org/wiki/Sodium
[24] Ronald Allen Knief, *Nuclear Energy Technology: Theory and Practice of Commercial Nuclear Power*, Hemisphere Publishing, Washington, 1981, pp. 566-570.

Chapter 18

[1] http://en.wikipedia.org/wiki/TRIGA
[2] http://triga.ga.com/
[3] https://commons.wikimedia.org/wiki/File:TrigaReactorCore.jpeg
[4] http://en.wikipedia.org/wiki/Cherenkov_radiation
[5] http://www.sckcen.be/en/Our-Research/Research-facilities/VENUS-zero-power-critical-facility
[6] http://www.psi.ch/
[7] http://www.ne.anl.gov/ne_web_photos/index.php?cid=15&pid=97
[8] http://irphep.inl.gov/
[9] http://proteus.web.psi.ch/GCFR-PROTEUS/index.html?forprint
[10] http://en.wikipedia.org/wiki/Advanced_Test_Reactor
[11] https://inlportal.inl.gov/portal/server.pt/gateway/PTARGS_0_1646_9670_0_0_18/atr.pdf
[12] *Advanced Test Reactor*, an information pamphlet published by the Idaho National Engineering Laboratory, undated
[13] http://en.wikipedia.org/wiki/File:Advanced_Test_Reactor.jpg
[14] http://creativecommons.org/licenses/by-sa/2.0/
[15] http://atrnsuf.inl.gov/
[16] https://secure.inl.gov/atrproposal/documents/ATRUsersGuide.pdf

Chapter 19

[1] http://www.world-nuclear.org/info/inf33.html
[2] http://en.wikipedia.org/wiki/United_States_naval_reactors
[3] http://www.world-nuclear.org/info/inf34.html
[4] http://www.nuscale.com/documents/Jun10NNNuScaleArticle_000.pdf
[5] S. M. Modro, et al., *Multi-Application Small Light-Water Reactor*, NERI Final Report, Idaho National Engineering and Environmental Laboratory, INEEL/EXT-04-01626, December 2003.
[6] http://www.nuscale.com/index.php
[7] Bruce Landrey and Jose N. Reyes, Jr., *NuScale's Passive Safety Approach*, Power Point presentation, September 2011, accessed by the link at http://www.nuscale.com/ot-Nuclear-Power-Presentations.php.
[8] http://www.gen4energy.com/
[9] http://www.nrc.gov/reactors/advanced/hyperion.html
[10] http://en.wikipedia.org/wiki/File:Hyperion_Power_Facility_Concept.jpg

[11] http://en.wikipedia.org/wiki/Lead-bismuth_eutectic
[12] http://en.wikipedia.org/wiki/Hyperion_Power_Generation
[13] http://www.ga.com/websites/ga/docs/em2/pdf/EM2_datasheet.pdf
[14] https://en.wikipedia.org/wiki/Energy_Multiplier_Module
[15] https://terrapower.com/
[16] https://en.wikipedia.org/wiki/TerraPower

Chapter 20

[1] http://en.wikipedia.org/wiki/Fusion_power
[2] http://en.wikipedia.org/wiki/ITER
[3] Francis F. Chen, *Introduction to Plasma Physics*, Plenum Press, New York, 1974.
[4] http://en.wikipedia.org/wiki/Interstellar_medium
[5] William H. Hayt, Jr., *Engineering Electromagnetics*, McGraw-Hill, New York, Third Edition, 1974.
[6] William K. Terry, "The Connection Between the Charged-Particle Current and the Displacement Current," *Am. J. Physics* 50 (8), 742 (1982).
[7] http://en.wikipedia.org/wiki/Michael_Faraday
[8] http://depts.washington.edu/rppl/papers/pst_7_1_001.pdf
[9] http://en.wikipedia.org/wiki/Tokamak
[10] http://en.wikipedia.org/wiki/Igor_Tamm
[11] http://en.wikipedia.org/wiki/Andrei_Sakharov
[12] http://en.wikipedia.org/wiki/Oleg_Lavrentiev
[13] George Schmidt, *Physics of High Temperature Plasmas: An Introduction*, Academic Press, New York, 1966, p. 96.
[14] http://www.iaea.org/programmes/ripc/physics/fec2000/pdf/iterp_14.pdf
[15] http://www.ipp.mpg.de/~Simon.Pinches/thesis/node29.html
[16] George Schmidt, *Physics of High Temperature Plasmas: An Introduction*, Academic Press, New York, 1966, pp. 217-221.
[17] http://en.wikipedia.org/wiki/Lev_Landau
[18] William Knox Terry, "Energy Confinement Considerations in Two Proposed Nuclear Fusion Reactors," PhD Dissertation, University of Washington, 1980.
[19] Francis F. Chen, *Introduction to Plasma Physics*, Plenum Press, New York, 1974, p. 299.
[20] http://adsabs.harvard.edu/abs/1981PhRvL..46.1337B
[21] Alan L. Hoffman, "Field-Reversed Configurations," *J. Fusion Energy* 17 (3), 201-205 (1998).
[22] Alan L. Hoffman, John T. Slough, and Dennis G. Harding, "Suppression of the n=2 rotational instability in field-reversed configurations," Proc. Fifth Symposium on the Physics and Technology of Compact Toroids, Mathematical Sciences Northwest, Seattle, Washington, November 16-18, 1982, A. L. Hoffman and R. D. Milroy, eds., p.51.
[23] http://en.wikipedia.org/wiki/Stellarator
[24] http://en.wikipedia.org/wiki/Reversed_field_pinch
[25] http://en.wikipedia.org/wiki/Spheromak
[26] https://www.llnl.gov/str/Hill.html
[27] https://www.llnl.gov/str/September05/Hill.html
[28] https://www.llnl.gov/str/Hill.html
[29] https://www.researchgate.net/publication/232061586_Sustained_Spheromak_Physics_

Experiment_SSPX_Design_and_physics_results
[30] http://www.pppl.gov/publications/pics/info_bull_nstx_0606.pdf
[31] https://en.wikipedia.org/wiki/National_Spherical_Torus_Experiment
[32] http://en.wikipedia.org/wiki/Inertial_confinement_fusion
[33] http://www.boston.com/bigpicture/2010/10/the_national_ignition_facility.html
[34] David Kramer, "High-energy-density science blooms at NIF," *Physics Today*, February 2017, pp. 33-35.
[35] http://en.wikipedia.org/wiki/File:NIF_beamline_diagram.png
[36] https://lasers.llnl.gov/about/how-nif-works
[37] https://www.llnl.gov/news/aroundthelab/2014/Feb/NR-14-02-06.html?utm_content=buffer07554&utm_medium=social&utm_source=twitter.com&utm_campaign=buffer#.U-4FKeNdW5J
[38] http://en.wikipedia.org/wiki/Joint_European_Torus
[39] http://en.wikipedia.org/wiki/Troubled_Asset_Relief_Program
[40] http://www.iter.org/newsline/-/2546
[41] http://en.wikipedia.org/wiki/Orders_of_magnitude_(magnetic_field)
[42] http://www.iter.org/default.aspx
[43] https://interestingengineering.com/chinas-anhei-tokamak-leads-the-way-to-fusion-energy
[44] http://en.wikipedia.org/wiki/DEMO

Chapter 21

[1] http://www.world-nuclear.org/info/inf75.html
[2] http://en.wikipedia.org/wiki/Uranium_mining
[3] http://en.wikipedia.org/wiki/Uranium_ore
[4] http://en.wikipedia.org/wiki/Uraninite
[5] http://en.wikipedia.org/wiki/Carnotite
[6] http://en.wikipedia.org/wiki/Pegmatite
[7] http://en.wikipedia.org/wiki/Nepheline_syenite
[8] http://large.stanford.edu/courses/2010/ph240/sagatov1/docs/isl.pdf
[9] http://en.wikipedia.org/wiki/Uranium_depletion
[10] http://www.nea.fr/science/docs/2009/nsc-doc2009-13.pdf
[11] http://en.wikipedia.org/wiki/Nuclear_reprocessing
[12] http://www.world-nuclear.org/info/inf98.html
[13] http://en.wikipedia.org/wiki/Enriched_uranium
[14] http://www.nrc.gov/materials/fuel-cycle-fac/ur-enrichment.html
[15] http://www.world-nuclear.org/info/inf28.html
[16] http://www.csmonitor.com/USA/2010/0528/Will-secret-technology-help-rogue-nations-get-nuclear-weapons
[17] http://en.wikipedia.org/wiki/Uranium_hexafluoride#Physical_Properties
[18] Samuel Glasstone and Alexander Sesonske, *Nuclear Reactor Engineering*, Van Nostrand Reinhold, New York, Third Edition, 1981, p. 480.
[19] cns.miis.edu/activities/media/ferguson_laser_enrichment.ppt
[20] http://www.nei.org/howitworks/factsheets/nuclearfuelproduction/
[21] http://en.wikipedia.org/wiki/Nuclear_fuel
[22] Samuel Glasstone and Alexander Sesonske, *Nuclear Reactor Engineering*, Van Nostrand Reinhold, New York, Third Edition, 1981, p. 481.

23 http://en.wikipedia.org/wiki/Sinter
24 http://www.freepatentsonline.com/3923933.html
25 http://en.wikipedia.org/wiki/Burnup
26 Anthony V. Nero, Jr., *A Guidebook to Nuclear Reactors*, University of California Press, 1979, p. 128.
27 http://www.janleenkloosterman.nl/papers/kuijper0601.pdf
28 http://www.ne.doe.gov/geniv/neGenIV9.html
29 James J. Duderstadt and Louis J. Hamilton, *Nuclear Reactor Analysis*, John Wiley & Sons, Inc., New York, 1976, pp. 596-600.
30 http://en.wikipedia.org/wiki/Nuclear_reprocessing
31 http://en.wikipedia.org/wiki/Actinide
32 http://en.wikipedia.org/wiki/Minor_actinides
33 http://www.osti.gov/bridge/product.biblio.jsp?osti_id=10170117
34 http://en.wikipedia.org/wiki/Reprocessed_uranium
35 http://en.wikipedia.org/wiki/Spent_nuclear_fuel
36 http://en.wikipedia.org/wiki/Uranium-236
37 http://www.ccnr.org/plute_bomb.html
38 http://www.ne.doe.gov/AFCI/neAFCI.html
39 http://www.moxproject.com/
40 https://www.aikenstandard.com/news/nnsa-document-details-one-year-of-mox-termina-tion-work/article_bb8051c4-d39f-11e8-9db9-ef482a88134c.html
41 https://www.aikenstandard.com/news/legal-team-deliberating-supreme-court-appeal-for-mox-case-s/article_15ba33b0-65f6-11e9-a07c-1ffb6d91d499.html
42 http://en.wikipedia.org/wiki/PUREX
43 http://www.wise-uranium.org/eftokc.html
44 http://en.wikipedia.org/wiki/Integral_Fast_Reactor
45 Terry Kammash, *Fusion Reactor Physics: Principles and Technology*, Ann Arbor Science, Ann Arbor, Michigan, 1976, pp. 52-53.
46 http://en.wikipedia.org/wiki/Helium-3
47 http://en.wikipedia.org/wiki/Aneutronic_fusion

Chapter 22

1 http://web.archive.org/web/20080514020437/http://www.ocrwm.doe.gov/documents/nwpa/css/nwpa.htm
2 http://en.wikipedia.org/wiki/Yucca_Mountain_nuclear_waste_repository
3 Environmental Protection Agency, "Risk Assessment Guidance for Superfund: Volume 1 - Human Health Evaluation Manual (Part B, Development of Risk-based Preliminary Remediation Goals)", 9285.7-01B, Washington, DC, December 1991.
4 http://www.osti.gov/bridge/purl.cover.jsp;jsessionid=854B9B492BD9C5E4524607FC-8C8BB350?purl=/6943981-sA7f1t/
5 http://en.wikipedia.org/wiki/Superfund
6 http://www.wipp.energy.gov/
7 http://en.wikipedia.org/wiki/Tuff
8 http://en.wikipedia.org/wiki/Yucca_Mountain
9 http://www.issues.org/18.3/p_carter.html
10 http://en.wikipedia.org/wiki/Ogallala_Aquifer

[11] http://etd.lib.ttu.edu/theses/available/etd-05122009-31295000269356/unrestricted/31295000269356.pdf

[12] http://en.wikipedia.org/wiki/Basalt_Waste_Isolation_Project

[13] http://reid.senate.gov/issues/upload/Termination-Language-for-the-Website.pdf

[14] http://www.frumforum.com/nuke-panel-keeps-yucca-project-alive

[15] http://www.world-nuclear-news.org/WR_Bill_to_liquidate_the_nuclear_waste_fund_2704092.html

[16] D. J. Meyers, *Site Characterization Plan Conceptual Design Report for BWIP High-Level Nuclear Waste Packages*, prepared by Gilbert/Commonwealth, Inc., for the Rockwell Hanford Operations Basalt Waste Isolation Project and the Department of Energy Richland Operations Office, Richland, Washington, SD-BWI-CDR-005, 16 April 1987.

[17] http://lobby.la.psu.edu/066_Nuclear_Repository/Agency_Activities/NRC/NRC_Conceptual_Design_Plan.htm

[18] "Technical Basis Document No. 14: Low Probability Seismic Events". *Office of Civilian Radioactive Waste Management*. U.S. Department of Energy. June 2004.

[19] Rodney C. Ewing and Frank N. von Hippel. Nuclear Waste Management in the United States – Starting Over, Science, Vol. 325, 10 July 2009, p. 152.

[20] http://en.wikipedia.org/wiki/Waste_Isolation_Pilot_Plant

[21] European Commission Joint Research Centre. "Strategic Cooperation Sets the Scene for Geological Disposal of Nuclear Waste in Europe." ScienceDaily 19 February 2010. 23 August 2010 <http://www.sciencedaily.com /releases/2010/02/100219163402.htm>.

[22] http://www.world-nuclear.org/info/inf21.html

[23] Gwyneth Cravens, *Power to Save the World: The Truth about Nuclear Energy*, Alfred A. Knopf, Hew York, 2007, pp. 280-288.

Chapter 23

[1] http://en.wikipedia.org/wiki/Nuclear_weapon

[2] http://en.wikipedia.org/wiki/Nuclear_weapon_design

[3] http://www.cdphe.state.co.us/HM/rf/

[4] http://www.washingtonpost.com/wp-dyn/content/article/2006/01/19/AR2006011903311.html

[5] https://en.wikipedia.org/wiki/Iran_nuclear_deal_framework

[6] https://www.nytimes.com/2019/07/07/world/middleeast/iran-nuclear-limits-breach.html

[7] http://www.fas.org/nuke/guide/dprk/nuke/index.html

[8] http://www.rertr.anl.gov/

[9] https://www.rertr.anl.gov/meeting_announcements/2019/

[10] http://en.wikipedia.org/wiki/N-Reactor

[11] A. M. Ougouag, W. K. Terry and H. D. Gougar, "Examination of the Potential for Diversion or Clandestine Dual Use of a Pebble-Bed Reactor to Produce Plutonium," Proceedings of HTR 2002, 1st International Topical Meeting on High Temperature Reactor Technology (HTR), Petten, The Netherlands, April 22-24, 2002.

[12] http://en.wikipedia.org/wiki/Integral_Fast_Reactor

[13] http://www.world-nuclear.org/info/inf69.html

[14] http://en.wikipedia.org/wiki/Strategic_Arms_Limitation_Talks

[15] http://www.iaea.org/Publications/Documents/Infcircs/Others/infcirc140.pdf

[16] http://en.wikipedia.org/wiki/Nuclear_Non-Proliferation_Treaty

[17] http://en.wikipedia.org/wiki/Nuclear_weapons_and_Israel
[18] http://en.wikipedia.org/wiki/International_Atomic_Energy_Agency
[19] http://en.wikipedia.org/wiki/Peaceful_nuclear_explosions
[20] http://www.cnn.com/2010/WORLD/meast/02/19/iran.nuclear.index.html?hpt=T2
[21] http://en.wikipedia.org/wiki/Nuclear_program_of_Iran
[22] http://en.wikipedia.org/wiki/Iran_hostage_crisis
[23] http://en.wikipedia.org/wiki/Bushehr_Nuclear_Power_Plant
[24] http://www.foxnews.com/politics/2017/04/19/iran-nuclear-deal-trump-administra-tion-says-tehran-complying-with-agreement.html
[25] http://www.reuters.com/article/us-iran-usa-sanctions-idUSKBN1971G5?feed-Type=RSS&feedName=worldNews
[26] http://en.wikipedia.org/wiki/September_11_attacks
[27] Reza Aslan, *How to Win a Cosmic War: God, Globalization, and the End of the War on Terror*, Random House, New York, 2009
[28] http://www.fas.org/sgp/crs/nuke/R40154.pdf
[29] http://en.wikipedia.org/wiki/Thorium_fuel_cycle
[30] http://en.wikipedia.org/wiki/THTR-300

Appendix II

[1] http://en.wikipedia.org/wiki/Nikola_Tesla
[2] http://en.wikipedia.org/wiki/Carl_Friedrich_Gauss
[3] http://en.wikipedia.org/wiki/Wilhelm_Eduard_Weber
[4] Graham Farmelo, *The Strangest Man: The Hidden Life of Paul Dirac, Mystic of the Atom*, Basic Books, New York, 2009
[5] https://en.wikipedia.org/wiki/Magnetic_monopole
[6] https://phys.org/news/2016-08-mysterious-magnetic-monopole.html
[7] http://en.wikipedia.org/wiki/Andr%C3%A9-Marie_Amp%C3%A8re
[8] http://en.wikipedia.org/wiki/Joseph_Henry
[9] William K. Terry, "The Connection Between the Charged-Particle Current and the Displacement Current," *Am. J. Physics* 50 (8), 742 (1982).
[10] http://en.wikipedia.org/wiki/Michael_Faraday

INDEX